Student Solutions Manual

Charles W. Haines
Rochester Institute of Technology

to accompany

Elementary Differential Equations, 7e and Elementary Differential Equations and Boundary Value Problems, 7e

William E. Boyce
Richard C. DiPrima

Rensselaer Polytechnic Institute

John Wiley & Sons, Inc.
New York • Chichester • Weinheim • Brisbane • Singapore • Toronto

To order books or for customer service call 1-800-CALL-WILEY (225-5945).

ISBN 0-471-39114-x

Printed in the United States of America

10 9 8 7 6 5 4 3 2 1

Printed and bound by Hamilton Printing Company

PREFACE

This supplement has been prepared for use in conjunction with the seventh editions of ELEMENTARY DIFFERNTIAL EQUATIONS AND BOUNDARY VALUE PROBLEMS and ELEMENTARY DIFFERENTIAL EQUATIONS, both by W.E. Boyce and R.C. DiPrima. The supplement contains a sampling of the problems from each section of the text. In most cases the complete details in determining the solutions are given while in the remainder of the problems helpful hints are provided. The problems chosen in each section represent, wherever possible, the variety of applications and types of examples that are covered in the written material of the text, thereby providing the student a complete set of examples from which to learn.

Students should be aware that following these solutions is very different from designing and constructing one's own solution. Using this supplemental resource appropriately for learning differential equations is outlined as follows:

 1. Make an honest attempt to solve the problem without using the guide.
 2. If needed, glance at the beginning of the solution in the guide and then try again to generate the complete solution. Continue using the guide for hints when you reach an impasse.
 3. Compare your final solution with the one provided to see whether yours is more or less efficient than the guide, since there is frequently more than one correct way to solve a problem.
 4. Ask yourself why that particular problem was assigned.

The use of a symbolic computational software package, in many cases, would greatly simplify finding the solution to a given problem, but the details given in this solutions manual are important for the student's understanding of the underlying mathematical principles and applications. In other cases, these software packages are essential for completing the given problem, as the calculations would be overwhelming using analytical techniques. In these cases, some steps or hints are given and then reference made to the use of an appropriate software package.

In order to simplify the text, the following abbreviations have been used:

D.E.	differential equation(s)
O.D.E.	ordinary differential equation(s)
P.D.E.	partial differential equation(s)
I.C.	initial condition(s)
I.V.P.	initial value problem(s)
B.C.	boundary condition(s)
B.V.P.	boundary value problem(s)

I wish to express my appreciation to Mrs. Susan A. Hickey for her excellent typing and proofreading of all stages of the manuscript. Dr. Josef Torok has also provided invaluable assistance with the preparation of the figures as well as assistance with many of the solutions involving the use of symbolic computational software.

Charles W. Haines
Professor of Mathematics and
     Mechanical Engineering
Rochester Institute of Technology
Rochester, New York
June 2000

CONTENTS

CHAPTER 1

<u>Section 1.1, Page 8</u>

2. For y > 3/2 we see that y' > 0
    and thus y(t) is increasing there.
    For y < 3/2 we have y' < 0 and thus
    y(t) is decreasing there.   Hence
    y(t) diverges from 3/2 as t→∞.

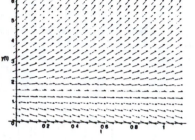

4. Observing the direction field,
    we see that for y>-1/2 we have
    y'<0, so the solution is
    decreasing here. Likewise,
    for y<-1/2 we have y'>0 and
    thus y(t) is increasing here.
    Since the slopes get closer to
    zero as y gets closer to -1/2,
    we conclude that y→-1/2 as t→∞.

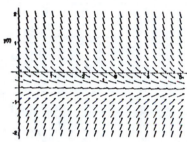

7. If all solutions approach 3, then 3 is the equilibrium
    solution and we want $\dfrac{dy}{dt}$ < 0 for y > 3 and $\dfrac{dy}{dt}$ > 0 for
    y < 3.   Thus $\dfrac{dy}{dt}$ = 3-y.

11. For y = 0 and y = 4 we have y' = 0 and thus y = 0 and
    y = 4 are equilibrium solutions. For y > 4, y' < 0 so if
    y(0) > 4 the solution approaches y = 4 from above. If
    0 < y(0) < 4, then y' > 0 and the solutions "grow" to y = 4
    as t→∞. For y(0) < 0 we see that y' < 0 and the solutions
    diverge from 0.

13. Since $y' = y^2$, y = 0 is the equilibrium solution and y' > 0
    for all y. Thus if y(0) > 0, solutions will diverge from
    0 and if y(0) < 0, solutions will aproach y = 0 as t→∞.

15a. Let q(t) be the number of grams of the substance in the
     water at any time. Then $\dfrac{dq}{dt}$ = 300(.01) − $\dfrac{300q}{1,000,000}$ =
     $300(10^{-2} - 10^{-6}q)$.

15b. The equilibrium solution occurs when q' = 0, or c = 10^4gm,
     independent of the amount present at t = 0 (all solutions
     approach the equilibrium solution).

16.   The D.E. expressing the evaporation is $\dfrac{dV}{dt} = -aS$, a > 0.

Now $V = \dfrac{4}{3}\pi r^3$ and $S = 4\pi r^2$, so $S = 4\pi\left(\dfrac{3}{4\pi}\right)^{2/3} V^{2/3}$.

Thus $\dfrac{dv}{dt} = -kV^{2/3}$, for k > 0.

21.

22.

24.

Section 1.2 Page 14

1b.   dy/dt = -2y+5 can be rewritten as $\dfrac{dy}{y-5/2} = -2dt$.   Thus

$\ln|y-5/2| = -2t+c_1$, or $y-5/2 = ce^{-2t}$. $y(0) = y_0$ yields

$c = y_0 - 5/2$, so $y = 5/2 + (y_0-5/2)e^{-2t}$.

If $y_0 > 5/2$, the solution starts above the equilibrium
solution and decreases exponentially and approaches 5/2
as t→∞.   Conversely, if y < 5/2, the solution starts
below 5/2 and grows exponentially and approaches 5/2 from

below as $t \to \infty$.

4a. Rewrite Eq.(ii) as $\dfrac{dy/dt}{y} = a$ and thus $\ln|y| = at + c_1$; or

$y = ce^{at}$.

4b. If $y = y_1(t) + k$, then $\dfrac{dy}{dt} = \dfrac{dy_1}{dt}$. Substituting both

these into Eq.(i) we get $\dfrac{dy_1}{dt} = a(y_1 + k) - b$. Since

$\dfrac{dy_1}{dt} = ay_1$, this leaves $ak - b = 0$ and thus $k = b/a$.

Hence $y = y_1(t) + b/a$ is the solution to Eq(i).

6b. From Eq.(11) we have $p = 900 + ce^{t/2}$. If $p(0) = p_0$, then
$c = p_0 - 900$ and thus $p = 900 + (p_0 - 900)e^{t/2}$. If
$p_0 < 900$, this decreases, so if we set $p = 0$ and solve
for T (the time of extinction) we obtain
$e^{T/2} = 900/(900-p_0)$, or $T = 2\ln[900/(900-p_0)]$.

8a. Use Eq.(26).

8b. Use Eq.(29).

10a. $\dfrac{dQ}{dt} = -rQ$ yields $\dfrac{dQ/dt}{Q} = -r$, or $\ln|Q| = -rt + c_1$. Thus

$Q = ce^{-rt}$ and $Q(0) = 100$ yields $c = 100$. Hence $Q = 100e^{-rt}$.
Setting $t = 1$, we have $82.04 = 100e^{-r}$, which yields
$r = .1980/\text{wk}$ or $r = .02828/\text{day}$.

13a. Rewrite the D.E. as $\dfrac{dQ/dt}{Q-CV} = \dfrac{-1}{CR}$, thus upon integrating and

simplifying we get $Q = De^{-t/CR} + CV$. $Q(0) = 0 \Rightarrow D = -CV$ and
thus $Q(t) = CV(1 - e^{-t/CR})$.

13b. $\lim\limits_{t \to \infty} Q(t) = CV$ since $\lim\limits_{t \to \infty} e^{-t/CR} = 0$.

13c. In this case $R\dfrac{dQ}{dt} + \dfrac{Q}{C} = 0$, $Q(t_1) = CV$. The solution of this

D.E. is $Q(t) = Ee^{-t/CR}$, so $Q(t_1) = Ee^{-t_1/CR} = CV$, or
$E = CVe^{t_1/CR}$. Thus $Q(t) = CVe^{t_1/CR}e^{-t/CR} = CVe^{-(t-t_1)/CR}$.

13a. CV = 20, CR = 2.5   13c. CV = 20, CR = 2.5 t_1 = 10

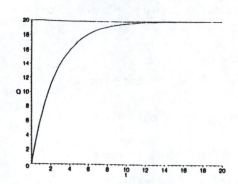
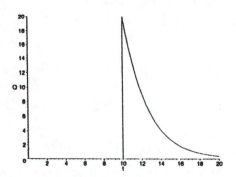

Section 1.3, Page 22

2.   The D.E. is second order since there is a second
     derivative of y appearing in the equation. The equation
     is nonlinear due to the y^2 term (as well as due to the y^2
     term multiplying the y″ term).

6.   This is a third order D.E. since the highest derivative is
     y‴ and it is linear since y and all its derivatives
     appear to the first power only. The terms t^2 and $\cos^2 t$ do
     not affect the linearity of the D.E.

8.   For $y_1(t) = e^{-3t}$ we have $y_1'(t) = -3e^{-3t}$ and $y_1''(t) = 9e^{-3t}$.
     Substitution of these into the D.E. yields
     $$9e^{-3t} + 2(-3e^{-3t}) - 3(e^{-3t}) = (9-6-3)e^{-3t} = 0.$$

14. Recall that if $u(t) = \int_0^t f(s)\,ds$, then $u'(t) = f(t)$.

16. Differentiating e^{rt} twice and substituting into the D.E.
     yields $r^2 e^{rt} - e^{rt} = (r^2-1)e^{rt}$. If $y = e^{rt}$ is to be a
     solution of the D.E. then the last quantity must be zero
     for all t. Thus $r^2-1 = 0$ since e^{rt} is never zero.

19. Differentiating t^r twice and substituting into the D.E.
     yields $t^2[r(r-1)t^{r-2}] + 4t[rt^{r-1}] + 2t^r = [r^2+3r+2]t^r$. If
     $y = t^r$ is to be a solution of the D.E., then the last term
     must be zero for all t and thus $r^2 + 3r + 2 = 0$.

22. The D.E. is second order since there are second partial
     derivatives of u(x,y) appearing. The D.E. is nonlinear
     due to the product of u(x,y) times u_x(or u_y).

26. Since $\dfrac{\partial u_1}{\partial t} = -\alpha^2 e^{-\alpha^2 t} \sin x$ and $\dfrac{\partial^2 u_1}{\partial x^2} = -e^{-\alpha^2 t} \sin x$ we have

$\alpha^2[-e^{-\alpha^2 t}\sin x] = -\alpha^2 e^{-\alpha^2 t}\sin x$, which is true for all t and x.

6

CHAPTER 2

Section 2.1, Page 38

1. $\mu(t) = \exp(\int 3dt) = e^{3t}$. Thus $e^{3t}(y'+3y) = e^{3t}(t+e^{-2t})$ or $\frac{d}{dt}(ye^{3t}) = te^{3t} + e^t$. Integration of both sides yields $ye^{3t} = \frac{1}{3}te^{3t} - \frac{1}{9}e^{3t} + e^t + c$, and division by e^{3t} gives the general solution. Note that $\int te^{3t}dt$ is evaluated by integration by parts, with $u = t$ and $dv = e^{3t}dt$.

2. $\mu(t) = e^{-2t}$.

3. $\mu(t) = e^t$.

4. $\mu(t) = \exp(\int \frac{dt}{t}) = e^{\ln t} = t$, so $(ty)' = 3t\cos 2t$, and integration by parts yields the general solution.

6. The equation must be divided by t so that it is in the form of Eq.(3): $y' + (2/t)y = (\sin t)/t$. Thus $\mu(t) = \exp(\int \frac{2dt}{t} = t^2$, and $(t^2 y)' = t\sin t$. Integration then yields $t^2 y = -t\cos t + \sin t + c$.

7. $\mu(t) = e^{t^2}$.

8. $\mu(t) = \exp(\int \frac{4tdt}{1+t^2}) = (1+t^2)^2$.

11. $\mu(t) = e^t$ so $(e^t y)' = 5e^t\sin 2t$. To integrate the right side you can integrate by parts (twice), use an integral table, or use a symbolic computational software program to find $e^t y = e^t(\sin 2t - 2\cos 2t) + c$.

13. $\mu(t) = e^{-t}$ and $y = 2(t-1)e^{2t} + ce^t$. To find the value for c, set $t = 0$ in y and equate to 1, the initial value of y. Thus $-2+c = 1$ and $c = 3$, which yields the solution of the given initial value problem.

15. $\mu(t) = \exp(\int \frac{2dt}{t}) = t^2$ and $y = t^2/4 - t/3 + 1/2 + c/t^2$. Setting $t = 1$ and $y = 1/2$ we have $c = 1/12$.

18. $\mu(t) = t^2$. Thus $(t^2 y)' = t\sin t$ and $t^2 y = -t\cos t + \sin t + c$. Setting $t = \pi/2$ and $y = 1$ yields $c = \pi^2/4 - 1$.

20. $\mu(t) = te^t$.

21b. $\mu(t) = e^{-t/2}$ so $(e^{-t/2}y)' = 2e^{-t/2}\cos t$. Integrating (see comments in #11) and dividing by $e^{-t/2}$ yields

$y(t) = -\dfrac{4}{5}\cos t + \dfrac{8}{5}\sin t + ce^{t/2}$. Thus $y(0) = -\dfrac{4}{5} + c = a$,

or $c = a + \dfrac{4}{5}$ and $y(t) = -\dfrac{4}{5}\cos t + \dfrac{8}{5}\sin t + (a + \dfrac{4}{5})e^{t/2}$.

If $(a + \dfrac{4}{5}) = 0$, then the solution is oscillatory for all

t, while if $(a + \dfrac{4}{5}) \neq 0$, the solution is unbounded as

$t \to \infty$. Thus $a_0 = -\dfrac{4}{5}$.

21a.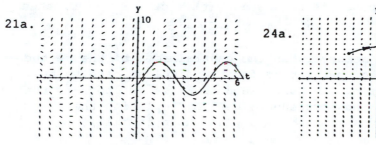

24a.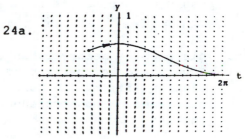

24b. $\mu(t) = \exp\displaystyle\int \dfrac{2dt}{t} = t^2$, so $(t^2 y)' = \sin t$ and

$y(t) = \dfrac{-\cos t}{t^2} + \dfrac{c}{t^2}$. Setting $t = -\dfrac{\pi}{2}$ yields

$\dfrac{4c}{\pi^2} = a$ or $c = \dfrac{a\pi^2}{4}$ and hence $y(t) = \dfrac{a\pi^2/4 - \cos t}{t^2}$, which

is unbounded as $t \to 0$ unless $a\pi^2/4 = 1$ or $a_0 = 4/\pi^2$.

24c. For $a = 4/\pi^2$ $y(t) = \dfrac{1 - \cos t}{t^2}$. To find the limit as

$t \to 0$ L'Hopital's Rule must be used:

$\displaystyle\lim_{t \to 0} y(t) = \lim_{t \to 0}\dfrac{\sin t}{2t} = \lim_{t \to 0}\dfrac{\cos t}{2} = \dfrac{1}{2}$.

28. $(e^{-t}y)' = e^{-t} + 3e^{-t}\sin t$ so

$e^{-t}y = -e^{-t} - 3e^{-t}(\dfrac{\sin t + \cos t}{2}) + c$ or

$y(t) = -1 - (\dfrac{3}{2})e^{-t}(\sin t + \cos t) + ce^{t}$.

If $y(t)$ is to remain bounded, we must have $c = 0$. Thus

$y(0) = -1 - \dfrac{3}{2} + c = y_0$ or $c = y_0 + \dfrac{5}{2} = 0$ and $y_0 = -\dfrac{5}{2}$.

30. $\mu(t) = e^{at}$ so the D.E. can be written as $(e^{at}y)' = be^{at}e^{-\lambda t} = be^{(a-\lambda)t}$. If $a \neq \lambda$, then integration and solution for y yields $y = [b/(a-\lambda)]e^{-\lambda t} + ce^{-at}$. Then $\lim\limits_{x \to \infty} y$ is zero since both λ and a are positive numbers.

    If $a = \lambda$, then the D.E. becomes $(e^{at}y)' = b$, which yields $y = (bt+c)/e^{\lambda t}$ as the solution. L'Hopital's Rule gives

    $$\lim_{t \to \infty} y = \lim_{t \to \infty} \frac{(bt+c)}{e^{\lambda t}} = \lim_{t \to \infty} \frac{b}{\lambda e^{\lambda t}} = 0.$$

32. There is no unique answer for this situation. One possible response is to assume $y(t) = ce^{-2t} + 3 - t$, then $y'(t) = -2ce^{-2t} - 1$ and thus $y' + 2y = 5 - 2t$.

35. This problem demonstrates the central idea of the method of variation of parameters for the simplest case. The solution (ii) of the homogeneous D.E. is extended to the corresponding nonhomogeneous D.E. by replacing the constant A by a function A(t), as shown in (iii).

36. Assume $y(t) = A(t)\exp(-\int(-2)dt) = A(t)e^{2t}$. Differentiating $y(t)$ and substituting into the D.E. yields $A'(t) = t^2$ since the terms involving A(t) add to zero. Thus $A(t) = t^3/3 + c$, which substituted into $y(t)$ yields the solution.

37. $y(t) = A(t)\exp(-\int\frac{dt}{t}) = A(t)/t$.

Section 2.2, Page 45

Problems 1 through 20 follow the pattern of the examples worked in this section. The first eight problems, however, do not have I.C. so the integration constant, c, cannot be found.

1. Write the equation in the form $ydy = x^2dx$. Integrating the left side with respect to y and the right side with respect to x yields

    $$\frac{y^2}{2} = \frac{x^3}{3} + C, \text{ or } 3y^2 - 2x^3 = c.$$

4. For $y \neq -3/2$ multiply both sides of the equation by $3 + 2y$ to get the separated equation $(3+2y)dy = (3x^2-1)dx$. Integration then yields

$3y + y^2 = x^3 - x + c.$

6.   We need $x \neq 0$ and $|y| < 1$ for this problem to be
      defined. Separating the variables we get $(1-y^2)^{-1/2}dy =$
      $x^{-1}dx$. Integrating each side yields $\arcsin y = \ln|x|+c$,
      so $y = \sin[\ln|x|+c]$, $x \neq 0$ (note that $|y| < 1$). Also,
      $y = \pm 1$ satisfy the D.E., since both sides are zero.

10a. Separating the variables we get $y\,dy = (1-2x)\,dx$, so
      $\dfrac{y^2}{2} = x - x^2 + c$. Setting $x = 0$ and $y = -2$ we have $2 = c$
      and thus $y^2 = 2x - 2x^2 + $ or $y = -\sqrt{2x - 2x^2 + 4}$. The
      negative square root must be used since $y(0) = -2$.

10c. Rewriting $y(x)$ as $-\sqrt{2(2-x)(x+1)}$, we see that y is
      defined for $-1 \leq x \leq 2$, However, since y' does not exist
      for $x = -1$ or $x = 2$, the solution is valid only for the
      open interval $-1 < x < 2$.

13.   Separate variables by factoring the denominator of the
      right side to get $y\,dy = \dfrac{2x}{1+x^2}dx$. Integration yields
      $y^2/2 = \ln(1+x^2)+c$ and use of the I.C. gives $c = 2$. Thus
      $y = \pm\,[2\ln(1+x^2)+4]^{1/2}$, but we must discard the plus
      square root because of the I.C. Since $1 + x^2 > 0$, the
      solution is valid for all x.

15.   Separating variables and integrating yields
      $y + y^2 = x^2 + c$. Setting $y = 0$ when $x = 2$ yields $c = -4$
      or $y^2 + y = x^2-4$. To solve for y complete the square on
      the left side by adding 1/4 to both sides. This yields
      $y^2 + y + \dfrac{1}{4} = x^2 - 4 + \dfrac{1}{4}$ or $(y + \dfrac{1}{2})^2 = x^2 - 15/4$. Taking
      the square root of both sides yields
      $y + \dfrac{1}{2} = \pm\sqrt{x^2 - 15/4}$, where the positive square root
      must be taken in order to satisfy the I.C. Thus
      $y = -\dfrac{1}{2} + \sqrt{x^2 - 15/4}$, which is defined for $x^2 \geq 15/4$ or
      $x \geq \sqrt{15}/2$. The possibility that $x < -\sqrt{15}/2$ is
      discarded due to the I.C.

17a. Separating variables gives $(2y-5)\,dy = (3x^2-e^x)\,dx$ and integration then gives $y^2 - 5y = x^3 - e^x + c$. Setting $x = 0$ and $y = 1$ we have $1 - 5 = 0 - 1 + c$, or $c = -3$. Thus $y^2 - 5y - (x^3 - e^x - 3) = 0$ and using the quadratic formula then gives

$$y(x) = \frac{5 \pm \sqrt{25 + 4(x^3 - e^x - 3)}}{2} = \frac{5}{2} - \sqrt{\frac{13}{4} + x^3 - e^x}.$$ The

negative square root is chosen due to the I.C.

17c. The interval of definition for y must be found numerically. Approximate values can be found by plotting

$$y_1(x) = \frac{13}{4} + x^3 \text{ and } y_2(x) = e^x \text{ and noting the values of } x$$

where the two curves cross.

19a. As above we start with $\cos 3y \, dy = -\sin 2x \, dx$ and integrate

to get $\frac{1}{3}\sin 3y = \frac{1}{2}\cos 2x + c$. Setting $y = \pi/3$ when

$x = \pi/2$ (from the I.C.) we find that $0 = -\frac{1}{2} + c$ or

$c = \frac{1}{2}$, so that $\frac{1}{3}\sin 3y = \frac{1}{2}\cos 2x + \frac{1}{2} = \cos^2 x$ (using the

appropriate trigonometric identity). To solve for y we must choose the branch that passes through the point $(\pi/2, \pi/3)$ and thus $3y = \pi - \arcsin(3\cos^2 x)$, or

$$y = \frac{\pi}{3} - \frac{1}{3}\arcsin(3\cos^2 x).$$

19c. The solution in part a is defined only for $0 \le 3\cos^2 x \le 1$, or $-\sqrt{1/3} \le \cos x \le \sqrt{1/3}$. Taking the indicated square roots and then finding the inverse cosine of each side yields $.9553 \le x \le 2.1863$, or $|x - \pi/2| \le 0.6155$, as the approximate interval.

21. We have $(3y^2 - 6y)dy = (1 + 3x^2)dx$ so that $y^3 - 3y^2 = x + x^3 - 2$, once the I.C. are used. From the D.E., the integral curve will have a vertical tangent when $3y^2 - 6y = 0$, or $y = 0, 2$. For $y = 0$ we have $x^3 + x - 2 = 0$, which is satisfied for $x = 1$, which is the only zero of the function $w = x^3 + x - 2$. Likewise, for $y = 2$, $x = -1$.

23. Separating variables gives $y^{-2}dy = (2 + x)dx$, so

$-y^{-1} = 2x + \dfrac{x^2}{2} + c.$ $y(0) = 1$ yields $c = -1$ and thus

$y = \dfrac{-1}{\dfrac{x^2}{2} + 2x - 1} = \dfrac{2}{2 - 4x - x^2}.$ This gives

$\dfrac{dy}{dx} = \dfrac{8 + 4x}{(2-4x-x^2)^2}$, so the minimum value is attained at

$x = -2$. Note that the solution is defined for $-2 - \sqrt{6} < x < -2 + \sqrt{6}$ (by finding the zeros of the denominator) and has vertical asymptotes at the end points of the interval.

25. Separating variables and integrating yields $3y + y^2 = \sin 2x + c$. $y(0) = -1$ gives $c = -2$ so that $y^2 + 3y + (2-\sin 2x) = 0$.   The quadratic formula then gives $y = -\dfrac{3}{2} + \sqrt{\sin 2x + 1/4}$, which is defined for $-.126 < x < 1.697$ (found by solving $\sin 2x = -.25$ for x and noting $x = 0$ is the initial point).   Thus we have $\dfrac{dy}{dx} = \dfrac{\cos 2x}{(\sin 2x + \dfrac{1}{4})}$, which yields $x = \pi/4$ as the only critical point in the above interval. Using the second derivative test or graphing the solution clearly indicates the critical point is a maximum.

27a. By sketching the direction field and using the D.E. we note that $y' < 0$ for $y > 4$ and y' approaches zero as y approaches 4. For $0 < y < 4$, $y' > 0$ and again approaches zero as y approaches 4. Thus $\lim\limits_{t\to\infty} y = 4$ if $y_0 > 0$. For $y_0 < 0$, $y' < 0$ for all y and hence y becomes negatively unbounded ($-\infty$) as t increases. If $y_0 = 0$, then $y' = 0$ for all t, so $y = 0$ for all t.

27b. Separating variables and using a partial fraction expansion we have $(\dfrac{1}{y} - \dfrac{1}{y-4})dy = \dfrac{4}{3}t\,dt$.   Hence $\ln\left|\dfrac{y}{y-4}\right| = \dfrac{2}{3}t^2 + c_1$ and thus $\left|\dfrac{y}{y-4}\right| = e^{c_1}e^{2t^2/3} = ce^{2t^2/3}$, where c is positive. For $y_0 = .5$ this becomes $\dfrac{y}{4-y} = ce^{2t^2/3}$ and thus $c = \dfrac{.5}{3.5} = \dfrac{1}{7}$. Using this value for c and solving for y yields $y(t) = \dfrac{4}{1 + 7e^{-2t^2/3}}$.

Setting this equal to 3.98 and solving for t yields
t = 3.29527.

29. Separating variables yields $\dfrac{cy+d}{ay+b}\,dy = dx$. If $a \neq 0$ and

   $ay+b \neq 0$ then $dx = (\dfrac{c}{a} + \dfrac{ad-bc}{a(ay+b)})\,dy$. Integration then
   yields the desired answer.

30c. If $v = y/x$ then $y = vx$ and $\dfrac{dy}{dx} = v + x\dfrac{dv}{dx}$ and thus the

   D.E. becomes $v + x\dfrac{dv}{dx} = \dfrac{v-4}{1-v}$. Subtracting v from both

   sides yields $x\dfrac{dv}{dx} = \dfrac{v^2-4}{1-v}$.

30d. The last equation in (c) separates into $\dfrac{1-v}{v^2-4}\,dv = \dfrac{1}{x}\,dx$. To

   integrate the left side use partial fractions to write
   $\dfrac{1-v}{v-4} = \dfrac{A}{v-2} + \dfrac{B}{v+2}$, which yields A = -1/4 and B = -3/4.

   Integration then gives $-\dfrac{1}{4}\ln|v-2| - \dfrac{3}{4}\ln|v+2| = \ln|x| - k$, or

   $\ln|x^4||v-2||v+2|^3 = 4k$ after manipulations using
   properties of the ln function.

31a. Simplifying the right side of the D.E. gives
   $dy/dx = 1 + (y/x) + (y/x)^2$ so the equation is homogeneous.

31b. The substitution y = vx leads to
   $v + x\dfrac{dv}{dx} = 1 + v + v^2$ or $\dfrac{dv}{1 + v^2} = \dfrac{dx}{x}$. Solving, we get
   $\arctan v = \ln|x| + c$. Substituting for v we obtain
   $\arctan(y/x) - \ln|x| = c$.

33b. Dividing the numerator and denominator of the right side
   by x and substituting y = vx we get $v + x\dfrac{dv}{dx} = \dfrac{4v - 3}{2-v}$

   which can be rewritten as $x\dfrac{dv}{dx} = \dfrac{v^2 + 2v - 3}{2 - v}$. Note that
   v = -3 and v = 1 are solutions of this equation. For
   $v \neq 1, -3$ separating variables gives

$$\frac{2-v}{(v+3)(v-1)}\,dv = \frac{1}{x}dx.$$ Applying a partial fraction

decomposition to the left side we obtain

$$[\frac{1}{4}\frac{1}{v-1} - \frac{5}{4}\frac{1}{v+3}]dv = \frac{dx}{x},$$ and upon integrating both sides

we find that $\frac{1}{4}\ln|v-1| - \frac{5}{4}\ln|v+3| = \ln|x| + c.$

Substituting for v and performing some algebraic
manipulations we get the solution in the implicit form
$|y-x| = c|y+3x|^{5}.$ v = 1 and v = -3 yield y = x and
y = -3x, respectively, as solutions also.

35b. As in Prob.33, substituting y = vx into the D.E. we get

$$v + x\frac{dv}{dx} = \frac{1+3v}{1-v},$$ or $x\frac{dv}{dx} = \frac{(v+1)^{2}}{1-v}.$ Note that v = -1 (or

y = -x) satisfies this D.E. Separating variables yields

$$\frac{1-v}{(v+1)^{2}}dv = \frac{dx}{x}.$$ Integrating the left side by parts we

obtain $\frac{v-1}{v+1} - \ln|v+1| = \ln|x| + c.$ Letting $v = \frac{y}{x}$ then

yields $\frac{y-x}{y+x} - \ln|\frac{y+x}{x}| = \ln|x| + c,$ or $\frac{y-x}{y+x} - \ln|y+x| = c.$

This answer differs from that in the text. The answer in
the text can be obtained by integrating the left side,
above, using partial fractions. By differentiating both
answers, it can be verified that indeed both forms satisfy
the D.E.

Section 2.3, Page 57

1.   Note that q(0) = 200 gm, where q(t) represents the amount of
     dye present at any time.

2.   Let S(t) be the amount of salt that is present at any time
     t, then S(0) = 0 is the original amount of salt in the tank,
     2 is the amount of salt entering per minute, and 2(S/120)
     is the amount of salt leaving per minute (all amounts
     measured in grams). Thus
     dS/dt = 2 - 2S/120, S(0) = 0.

3.   We must first find the amount of salt that is present after
     10 minutes. For the first 10 minutes (if we let Q(t) be the
     amount of salt in the tank):
      = (2) - 2, Q(0) = 0. This I.V.P. has the solution: Q(10)
     = 50(1-)   9.063 lbs. of salt in the tank after the first 10

minutes. At this point no more salt is allowed to enter, so
the new I.V.P. (letting P(t) be the amount of salt in the
tank) is:
  = (0)(2) – 2, P(0) = Q(10) = 9.063. The solution of this
problem is P(t) = 9.063, which yields P(10) = 7.42 lbs.

4.. Salt flows into the tank at the rate of (1)(3) lb/min. and
it flows out of the tank at the rate of (2) lb/min. since
the volume of water in the tank at any time t is 200 +
(1)(t) gallons (due to the fact that water flows into the
tank faster than it flows out). Thus the I.V.P. is dQ/dt =
3 – Q(t), Q(0) = 100.

7a. Set = 0 in Eq.(16) (or solve Eq.(15) with S(0) = 0).

7b. Set r = .075, t = 40 and S(t) = $1,000,000 in the answer to
(a) and then solve for k.

7c. Set k = $2,000, t = 40 and S(t) = $1,000,000 in the answer
to (a) and then solve numerically for r.

9. The rate of accumulation due to interest is .1S and the rate
of decrease is k dollars per year and thus
dS/dt = .1S – k, S(0) = $8,000. Solving this for S(t)
yields S(t) = 8000 – 10k(–1). Setting S = 0 and substitution
of t = 3 gives k = $3,086.64 per year. For 3 years this
totals $9,259.92, so $1,259.92 has been paid in interest.

10. Since we are assuming continuity, either convert the monthly
payment into an annual payment or convert the yearly
interest rate into a monthly interest rate for 240 months.
Then proceed as in Prob. 9.

12a. Using Eq. (15) we have – S = 800(1) or
     S (800t), S(0) 100,000. Using an integrating factor and
integration by parts (or using a D.E. solver) we get S(t)
t c. Using the I.C. yields c . Substituting this value
into S, setting S(t) 0, and solving numerically for t
yields t 135.363 months.

16a. This problem can be done numerically using appropriate D.E.
solver software. Analytically we have
     (.1.2sint)dt by separating variables and thus
y(t) cexp(.1t.2cost). y(0) 1 gives c , so
y(t) exp(.2.1t.2cost). Setting y 2 yields
ln2 .2 .1 .2cos, which can be solved numerically to give
 2.9632. If y(0) , then as above,
y(t) exp(.2.1t.2cost). Thus if we set y 2 we get the

same numerical equation for and hence the doubling time has
not changed.

18. From Eq.(26) we have 19 = 7 + (20-7) or
    k = -ln = ln(13/12). Hence if (T) = 15 we get: 15 = 7 +
    13. Solving for T yields
    T = ln(8/13)/-ln(13/12) = ln(13/8)/ln(13/12) min.

19. Hint: let Q(t) be the quantity of carbon monoxide in the
    room at any time t. Then the concentration is given by x(t)
    = Q(t)/1200.

20a. The required I.V.P. is dQ/dt = kr + P - r,
    Q(0) = V. Since c = Q(t)/V, the I.V.P. may be rewritten
    Vc(t) = kr + P - rc, c(0) = , which has the solution c(t) =
    k + + (- k -).

20b. Set k = 0, P = 0, t = T and c(T) = .5 in the solution found
    in (a).

21a.If we measure x positively upward from the ground, then
    Eq.(4) of Section 1.1 becomes m = -mg, since there is no air
    resistance. Thus the I.V.P. for v(t) is
    dv/dt = -g, v(0) = 20. Hence
     = v(t) = 20 - gt and x(t) = 20t - (g/2) + c. Since x(0) =
    30, c = 30 and $x(t) = 20t - (g/2)t^2 + 30$. At the maximum
    height $v(t_m) = 0$ and thus
    t_m = 20/9.8 = 2.04 sec., which when substituted in the
    equation for x(t) yields the maximum height.

21b. At the ground $x(t_g)$ = 0 and thus $20t_g - 4.9t_g^2 + 30 = 0$.

22. The I.V.P. in this case is $m\frac{dv}{dt} = -\frac{1}{30}v - mg$, v(0) = 20,
    where the positive direction is measured upward.

24a. The I.V.P. is $m\frac{dv}{dt}$ = mg - .75v, v(0) = 0 and v is
    measured positively downward. Since m = 180/32, the D.E.
    becomes $\frac{dv}{dt} = 32 - \frac{2}{15}v$ and thus $v(t) = 240(1-e^{-2t/15})$ so
    that v(10) = 176.7 ft/sec.

24b. Integration of v(t) as found in (a) yields
    $x(t) = 240t + 1800(e^{-2t/15}-1)$ where x is measured
    positively down from the altitude of 5000 feet. Set

t = 10 to find the distance traveled when the parachute opens.

24c. After the parachute opens the I.V.P. is $m\dfrac{dv}{dt}$ = mg-12v,

v(0) = 176.7, which has the solution
v(t) = $161.7e^{-32t/15}$ + 15 and where t = 0 now represents the time the parachute opens. Letting t→∞ yields the limiting velocity.

24d. Integrate v(t) as found in (c) to find
x(t) = 15t - $75.8e^{-32t/15}$ + C_2. C_2 = 75.8 since x(0) = 0, x now being measured from the point where the parachute opens. Setting x = 3925.5 will then yield the length of time the skydiver is in the air after the parachute opens.

26a. Again, if x is measured positively upward, then Eq.(4) of Sect.1.1 becomes $m\dfrac{dv}{dt}$ = -mg - kv.

26b. From part (a) v(t) = $-\dfrac{mg}{k}$ + $[v_0 + \dfrac{mg}{k}]e^{-kt/m}$. As k → 0 this has the indeterminant form of -∞ + ∞. Thus rewrite v(t) as v(t) = $[-mg + (v_0 k + mg)e^{-kt/m}]/k$ which has the indeterminant form of 0/0, as k → 0 and hence L'Hopital's Rule may be applied with k as the variable.

27a. The equation of motion is m(dv/dt) = w-R-B which, in this problem, is
$\dfrac{4}{3}\pi a^3 \rho$(dv/dt) = $\dfrac{4}{3}\pi a^3 \rho g$ - 6πμav - $\dfrac{4}{3}\pi a^3 \rho' g$. The limiting velocity occurs when dv/dt = 0.

27b. Since the droplet is motionless, v = dv/dt = 0, we have the equation of motion 0 = $(\dfrac{4}{3})\pi a^3 \rho g$ - Ee - $(\dfrac{4}{3})\pi a^3 \rho' g$, where ρ is the density of the oil and ρ' is the density of air. Solving for e yields the answer.

28. All three parts can be answered from one solution if k represents the resistance and if the method of solution of Example 4 is used. Thus we have
$m\dfrac{dv}{dt}$ = $mv\dfrac{dv}{dx}$ = mg - kv, v(0) = 0, where we have assumed the velocity is a function of x. The solution of this

I.V.P. involves a logarithmic term, and thus the answers to parts (a) and (c) must be found using a numerical procedure.

29b. Note that 32 ft/sec^2 = 78,545 m/hr^2.

30. This problem is the same as Example 4 through Eq.(29).

In this case the I.C. is $v(\xi R) = v_o$, so $c = \dfrac{v_0^2}{2} - \dfrac{gR}{1+\xi}$.

The escape velocity, v_e, is found by noting that $v_o^2 \geq \dfrac{2gR}{1+\xi}$

in order for v^2 to always be positive. From Example 4, the escape velocity for a surface launch is $v_e(0) = \sqrt{2gR}$. We want the escape velocity of $x_o = \xi R$ to have the relation $v_e(\xi R) = .85v_e(0)$, which yields

$\xi = (0.85)^{-2} - 1 \cong 0.384$. If R = 4000 miles then $x_o = \xi R = 1536$ miles.

31b. From part a) $\dfrac{dx}{dt} = v = u\cos A$ and hence

$x(t) = (u\cos A)t + d_1$. Since $x(0) = 0$, we have $d_1 = 0$ and

$x(t) = (u\cos A)t$. Likewise $\dfrac{dy}{dt} = -gt + u\sin A$ and

therefore $y(t) = -gt^2/2 + (u\sin A)t + d_2$. Since $y(0) = h$

we have $d_2 = h$ and $y(t) = -gt^2/2 + (u\sin A)t + h$.

31d. Let t_w be the time the ball reaches the wall. Then

$x(t_w) = L = (u\cos A)t_w$ and thus $t_w = \dfrac{L}{u\cos A}$. For the ball

to clear the wall $y(t_w) \geq H$ and thus (setting

$t_w = \dfrac{L}{u\cos A}$, g = 32 and h = 3 in y) we get

$\dfrac{-16L^2}{u^2\cos^2 A} + L\tan A + 3 \geq H$.

31e. Setting L = 350 and H = 10 we get $\dfrac{-161.98}{\cos^2 A} + 350\dfrac{\sin A}{\cos A} \geq 7$

or $7\cos^2 A - 350\cos A\sin A + 161.98 \leq 0$. This can be solved numerically or by plotting the left side as a function of A and finding where the zero crossings are.

31f. Setting L = 350, and H = 10 in the answer to part d

yields $\dfrac{-16(350)^2}{u^2\cos^2 A}$ + 350tanA = 7, where we have chosen the
equality sign since we want to just clear the wall.

Solving for u^2 we get $u^2 = \dfrac{1,960,000}{175\sin 2A - 7\cos^2 A}$.   Now u will
have a minimum when the denominator has a maximum.   Thus
350cos2A + 7sin2A = 0, or tan2A = −50, which yields
A = .7954 rad. and u = 106.89 ft./sec.

Section 2.4, Page 72

1. If the equation is written in the form of Eq.(1), then
    p(t) = (lnt)/(t-3) and g(t) = 2t/(t-3). These are defined
    and continuous on the intervals (0,3) and (3,∞), but
    since the initial point is t = 1, the solution will be
    continuous on 0 < t < 3.

4. p(t) = 2t/(2-t)(2+t) and g(t) = $3t^2$/(2-t)(2+t).

8. Theorem 2.4.2 guarantees a unique solution to the D.E.
    through any point (t_0, y_0) such that $t_0^2 + y_0^2 < 1$ since
    $\dfrac{\partial f}{\partial y}$ = $-y(1-t^2-y^2)^{1/2}$ is defined and continuous only for
    $1-t^2-y^2 > 0$.   Note also that f = $(1-t^2-y^2)^{1/2}$ is defined
    and continuous in this region as well as on the boundary
    $t^2+y^2 = 1$.   The boundary can't be included in the final
    region due to the discontinuity of $\dfrac{\partial f}{\partial y}$ there.

11. In this case f = $\dfrac{1+t^2}{y(3-y)}$ and $\dfrac{\partial f}{\partial y}$ = $\dfrac{1+t^2}{y(3-y)^2}$ - $\dfrac{1+t^2}{y^2(3-y)}$,
    which are both continuous everywhere except for y = 0 and
    y = 3.

13. The D.E. may be written as ydy = -4tdt so that
    $\dfrac{y^2}{2}$ = $-2t^2+c$, or $y^2 = c-4t^2$.   The I.C. then yields
    y_0^2 = c, so that $y^2 = y_0^2 - 4t^2$ or y = $\pm\sqrt{y_0^2-4t^2}$, which is
    defined for $4t^2 < y_0^2$ or $|t| < |y_0|/2$.   Note that $y_0 \neq 0$
    since Theorem 2.4.2 does not hold there.

17. From the direction field and the given D.E. it is noted
    that for t > 0 and y < 0 that y′ < 0, so y → −∞ for
    y_0 < 0.   Likewise, for 0 < y_0 < 3, y′ > 0 and y′ → 0 as

$y \to 3$, so $y \to 3$ for $0 < y_0 < 3$ and for $y_0 > 3$, $y' < 0$ and again $y' \to 0$ as $y \to 3$, so $y \to 3$ for $y_0 > 3$. For $y_0 = 3$, $y' = 0$ and $y = 3$ for all t and for $y_0 = 0$, $y' = 0$ and $y = 0$ for all t.

22a. For $y_1 = 1-t$, $y_1' = -1 = \dfrac{-t+[t^2+4(1-t)]^{1/2}}{2}$

$$= \frac{-t+[(t-2)^2]^{1/2}}{2}$$

$$= \frac{-t+|t-2|}{2} = -1 \text{ if}$$

$(t-2) \geq 0$, by the definition of absolute value. Setting $t = 2$ in y_1 we get $y_1(2) = -1$, as required.

22b. By Theorem 2.4.2 we are guaranteed a unique solution only where $f(t,y) = \dfrac{-t+(t^2+4y)^{1/2}}{2}$ and $f_y(t,y) = (t^2+4y)^{-1/2}$ are continuous. In this case the initial point $(2,-1)$ lies in the region $t^2 + 4y \leq 0$, in which case $\dfrac{\partial f}{\partial y}$ is not continuous and hence the theorem is not applicable and there is no contradiction.

22c. If $y = y_2(t)$ then we must have $ct + c^2 = -t^2/4$, which is not possible since c must be a constant.

23a. To show that $\phi(t) = e^{2t}$ is a solution of the D.E., take its derivative and substitute into the D.E.

24. $[c\,\phi(t)]' + p(t)[c\phi(t)] = c[\phi'(t) + p(t)\phi(t)] = 0$ since $\phi(t)$ satisfies the given D.E.

25. $[y_1(t) + y_2(t)]' + p(t)[y_1(t) + y_2(t)] =$ $y_1'(t) + p(t)y_1(t) + y_2'(t) + p(t)y_2(t) = 0 + g(t)$.

27a. For $n = 0,1$, the D.E. is linear and Eqs.(3) and (4) apply.

27b. Let $v = y^{1-n}$ then $\dfrac{dv}{dt} = (1-n)y^{-n}\dfrac{dy}{dt}$ so $\dfrac{dy}{dt} = \dfrac{1}{1-n}y^n\dfrac{dv}{dt}$, which makes sense when $n \neq 0,1$. Substituting into the D.E. yields $\dfrac{y^n}{1-n}\dfrac{dv}{dt} + p(t)y = q(t)y^n$ or

$v' + (1-n)p(t)y^{1-n} = (1-n)q(t)$. Setting $v = y^{1-n}$ then yields a linear D.E. for v.

28. $n = 3$ so $v = y^{-2}$ and $\dfrac{dv}{dt} = -2y^{-3}\dfrac{dy}{dt}$ or $\dfrac{dy}{dt} = -\dfrac{1}{2}y^3\dfrac{dv}{dt}$.

Substituting this into the D.E. gives

$-\dfrac{1}{2}y^3\dfrac{dv}{dt} + \dfrac{2}{t}y = \dfrac{1}{t^2}y^3$. Simplifying and setting

$y^{-2} = v$ then gives the linear D.E.

$v' - \dfrac{4}{t}v = -\dfrac{2}{t^2}$, where $\mu(t) = \dfrac{1}{t^4}$ and

$v(t) = ct^4 + \dfrac{2}{5t} = \dfrac{2+5ct^5}{5t}$. Thus $y = \pm[5t/(2+5ct^5)]^{1/2}$.

29. $n = 2$ so $v = y^{-1}$ and $\dfrac{dv}{dt} = -y^2\dfrac{dv}{dt}$. Thus the D.E.

becomes $-y^2\dfrac{dv}{dt} - ry = -ky^2$ or $\dfrac{dv}{dt} + rv = k$. Hence

$\mu(t) = e^{rt}$ and $v = k/r + ce^{-rt}$. Setting $v = 1/y$ then yields the solution.

32. Since g(t) is continuous on the interval $0 \le t \le 1$ we may solve the I.V.P.

$y_1' + 2y_1 = 1$, $y_1(0) = 0$ on that interval to obtain

$y_1 = 1/2 - (1/2)e^{-2t}$, $0 \le t \le 1$. g(t) is also continuous for $1 < t$; and hence we may solve $y_2' + 2y_2 = 0$ to obtain

$y_2 = ce^{-2t}$, $1 < t$. The solution y of the original I.V.P. must be continuous (since its derivative must exist) and hence we need c in y_2 so that y_2 at 1 has the same value as y_1 at 1. Thus

$ce^{-2} = 1/2 - e^{-2}/2$ or $c = (1/2)(e^2-1)$ and we obtain

$$y = \begin{cases} 1/2 - (1/2)e^{-2t} & 0 \le t \le 1 \\ 1/2(e^2-1)e^{-2t} & 1 \le t \end{cases} \qquad \text{and}$$

$$y' = \begin{cases} e^{-2t} & 0 \le t \le 1 \\ (1-e^2)e^{-2t} & 1 < t. \end{cases}$$

Evaluating the two parts of y' at $t_0 = 1$ we see that they are different, and hence y' is not continuous at $t_0 = 1$.

Section 2.5, Page 84

Problems 1 through 13 follow the pattern illustrated in
Fig.2.5.3 and the discussion following Eq.(11).

3.   The critical points are found by setting $\frac{dy}{dt}$ equal to

zero. Thus y = 0,1,2 are the critical points. The graph
of y(y-1)(y-2) is positive for 0 < y < 1 and 2 < y and
negative for 1 < y < 2. Thus y(t) is increasing

($\frac{dy}{dt}$ > 0) for 0 < y < 1 and 2 < y and decreasing

($\frac{dy}{dt}$ < 0) for 1 < y < 2. Therefore 0 and 2 are unstable

critical points while 1 is an asymptotically stable
critical point.

6. $\frac{dy}{dt}$ is zero only when arctany is zero (ie, y = 0). $\frac{dy}{dt}$ > 0

for y < 0 and $\frac{dy}{dt}$ < 0 for y > 0. Thus y = 0 is an

asymptotically stable critical point.

7c.   Separate variables to get $\frac{dy}{(1-y)^2}$ = kt.   Integration

yields $\frac{1}{1-y}$ = kt + c, or y = 1 - $\frac{1}{kt + c}$ = $\frac{kt + c - 1}{kt + c}$.

Setting t = 0 and y(0) = y_0 yields y_0 = $\frac{c-1}{c}$ or

c = $\frac{1}{1-y_0}$.   Hence y(t) = $\frac{(1-y_0)kt + y_0}{(1-y_0)kt + 1}$. Note that for

y_0 < 1 y $\rightarrow$ $(1-y_0)k/(1-y_0)k$ = 1 as t $\rightarrow$ ∞. For y_0 > 1
notice that the denominator will have a zero for some
value of t, depending on the values chosen for y_0 and k.
Thus the solution has a discontinuity at that point.

9.   Setting $\frac{dy}{dt}$ = 0 we find y = 0, ± 1 are the critical

points. Since $\frac{dy}{dt}$ > 0 for y < -1 and y > 1 while $\frac{dy}{dt}$ < 0

for -1 < y < 1 we may conclude that y = -1 is
asymptotically stable, y = 0 is semistable, and y = 1 is
unstable.

11. $y = b^2/a^2$ and $y = 0$ are the only critical points. For
     $0 < y < b^2/a^2$, $\dfrac{dy}{dt} < 0$ and thus $y = 0$ is asymptotically
     stable. For $y > b^2/a^2$, $dy/dt > 0$ and thus
     $y = b^2/a^2$ is unstable.

14. If $f'(y_1) < 0$ then the slope of f is negative at y_1 and
     thus $f(y) > 0$ for $y < y_1$ and $f(y) < 0$ for $y > y_1$ since
     $f(y_1) = 0$. Hence y_1 is an asysmtotically stable critical
     point. A similar argument will yield the result for
     $f'(y_1) > 0$.

16b. By taking the derivative of $y \ln(K/y)$ it can be shown that
     the graph of $\dfrac{dy}{dt}$ vs y has a maximum point at $y = K/e$. Thus
     $\dfrac{dy}{dt}$ is positive and increasing for $0 < y < K/e$ and thus y(t)
     is concave up for that interval. Similarly $\dfrac{dy}{dt}$ is positive
     and decreasing for $K/e < y < K$ and thus y(t) is concave down
     for that interval.

16c. $\ln(K/y)$ is very large for small values of y and thus
     $(ry)\ln(K/y) > ry(1 - y/K)$ for small y. Since $\ln(K/y)$ and
     $(1 - y/K)$ are both strictly decreasing functions of y and
     since $\ln(K/y) = (1 - y/K)$ only for $y = K$, we may conclude
     that $\dfrac{dy}{dt} = (ry)\ln(K/y)$ is never less than
     $\dfrac{dy}{dt} = ry(1 - y/K)$.

17a. If $u = \ln(y/K)$ then $y = Ke^u$ and $\dfrac{dy}{dt} = Ke^u\dfrac{du}{dt}$ so that the
     D.E. becomes $du/dt = -ru$.

18a. The D.E. is $dV/dt = k - \alpha\pi r^2$. The volume of a cone of
     height L and radius r is given by $V = \pi r^2 L/3$ where
     $L = hr/a$ from symmetry. Solving for r yields the desired
     solution.

18b. Equilibrium is given by $k - \alpha\pi r^2 = 0$.

18c. The equilibrium height must be less than h.

20b. Use the results of Problem 14.

20d. Differentiate Y with respect to E.

21a. Set $\dfrac{dy}{dt} = 0$ and solve for y using the quadratic formula.

21b. Use the results of Problem 14.

21d. If h > rK/4 there are no critical points (see part a) and $\dfrac{dy}{dt} < 0$ for all t.

24a. If z = x/n then $dz/dt = \dfrac{1}{n}\dfrac{dx}{dt} - \dfrac{x}{n^2}\dfrac{dn}{dt}$. Use of Equations (i) and (ii) then gives the I.V.P. (iii).

24b. Separate variables to get $\dfrac{vdz}{z(1-vz)} = -\beta dt$. Using partial fractions this becomes $\dfrac{dz}{z} + \dfrac{dz}{1-z} = -\beta dt$. Integration and solving for z yields the answer.

24c. Find z(20).

26a. Plot dx/dt vs x and observe that x = p and x = q are critical points. Also note that dx/dt > 0 for x < min(p,q) and x > max(p,q) while dx/dt < 0 for x between min(p,q) and max(p,q). Thus x = min(p,q) is an asymptotically stable point while x = max(p,q) is unstable. To solve the D.E., separate variables and use partial fractions to obtain $\dfrac{1}{q-p}[\dfrac{dx}{q-x} - \dfrac{dx}{p-x}] = \alpha dt$. Integration and solving for x yields the solution.

26b. x = p is a semistable critical point and since $\dfrac{dx}{dt} > 0$, x(t) is an increasing function. Thus for x(0) = 0, x(t) approaches p as t → ∞. To solve the D.E., separate variables and integrate.

Section 2.6, Page 95

3. M(x,y) = $3x^2-2xy+2$ and N(x,y) = $6y^2-x^2+3$, so $M_y = -2x = N_x$ and thus the D.E. is exact. Integrating M(x,y) with

respect to x we get $\psi(x,y) = x^3 - x^2y + 2x + H(y)$.
Taking the partial derivative of this with respect to y
and setting it equal to $N(x,y)$ yields $-x^2+h'(y) =$
$6y^2-x^2+3$, so that $h'(y) = 6y^2 + 3$ and $h(y) = 2y^3 + 3y$.
Substitute this $h(y)$ into $\psi(x,y)$ and recall that the
equation which defines $y(x)$ implicitly is $\psi(x,y) = c$.
Thus $x^3 - x^2y + 2x + 2y^3 + 3y = c$ is the equation that
yields the solution.

5. Writing the equation in the form $M(x,y)dx + N(x,y)dy = 0$
    gives $M(x,y) = ax + by$ and $N(x,y) = bx + cy$. Now
    $M_y = b = N_x$ and the equation is exact. Integrating

    $M(x,y)$ with respect to x yields $\psi(x,y) = (a/2)x^2 + bxy +$
    $h(y)$. Differentiating ψ with respect to y (x constant)
    and setting $\psi_y(x,y) = N(x,y)$ we find that $h'(y) = cy$ and

    thus $h(y) = (c/2)y^2$. Hence the solution is given by
    $(a/2)x^2 + bxy + (c/2)y^2 = k$.

7. $M_y(x,y) = e^xcosy - 2sinx = N_x(x,y)$ and thus the D.E. is
    exact. Integrating $M(x,y)$ with respect to x gives
    $\psi(x,y) = e^xsiny + 2ycosx + h(y)$. Finding $\psi_y(x,y)$ from
    this and setting that equal to $N(x,y)$ yields $h'(y) = 0$
    and thus $h(y)$ is a constant. Hence an implicit solution
    of the D.E. is $e^xsiny + 2ycosx = c$. The solution $y = 0$
    is also valid since it satisfies the D.E. for all x.

9. If you try to find $\psi(x,y)$ by integrating $M(x,y)$ with
    respect to x you must integrate by parts. Instead find
    $\psi(x,y)$ by integrating $N(x,y)$ with respect to y to obtain
    $\psi(x,y) = e^{xy}cos2x - 3y + g(x)$. Now find $g(x)$ by
    differentiating $\psi(x,y)$ with respect to x and set that
    equal to $M(x,y)$, which yields $g'(x) = 2x$ or $g(x) = x^2$.

12. As long as $x^2 + y^2 \neq 0$, we can simplify the equation by
    multiplying both sides by $(x^2 + y^2)^{3/2}$. This gives the
    exact equation $xdx + ydy = 0$. The solution to this
    equation is given implicitly by $x^2 + y^2 = c$. If you
    apply Theorem 2.6.1 and its construction without the
    simplification, you get $(x^2 + y^2)^{-1/2} = C$ which can be
    written as $x^2 + y^2 = c$ under the same assumption required
    for the simplification.

14. $M_y = 1$ and $N_x = 1$, so the D.E. is exact. Integrating

M(x,y) with respect to x yields

$\psi(x,y) = 3x^3 + xy - x + h(y)$. Differentiating this with respect to y and setting $\psi_y(x,y) = N(x,y)$ yields

$h'(y) = -4y$ or $h(y) = -2y^2$. Thus the implicit solution is $3x^3 + xy - x - 2y^2 = c$. Setting $x = 1$ and $y = 0$ gives $c = 2$ so that $2y^2 - xy + (2+x-3x^3) = 0$ is the implicit solution satisfying the given I.C. Use the quadratic formula to find $y(x)$, where the negative square root is used in order to satisfy the I.C. The solution will be valid for $24x^3 + x^2 - 8x - 16 > 0$.

15. We want $M_y(x,y) = 2xy + bx^2$ to be equal to $N_x(x,y) = 3x^2 + 2xy$. Thus we must have $b = 3$. This gives $\psi(x,y) = \frac{1}{2}x^2y^2 + x^3y + h(y)$ and consequently $h'(y) = 0$. After multiplying through by 2, the solution is given implicitly by $x^2y^2 + 2x^3y = c$.

19. $M_y(x,y) = 3x^2y^2$ and $N_x(x,y) = 1 + y^2$ so the equation is not exact by Theorem 2.6.1. Multiplying by the integrating factor $\mu(x,y) = 1/xy^3$ we get

$x + \dfrac{(1+y^2)}{y^3}y' = 0$, which is an exact equation since $M_y = N_x = 0$ (it is also separable). In this case $\psi = \frac{1}{2}x^2 + h(y)$ and $h'(y) = y^{-3} + y^{-1}$ so that $x^2 - y^{-2} + 2\ln|y| = c$ gives the solution implicitly.

22. Multiplication of the given D.E. (which is not exact) by $\mu(x,y) = xe^x$ yields $(x^2 + 2x)e^x\sin y\, dx + x^2e^x\cos y\, dy$, which is exact since $M_y(x,y) = N_x(x,y) = (x^2+2x)e^x\cos y$. To solve this exact equation it's easiest to integrate $N(x,y) = x^2e^x\cos y$ with respect to y to get $\psi(x,y) = x^2e^x\sin y + g(x)$. Solving for $g(x)$ yields the implicit solution.

23. This problem is similar to the derivation leading up to Eq.(26). Assuming that μ depends only on y, we find from Eq.(25) that $\mu' = Q\mu$, where $Q = (N_x - M_y)/M$ must depend on y alone. Solving this last D.E. yields $\mu(y)$ as given. This method provides an alternative approach to Problems 27 through 30.

25. The equation is not exact so we must attempt to find an integrating factor. Since $\frac{1}{N}(M_y-N_x) = \frac{3x^2 + 2x + 3y^2 - 2x}{x^2 + y^2} = 3$ is a function of x alone there is an integrating factor depending only on x, as shown in Eq.(26). Then $d\mu/dx = 3\mu$, and the integrating factor is $\mu(x) = e^{3x}$. Hence the equation can be solved as in Example 4.

26. An integrating factor can be found which is a function of x only, yielding $\mu(x) = e^{-x}$. Alternatively, you might recognize that $y' - y = e^{2x} - 1$ is a linear first order equation which can be solved as in Section 2.1.

27. Using the results of Problem 23, it can be shown that $\mu(y) = y$ is an integrating factor. Thus multiplying the D.E. by y gives $ydx + (x - ysiny)dy = 0$, which can be identified as an exact equation. Alternatively, one can rewrite the last equation as $(ydx + xdy) - ysiny\,dy = 0$. The first term is $d(xy)$ and the last can be integrated by parts. Thus we have $xy + ycosy - siny = c$.

29. Multiplying by siny we obtain $e^x siny\,dx + e^x cosy\,dy + 2y\,dy = 0$, and the first two terms are just $d(e^x siny)$. Thus, $e^x siny + y^2 = c$.

31. Using the results of Problem 24, it can be shown that $\mu(xy) = xy$ is an integrating factor. Thus, multiplying by xy we have $(3x^2y + 6x)dx + (x^3 + 3y^2)dy = 0$, which can be identified as an exact equation. Alternatively, we can observe that the above equation can be written as $d(x^3y) + d(3x^2) + d(y^3) = 0$, so that $x^3y + 3x^2 + y^3 = c$.

Section 2.7, Page 103

1d. The exact solution to this I.V.P. is $y = \phi(t) = t + 2 - e^{-t}$.

3a. The Euler formula is $y_{n+1} = y_n + h(2y_n - t_n + 1/2)$ for $n = 0,1,2,3$ and with $t_0 = 0$ and $y_0 = 1$. Thus
$y_1 = y_0 + .1(2y_0 - t_0 + 1/2) = 1.25$,
$y_2 = 1.25 + .1[2(1.25) - (.1) + 1/2] = 1.54$,
$y_3 = 1.54 + .1[2(1.54) - (.2) + 1/2] = 1.878$, and
$y_4 = 1.878 + .1[2(1.878) - (.3) + 1/2] = 2.2736$.

3b.   Use the same formula as in Problem 3a, except now h = .05
      and n = 0,1...7. Notice that only results for n = 1,3,5
      and 7 are needed to compare with part a.

3c.   Again, use the same formula as above with h = .025 and
      n = 0,1...15. Notice that only results for n = 3,7,11 and
      15 are needed to compare with parts a and b.

3d. $y' = 1/2 - t + 2y$ is a first order linear D.E. Rewrite
      the equation in the form $y' - 2y = 1/2 - t$ and multiply
      both sides by the integrating factor e^{-2t} to obtain
      $(e^{-2t}y)' = (1/2 - t)e^{-2t}$. Integrating the right side by
      parts and multiplying by e^{2t} we obtain $y = ce^{-2t} + t/2$.
      The I.C. $y(0) = 1 \rightarrow c = 1$ and hence the solution of the
      I.V.P. is $y = \phi(x) = e^{2t} + t/2$. Thus $\phi(0.1) = 1.2714$,
      $\phi(0.2) = 1.59182$, $\phi(0.3) = 1.97212$, and $\phi(0.4) = 2.42554$.

4d.   The exact solution to this I.V.P. is
      $y = \phi(t) = (6\cos t + 3\sin t - 6e^{-2t})/5$.

6.   For y(0) > 0 the
      solutions appear to
      converge. For y(0)<0
      the solutions diverge.

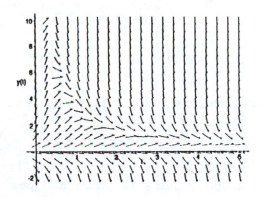

9.   All solutions seem
      to diverge.

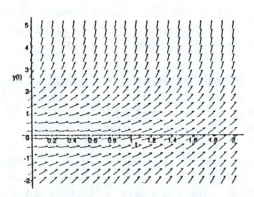

13a. The Euler formula is

$$y_{n+1} = y_n + h(\frac{4-t_n y_n}{1 + y_n^2}), \text{ where } t_0 = 0 \text{ and}$$

$y_0 = y(0) = -2.$ Thus, for $h = .1$, we get

$y_1 = -2 + .1(4/5) = -1.92$

$$y_2 = -1.92 + .1(\frac{4-.1(-1.92)}{1 + (1.92)^2}) = -1.83055$$

$$y_3 = -1.83055 + .1(\frac{4-.2(-1.83055)}{1+(1.83055)^2}) = -1.7302$$

$$y_4 = -1.7302 + .1(\frac{4-.3(-1.7302)}{1 + (1.7302)^2}) = -1.617043$$

$$y_5 = -1.617043 + .1(\frac{4 - .4(-1.617043)}{1 + (1.617043)^2}) = -1.488494.$$

Thus, $y(.5) \cong -1.488494.$

15a. The Euler formula is

$$y_{n+1} = y_n + .1 \ (\frac{3t_n^2}{3y_n^2 -4}), \text{ where } t_0 = 1 \text{ and } y_0 = 0. \text{ Thus}$$

$$y_1 = 0 + .1 \ (\frac{3}{-4}) = -.075 \text{ and}$$

$$y_2 = -.075 + .1 \ (\frac{3(1.1)^2}{3(.075)^2-4}) = .166134.$$

15c. There are two factors that explain the large differences. From the limit, the slope of y, y', becomes very "large" for values of y near -1.155. Also, the slope changes sign at $y = -1.155$. Thus for part a, $y(1.7) \cong y_7 = -1.178$, which is close to -1.155 and the slope y' here is large and positive, creating the large change in $y_8 \cong y(1.8)$. For part b, $y(1.65) \cong -1.125$, resulting in a large negative slope, which yields $y(1.70) \cong -3.133$. The slope at this point is now positive and the remainder of the solutions "grow" to -3.098 for the approcimation to $y(1.8)$.

16. For the four step sizes given, the approximte values for $y(.8)$ are 3.5078, 4.2013, 4.8004 and 5.3428. Thus, since these changes are still rather "large", it is hard to give an estimate other than $y(.8)$ is at least 5.3428. By using $h = .005$, .0025 and .001, we find further approximate values of $y(.8)$ to be 5.576, 5.707 and 5.790. Thus a better estimate now is for $y(.8)$ to be between 5.8 and 6. No reliable estimate is obtainable for $y(1)$, which is consistent with the direction field of Prob.9.

18. It is helpful, in understanding this problem, to also calculate $y'(t_n) = y_n(.1 y_n^2 - t_n)$. For $\alpha = 2.38$ this term remains positive and grows very large for $t_n > 2$. On the other hand, for $\alpha = 2.37$ this term decreases and eventually becomes negative for $t_n \cong 1.6$ (for $h = .01$). For $\alpha = 2.37$ and $h = .1$, .05 and .01, $y(2.00)$ has the approximations of 4.48, 4.01 and 3.50 respectively. A small step size must be used, due to the sensitivety of the slope field, given by $y_n (.1y_n^2 - t_n)$.

22. Using Eq.(8) we have $y_{n+1} = y_n + h(2y_n - 1) = (1+2h)y_n - h$. Setting $n + 1 = k$ (and hence $n = k-1$) this becomes $y_k = (1 + 2h)y_{k-1} - h$, for $k = 1,2,\dots$. Since $y_0 = 1$, we have $y_1 = 1 + 2h - h = 1 + h = (1 + 2h)/2 + 1/2$, and hence $y_2 = (1 + 2h)y_1 - h = (1 + 2h)^2/2 + (1 + 2h)/2 - h$
$$= (1 + 2h)^2/2 + 1/2;$$
$y_3 = (1 + 2h)y_2 - h = (1 + 2h)^3/2 + (1 + 2h)/2 - h$
$$= (1 + 2h)^3/2 + 1/2.$$ Continuing in this fashion (or using induction) we obtain $y_k = (1 + 2h)^k/2 + 1/2$. For fixed $x > 0$ choose $h = x/k$. Then substitute for h in the last formula to obtain $y_k = (1 + 2x/k)^k/2 + 1/2$. Letting $k \to \infty$ we find (See hint for Problem 20d.) $y(x) = y_k \to e^{2x}/2 + 1/2$, which is the exact solution.

Section 2.8, Page 113

1. Let $s = t-1$ and $w(s) = y(t(s)) - 2$, then when $t = 1$ and $y = 2$ we have $s = 0$ and $w(0) = 0$. Also,
$$\frac{dw}{ds} = \frac{dw}{dt} \cdot \frac{dt}{ds} = \frac{d}{dt}(y-2)\frac{dt}{ds} = \frac{dy}{dt} \text{ and hence}$$
$$\frac{dw}{ds} = (s+1)^2 + (w+2)^2,$$ upon substitution into the given D.E.

4a. Following Ex. 1 of the text, from Eq.(7) we have
$$\phi_{n+1}(t) = \int_0^t f(s,\phi(s))ds,$$ where $f(t,\phi) = -1 - \phi$ for this problem. Thus if $\phi_0(t) = 0$, then $\phi_1(t) = -\int_0^t ds = -t;$
$$\phi_2(t) = -\int_0^t (1-s)ds = -t + \frac{t^2}{2};$$
$$\phi_3(t) = -\int_0^t (1-s + \frac{s^2}{2})ds = -t + \frac{t^2}{2} - \frac{t^3}{2\cdot3};$$

$$\phi_4(t) = -\int_0^t (1 - s + \frac{s^2}{2} - \frac{s^3}{3!})ds = -t + \frac{t^2}{2} - \frac{t^3}{3!} + \frac{t^4}{4!}.$$

Based upon these we hypothesize that: $\phi_n(t) = \sum_{k=1}^{n} \frac{(-1)^k t^k}{k!}$

and use mathematical induction to verify this form for $\phi_n(t)$. Using Eq.(7) again we have:

$$\phi_{n+1}(t) = -\int_0^t [1 + \phi_n(s)] ds = -t - \sum_{k=1}^{n} \frac{(-1)^k t^{k+1}}{(k+1)!}$$

$$= \sum_{k=0}^{n} \frac{(-1)^{k+1} t^{k+1}}{(k+1)!} = \sum_{i=1}^{n+1} \frac{(-1)^i t^i}{i!}, \text{ where } i = k+1. \text{ Since this}$$

is the same form for $\phi_{n+1}(t)$ as derived from $\phi_n(t)$ above, we have verified by mathematical induction that $\phi_n(t)$ is as given.

4c. From part a, let $\phi(t) = \lim_{n \to \infty} \phi_n(t) = \sum_{k=1}^{\infty} \frac{(-1)^k t^k}{k!}$

$$= -t + \frac{t^2}{2} - \frac{t^3}{3!} + \ldots .$$

Since this is a power series, recall from calculus that:

$$e^{at} = \sum_{k=0}^{\infty} \frac{a^k t^k}{k!} = 1 + at + \frac{a^2 t^2}{2} + \frac{a^3 t^3}{3!} + \ldots . \text{ If we let}$$

$a = -1$, then we have $e^{-t} = 1 - t + \frac{t^2}{2} - \frac{t^3}{3!} + \ldots = 1 + \phi(t)$.

Hence $\phi(t) = e^{-t} - 1$.

7. As in Prob.4,

$$\phi_1(t) = \int_0^t (s\phi_0(s) + 1)ds = s \Big|_0^t = t$$

$$\phi_2(t) = \int_0^t (s^2 + 1)ds = (\frac{s^3}{3} + s) \Big|_0^t = t + \frac{t^3}{3}$$

$$\phi_3(t) = \int_0^t (s^2 + \frac{s^4}{3} + 1)ds = (\frac{s^3}{3} + \frac{s^5}{3 \cdot 5} + s) \Big|_0^t = t + \frac{t^3}{3} + \frac{t^5}{3 \cdot 5}.$$

Based upon these we hypothesize that:

$$\phi_n(t) = \sum_{k=1}^{n} \frac{t^{2k-1}}{1 \cdot 3 \cdot 5 \cdots (2k-1)} \quad \text{and use mathematical induction}$$

to verify this form for $\phi_n(t)$. Using Eq.(7) again we have:

$$\phi_{n+1}(t) = \int_0^t \left(\sum_{k=1}^{n} \frac{s^{2k}}{1\cdot3\cdot5\cdots(2k-1)} + 1 \right) ds$$

$$= \sum_{k=1}^{n} \frac{t^{2k+1}}{1\cdot3\cdot5\cdots(2k+1)} + t$$

$$= \sum_{k=0}^{n} \frac{t^{2k+1}}{1\cdot3\cdot5\cdots(2k+1)}$$

$$= \sum_{i=1}^{n+1} \frac{t^{2i-1}}{1\cdot3\cdot5\cdots(2i-1)}, \text{ where } i = k+1. \text{ Since this is}$$

the same form for $\phi_{n+1}(t)$ as derived from $\phi_n(t)$ above, we have verified by mathematical induction that $\phi_n(t)$ is as given.

11. Recall that $\sin x = x - \dfrac{x^3}{3!} + \dfrac{x^5}{5!} + O(x^7)$. Thus, for

$$\phi_2(t) = t - \frac{t^2}{2!} + \frac{t^4}{4!} - \frac{t^6}{6!} + O(t^7) \text{ we have}$$

$$\sin[\phi_2(t)] = \left(t - \frac{t^2}{2!} + \frac{t^4}{4!} - \frac{t^6}{6!} \right) - \frac{(t - \frac{t^2}{2!})^3}{3!} + \frac{t^5}{5!} + O(t^7).$$

Section 2.9, Page 124

2. Using the given difference equation we have for n=0, $y_1 = y_0/2$; for n=1, $y_2 = 2y_1/3 = y_0/3$; and for n=2, $y_3 = 3y_2/4 = y_0/4$. Thus we guess that $y_n = y_0/(n+1)$, and the given equation then gives $y_{n+1} = \dfrac{n+1}{n+2} y_n = y_0/(n+2)$, which, by mathematical induction, verifies $y_n = y_0/(n+1)$ as the solution for all n.

5. From the given equation we have $y_1 = .5y_0+6$.

$$y_2 = .5y_1 + 6 = (.5)^2 y_0 + 6\left(1 + \frac{1}{2}\right) \text{ and}$$

$$y_3 = .5y_2 + 6 = (.5)^3 y_0 + 6\left(1 + \frac{1}{2} + \frac{1}{4}\right). \text{ In general, then}$$

$$y_n = (.5)^n y_0 + 6(1 + \frac{1}{2} + \cdots + \frac{1}{2^{n-1}})$$

$$= (.5)^n y_0 + 6(\frac{1 - (1/2)^n}{1 - 1/2})$$

$$= (.5)^n y_0 + 12 - (.5)^n 12$$

$$= (.5)^n (y_0 - 12) + 12.$$ Mathematical induction can now be used to prove that this is the correct solution.

10. The governing equation is $y_{n+1} = \rho y_n - b$, which has the

solution $y_n = \rho^n y_0 - \dfrac{1-\rho^n}{1-\rho} b$ (Eq.(14) with a negative b).

Setting $y_{360} = 0$ and solving for b we obtain

$b = \dfrac{(1-\rho)\rho^{360} y_0}{1-\rho^{360}}$, where $\rho = 1.0075$ for part a.

13. You must solve Eq.(14) numerically for ρ when n = 240, $y_{240} = 0$, b = -\$900 and y_0 = \$95,000.

14. Substituting $u_n = \dfrac{\rho-1}{\rho} + v_n$ into Eq.(21) we get

$$\frac{\rho-1}{\rho} + v_{n+1} = \rho(\frac{\rho-1}{\rho} + v_n)(1 - \frac{\rho-1}{\rho} - v_n) \text{ or}$$

$$v_{n+1} = -\frac{\rho-1}{\rho} + (\rho-1 + \rho v_n)(\frac{1}{\rho} - v_n)$$

$$= \frac{1-\rho}{\rho} + \frac{\rho-1}{\rho} - (\rho-1)v_n + v_n - \rho v_n^2 = (2-\rho)v_n - \rho v_n^2.$$

15a. For $u_0 = .2$ we have $u_1 = 3.2u_0(1-u_0) = .512$ and $u_2 = 3.2u_1(1-u_1) = .7995392$. Likewise $u_3 = .51288406$, $u_4 = .7994688$, $u_5 = .51301899$, $u_6 = .7994576$ and $u_7 = .5130404$. Continuing in this fashion, $u_{14} = u_{16} = .79945549$ and $u_{15} = u_{17} = .51304451$.

17. For both parts of this problem a computer spreadsheet was used and an initial value of u_0 = .2 was chosen.
Different initial values or different computer programs may need a slightly different number of iterations to reach the limiting value.

17a. The limiting value of .65517 (to 5 decimal places) is reached after approximately 100 iterations for ρ = 2.9. The limiting value of .66102 (to 5 decimal places) is reached after approximately 200 iterations for ρ = 2.95.

The limiting value of .66555 (to 5 decimal places) is reached after approximately 910 iterations for ρ = 2.99.

17b. The solution oscillates between .63285 and .69938 after approximately 400 iterations for ρ = 3.01. The solution oscillates between .59016 and .73770 after approximately 130 iterations for ρ = 3.05. The solution oscillates between .55801 and .76457 after approximately 30 iterations for ρ = 3.1. For each of these cases additional iterations verified the oscillations were correct to five decimal places.

18. For an initial value of .2 and ρ = 3.448 we have the solution oscillating between .4403086 and .8497146. After approximately 3570 iterations the eighth decimal place is still not fixed, though. For the same initial value and ρ = 3.45 the solution oscillates between the four values: .43399155, .84746795, .44596778 and .85242779 after 3700 iterations.. For ρ = 3.449, the solution is still varying in the fourth decimal place after 3570 iterations, but there appear to be four values.

Miscellaneous Problems, Page 126

Before trying to find the solution of a D.E. it is necessary to know its type. The student should first classify the D.E. before reading this section, which indentifies the type of each equation in Problems 1 through 32.

1. Linear

2. Homogeneous

3. Exact

4. Linear equation in x(y)

5. Exact

6. Linear

7. Letting $u = x^2$ yields $\dfrac{dy}{dx} = 2x\dfrac{dy}{du}$ and thus

$\dfrac{du}{dy} - 2yu = 2y^3$ which is linear in u(y).

8. Linear

9. Exact

10. Integrating factor depends on x only

11. Exact

12. Linear   13. Homogeneous

14. Exact or homogeneous   15. Separable

16. Homogeneous   17. Linear

18. Linear or homogeneous   19. Integrating factor
                                         depends on x only

20. Separable   21. Homogeneous

22. Separable   23. Bernoulli equation

24. Separable   25. Exact

26. Integrating factor   27. Integrating factor
      depends on x only   depends on x only

28. Exact   29. Homogeneous

30. Linear equation in x(y) 31. Separable

32. Integrating factor depends on y only.

CHAPTER 3

3. Assume $y = e^{rt}$, which gives $y' = re^{rt}$ and $y'' = r^2 e^{rt}$.
    Substitution into the D.E. yields $(6r^2 - r - 1)e^{rt} = 0$.
    Since $e^{rt} \neq 0$, we have the characteristic equation
    $6r^2 - r - 1 = 0$, or $(3r+1)(2r-1) = 0$. Thus
    $r = -1/3,\ 1/2$ and $y = c_1 e^{t/2} + c_2 e^{-t/3}$.

5. The characteristic equation is $r^2 + 5r = 0$, so the roots
    are $r_1 = 0$, and $r_2 = -5$. Thus
    $$y = c_1 e^{0t} + c_2 e^{-5t} = c_1 + c_2 e^{-5t}.$$

7. The characteristic equation is $r^2 - 9r + 9 = 0$ so that
    $r = (9 \pm \sqrt{81-36})/2 = (9 \pm 3\sqrt{5})/2$ using the quadratic
    formula. Hence
    $$y = c_1 \exp[(9+3\sqrt{5})t/2] + c_2 \exp[(9-3\sqrt{5})t/2].$$

10. Substituting $y = e^{rt}$ in the D.E.
    we obtain the characteristic
    equation $r^2 + 4r + 3 = 0$, which
    has the roots $r_1 = -1$, $r_2 = -3$.
    Thus $y = c_1 e^{-t} + c_2 e^{-3t}$ and
    $y' = -c_1 e^{-t} - 3c_2 e^{-3t}$.
    Substituting $t = 0$ we then have
    $c_1 + c_2 = 2$ and $-c_1 - 3c_2 = -1$,
    yielding $c_1 = 5/2$ and
    $c_2 = -1/2$. Thus $y = \dfrac{5}{2}e^{-t} - \dfrac{1}{2}e^{-3t}$
    and hence $y \to 0$ as $t \to \infty$.

15. The characteristic equation is $r^2 + 8r - 9 = 0$, so that
    $r_1 = 1$ and $r_2 = -9$ and the general solution is
    $y = c_1 e^{t} + c_2 e^{-9t}$. Since the I.C. are given at $t = 1$, it
    is convenient to write the general solution in the form
    $y = k_1 e^{(t-1)} + k_2 e^{-9(t-1)}$. Note that
    $c_1 = k_1 e^{-1}$ and $c_2 = k_2 e^{9}$. The advantage of the latter
    form of the general solution becomes clear when we apply

the I.C. $y(1) = 1$ and $y'(1) = 0$. This latter form of y
gives $y' = k_1 e^{(t-1)} - 9k_2 e^{-9(t-1)}$ and thus setting $t = 1$ in
y and y' yields the equations $k_1 + k_2 = 1$ and
$k_1 - 9k_2 = 0$. Solving for k_1 and k_2 we find that
$y = (9e^{(t-1)} + e^{-9(t-1)})/10$. Since $e^{(t-1)}$ has a positive
exponent for $t > 1$, $y \to \infty$ as $t \to \infty$.

17. Comparing the given solution to Eq(17), we see that $r_1 = 2$
and $r_2 = -3$ are the two roots of the characteristic

equation. Thus we have $(r-2)(r+3) = 0$, or $r^2 + r - 6 = 0$
as the characteristic equation. Hence the given solution
is for the D.E. $y'' + y' - 6y = 0$.

19. The roots of the characteristic equation are $r = 1, -1$
and thus the general solution is $y(t) = c_1 e^t + c_2 e^{-t}$.

$y(0) = c_1 + c_2 = \dfrac{5}{4}$ and $y'(0) = c_1 - c_2 = -\dfrac{3}{4}$, yielding

$y(t) = \dfrac{1}{4}e^t + e^{-t}$. From this $y'(t) = \dfrac{1}{4}e^t - e^{-t} = 0$ or

$e^{2t} = 4$ or $t = \ln 2$. The second derivative test or a
graph of the solution indicates this is a minimum point.

21. The general solution is $y = c_1 e^{-t} + c_2 e^{2t}$. Using the I.C.
we obtain $c_1 + c_2 = \alpha$ and $-c_1 + 2c_2 = 2$, so adding the two
equations we find $3c_2 = \alpha + 2$. If y is to approach zero
as $t \to \infty$, c_2 must be zero. Thus $\alpha = -2$.

24. The roots of the characteristic equation are given by
$r = -2$, $\alpha - 1$ and thus $y(t) = c_1 e^{-2t} + c_2 e^{(\alpha-1)t}$. Hence,
for $\alpha < 1$, all solutions tend to zero as $t \to \infty$. For
$\alpha > 1$, the second term becomes unbounded, but not the
first, so there are no values of α for which all
solutions become unbounded.

25a. The characteristic equation is $2r^2 + 3r - 2 = 0$, so
$r_1 = -2$ and $r_2 = 1/2$ and $y = c_1 e^{-2t} + c_2 e^{t/2}$. The I.C.

yield $c_1 + c_2 = 2$ and $-2c_1 + \dfrac{1}{2}c_2 = -\beta$ so that

$c_1 = (1 + 2\beta)/5$ and $c_2 = (4-2\beta)/5$.

25c. From part (a), if $\beta = 2$ then $y(t) = e^{-2t}$ and the solution
    simply decays to zero. For $\beta > 2$, the solution becomes
    unbounded negatively, and again there is no minimum
    point.

27. The second solution must decay faster than e^{-t}, so choose
    e^{-2t} or e^{-3t} etc. as the second solution. Then proceed as
    in Problem 17.

28. Let $v = y'$, then $v' = y''$ and thus the D.E. becomes
    $t^2 v' + 2tv - 1 = 0$ or $t^2 v' + 2tv = 1$. The left side is
    recognized as $(t^2 v)'$ and thus we may integrate to obtain
    $t^2 v = t + c$ (otherwise, divide both sides of the D.E. by
    t^2 and find the integrating factor, which is just t^2 in
    this case). Solving for $v = dy/dt$ we find
    $dy/dt = 1/t + c/t^2$ so that $y = \ln t + c_1/t + c_2$.

30. Set $v = y'$, then $v' = y''$ and thus the D.E. becomes
    $v' + tv^2 = 0$. This equation is separable and has the
    solution $-v^{-1} + t^2/2 = c$ or $v = y' = -2/(c_1 - t^2)$ where
    $c_1 = 2c$. We must consider separately the cases $c_1 = 0$,
    $c_1 > 0$ and $c_1 < 0$. If $c_1 = 0$, then $y' = 2/t^2$ or
    $y = -2/t + c_2$. If $c_1 > 0$, let $c_1 = k^2$. Then
    $y' = -2/(k^2 - t^2) = -(1/k)[1/(k-t) + 1/(k+t)]$, so that
    $y = (1/k)\ln|(k-t)/(k+t)| + c_2$. If $c_1 < 0$, let $c_1 = -k^2$.
    Then $y' = 2/(k^2 + t^2)$ so that $y = (2/k)\tan^{-1}(t/k) + c_2$.
    Finally, we note that $y =$ constant is also a solution of
    the D.E.

34. Following the procedure outlined, let $v = dy/dt$ and
    $y'' = dv/dt = v\, dv/dy$. Thus the D.E. becomes
    $yv\, dv/dy + v^2 = 0$, which is a separable equation with the
    solution $v = c_1/y$. Next let $v = dy/dt = c/y$, which again
    separates to give the solution $y^2 = c_1 t + c_2$.

37. Again let $v = y'$ and $v' = v\, dv/dy$ to obtain
    $2y^2 v\, dv/dy + 2yv^2 = 1$. This is an exact equation with
    solution $v = \pm y^{-1}(y + c_1)^{1/2}$. To solve this equation,

we write it in the form $\pm\, ydy/(y+c_1)^{1/2} = dt$.   On
observing that the left side of the equation can be
written as $\pm[(y+c_1) - c_1]dy/(y+c_1)^{1/2}$ we integrate and
find $\pm\, (2/3)(y-2c_1)(y+c_1)^{1/2} = t + c_2$.

39.   If $v = y'$, then $v' = vdv/dy$ and the D.E. becomes
$v\, dv/dy + v^2 = 2e^{-y}$.   Dividing by v we obtain
$dv/dy + v = 2v^{-1}e^{-y}$, which is a Bernoulli equation (see
Prob.27, Section 2.4).   Let $w(y) = v^2$, then $dw/dy = 2v$
dv/dy and the D.E. then becomes $dw/dy + 2w = 4e^{-y}$, which
is linear in w.   Its solution is $w = v^2 = ce^{-2y} + 4e^{-y}$.
Setting $v = dy/dt$, we obtain a separable equation in y
and t, which is solved to yield the solution.

40.   Since both t and y are missing, either approach used
above will work.   In this case it's easier to use the
approach of Problems 28-33, so let $v = y'$ and thus $v' = y''$
and the D.E. becomes $vdv/dt = 2$.

43.   The variable y is missing.   Let $v = y'$, then $v' = y''$ and
the D.E. becomes $vv' - t = 0$.   The solution of the
separable equation is $v^2 = t^2 + c_1$.   Substituting $v = y'$
and applying the I.C. $y'(1) = 1$, we obtain $y' = t$.   The
positive square root was chosen because $y' > 0$ at $t = 1$.
Solving this last equation and applying the I.C. $y(1) = 2$,
we obtain $y = t^2/2 + 3/2$.

Section 3.2, Page 145

2. $W(\cos t, \sin t) = \begin{vmatrix} \cos t & \sin t \\ -\sin t & \cos t \end{vmatrix} = \cos^2 t + \sin^2 t = 1$.

4. $W(x, xe^x) = \begin{vmatrix} x & xe^x \\ 1 & e^x + xe^x \end{vmatrix} = xe^x + x^2e^x - xe^x = x^2e^x$.

8.   Dividing by $(t-1)$ we have $p(t) = -3t/(t-1)$, $q(t) = 4/(t-1)$
and $g(t) = \sin t/(t-1)$, so the only point of discontinuity is
$t = 1$.   By Theorem 3.2.1, the largest interval is $-\infty < t < 1$,
since the initial point is $t_0 = -2$.

12. $p(x) = 1/(x-2)$ and $q(x) = \tan x$, so $x = \pi/2, 2, 3\pi/2...$ are points of discontinuity. Since $t_0 = 3$, the interval specified by Theorem 3.2.1 is $2 < x < 3\pi/2$.

14. For $y = t^{1/2}$, $y' = \frac{1}{2}t^{-1/2}$ and $y'' = -\frac{1}{4}t^{-3/2}$. Thus

$yy'' + (y')^2 = -\frac{1}{4}t^{-1} + \frac{1}{4}t^{-1} = 0$. Similarly $y = 1$ is also

a solution. If $y = c_1(1) + c_2t^{1/2}$ is substituted in the

D.E. you will get $-c_1c_2/4t^{3/2}$, which is zero only if

$c_1 = 0$ or $c_2 = 0$. Thus the linear combination of two

solutions is not, in general, a solution. Theorem 3.2.2

is not contradicted however, since the D.E. is not

linear.

15. $y = \phi(t)$ is a solution of the D.E. so $L[\phi](t) = g(t)$.

Since L is a linear operator,

$L[c\phi](t) = cL[\phi](t) = cg(t)$. But, since $g(t) \neq 0$,

$cg(t) = g(t)$ if and only if $c = 1$. This is not a

contradiction of Theorem 3.2.2 since the linear D.E. is

not homogeneous.

18. $W(f,g) = \begin{vmatrix} t & g \\ 1 & g' \end{vmatrix} = tg' - g = t^2e^t$, or $g' - \frac{1}{t}g = te^t$. This

has an integrating factor of $\frac{1}{t}$ and thus $\frac{1}{t}g' - \frac{1}{t^2}g = e^t$

or $(\frac{1}{t}g)' = e^t$. Integrating and multiplying by t we

obtain $g(t) = te^t + ct$.

21. From Section 3.1, e^t and e^{-2t} are two solutions, and

since $W(e^t, e^{-2t}) \neq 0$ they form a fundamental set of

solutions. To find the fundamental set specified by

Theorem 3.2.5, let $y(t) = c_1e^t + c_2e^{-2t}$, where c_1 and c_2

satisfy

$c_1 + c_2 = 1$ and $c_1 - 2c_2 = 0$ for y_1. Solving, we find

$y_1 = \frac{2}{3}e^t + \frac{1}{3}e^{-2t}$. Likewise, c_1 and c_2 satisfy

$c_1 + c_2 = 0$ and $c_1 - 2c_2 = 1$ for y_2, so that

$y_2 = \frac{1}{3}e^t - \frac{1}{3}e^{-2t}$.

25. For $y_1 = x$, we have $x^2(0) - x(x+2)(1) + (x+2)(x) = 0$ and
    for $y_2 = xe^x$ we have $x^2(x+2)e^x - x(x+2)(x+1)e^x + (x+2)xe^x = 0$.
    From Problem 4, $W(x, xe^x) = x^2 e^x \neq 0$ for $x > 0$, so y_1 and
    y_2 form a fundamental set of solutions.

27. Suppose that
    $P(x)y'' + Q(x)y' + R(x)y = [P(x)y']' + [f(x)y]'$. On
    expanding the right side and equating coefficients, we
    find $f'(x) = R(x)$ and $P'(x) + f(x) = Q(x)$. These two
    conditions on f can be satisfied if
    $R(x) = Q'(x) - P''(x)$ which gives the necessary condition
    $P''(x) - Q'(x) + R(x) = 0$.

30. We have $P(x) = x$, $Q(x) = -\cos x$, and $R(x) = \sin x$ and the
    condition for exactness is satisfied. Also, from Problem
    27, $f(x) = Q(x) - P'(x) = -\cos x - 1$, so the D.E. becomes
    $(xy')' - [(1 + \cos x)y]' = 0$. Hence
    $xy' - (1 + \cos x)y = c_1$. This is a first order linear
    D.E. and the integrating factor (after dividing by x) is
    $\mu(x) = \exp[-\int x^{-1}(1 + \cos x)dx]$. The general solution is
    $y = [\mu(x)]^{-1}[c_1 \int_{x_0}^{x} t^{-1} \mu(t)dt + c_2]$.

32. We want to choose $\mu(x)$ and $f(x)$ so that $\mu(x)P(x)y'' + \mu(x)Q(x)y' + \mu(x)R(x)y = [\mu(x)P(x)y']' + [f(x)y]'$.
    Expand the right side and equate coefficients of y'', y'
    and y. This gives $\mu'(x)P(x) + \mu(x)P'(x) + f(x) = \mu(x)Q(x)$ and $f'(x) = \mu(x)R(x)$. Differentiate the first
    equation and then eliminate $f'(x)$ to obtain the adjoint
    equation $P\mu'' + (2P' - Q)\mu' + (P'' - Q' + R)\mu = 0$.

34. $P = 1 - x^2$, $Q = -2x$ and $R = \alpha(\alpha+1)$. Thus
    $2P' - Q = -4x + 2x = -2x = Q$ and
    $P'' - Q' + R = -2 + 2 + \alpha(\alpha+1) = \alpha(\alpha+1) = R$.

36. Write the adjoint D.E. given in Problem 32 as
    $\hat{P}\mu'' + \hat{Q}\mu' + \hat{R}\mu = 0$ where $\hat{P} = P$, $\hat{Q} = 2P' - Q$, and
    $\hat{R} = P'' - Q' + R$. The adjoint of this equation, namely
    the adjoint of the adjoint, is
    $\hat{P}y'' + (2\hat{P}' - \hat{Q})y' + (\hat{P}'' - \hat{Q}' + \hat{R})y = 0$. After
    substituting for $\hat{P}$, $\hat{Q}$, and $\hat{R}$ and simplifying, we obtain
    $Py'' + Qy' + Ry = 0$. This is the same as the original
    equation.

37. From Problem 32 the adjoint of $Py'' + Qy' + Ry = 0$ is
     $P\mu'' + (2P' - Q)\mu' + (P'' - Q' + R)\mu = 0$. The two equations
     are the same if $2P' - Q = Q$ and $P'' - Q' + R = R$.   This
     will be true if $P' = Q$.   Hence the original D.E. is self-
     adjoint if $P' = Q$.   For Problem 33, $P(x) = x^2$ so
     $P'(x) = 2x$ and $Q(x) = x$.   Hence the Bessel equation of
     order ν is not self-adjoint.   In a similar manner we find
     that Problems 34 and 35 are self-adjoint.

Section 3.3, Page 152

2.   Since $\cos3\theta = 4\cos^3\theta - 3\cos\theta$ we have
     $\cos3\theta - (4\cos^3\theta - 3\cos\theta) = 0$ for all θ.   From Eq.(1) we have
     $k_1 = 1$ and $k_2 = -1$ and thus $\cos3\theta$ and $4\cos^3\theta - 3\cos\theta$ are
     linearly dependent.

6. $W(t, t^{-1}) = \begin{vmatrix} t & t^{-1} \\ 1 & -t^{-2} \end{vmatrix} = -2/t \neq 0$.

7.   For $t>0$ $g(t) = t$ and hence $f(t) - 3g(t) = 0$ for all t.
     Therefore f and g are linearly dependent on $0<t$.   For $t<0$
     $g(t) = -t$ and $f(t) + 3g(t) = 0$, so again f and g are
     linearly dependent on $t<0$.   For any interval that
     includes the origin, such as $-1<t<2$, there is no c for
     which $f(t) + cg(t) = 0$ for all t, and hence f and g are
     linearly independent on this interval.

12. The D.E. is linear and homogeneous.   Hence, if y_1 and y_2
     are solutions, then $y_3 = y_1 + y_2$ and $y_4 = y_1 - y_2$ are
     solutions. $W(y_3, y_4) = y_3 y_4' - y_3' y_4 = (y_1 + y_2)(y_1' - y_2') -$
     $(y_1' + y_2')(y_1 - y_2) = -2(y_1 y_2' - y_1' y_2) = -2W(y_1, y_2)$, is not
     zero since y_1 and y_2 are linearly independent solutions.
     Hence y_3 and y_4 form a fundamental set of solutions.
     Conversely, solving the first two equations for y_1 and
     y_2, we have $y_1 = (y_3 + y_4)/2$ and $y_2 = (y_3 - y_4)/2$, so y_1 and
     y_2 are solutions. Finally, from above we have
     $W(y_1 y_2) = -W(y_3, y_4)/2$.

15. Writing the D.E. in the form of Eq.(7), we have
     $p(t) = -(t+2)/t$.   Thus Eq.(8) yields

$$W(t) = c \exp[-\int \frac{-(t+2)}{t} \, dt] = ct^2 e^t.$$

20. From Eq.(8) we have $W(y_1, y_2) = c \exp[-\int p(t) \, dt]$, where $p(t) = 2/t$ from the D.E. Thus $W(y_1, y_2) = c/t^2$. Since $W(y_1, y_2)(1) = 2$ we find $c = 2$ and thus $W(y_1, y_2)(5) = 2/25$.

24. Let c be the point in I at which both y_1 and y_2 vanish. Then $W(y_1, y_2)(c) = y_1(c)y_2'(c) - y_1'(c)y_2(c) = 0$. Hence, by Theorem 3.3.3 the functions y_1 and y_2 cannot form a fundamental set.

26. Suppose that y_1 and y_2 have a point of inflection at t_0 and either $p(t_0) \neq 0$ or $q(t_0) \neq 0$. Since $y_1''(t_0) = 0$ and $y_2''(t_0) = 0$ it follows from the D.E. that $p(t_0)y_1'(t_0) + q(t_0)y_1(t_0) = 0$ and $p(t_0)y_2'(t_0) + q(t_0)y_2(t_0) = 0$. If $p(t_0) = 0$ and $q(t_0) \neq 0$ then $y_1(t_0) = y_2(t_0) = 0$, and $W(y_1, y_2)(t_0) = 0$ so the solutions cannot form a fundamental set. If $p(t_0) \neq 0$ and $q(t_0) = 0$ then $y_1'(t_0) = y_2'(t_0) = 0$ and $W(y_1, y_2)(t_0) = 0$, so again the solutions cannot form a fundamental set. If $p(t_0) \neq 0$ and $q(t_0) = 0$ then $y_1'(t_0) = q(t_0)y_1(t_0)/p(t_0)$ and $y_2'(t_0) = q(t_0)y_2(t_0)/p(t_0)$ and thus
$$\begin{aligned} W(y_1, y_2)(t_0) &= y_1(t_0)y_2'(t_0) - y_1'(t_0)y_2(t_0) \\ &= y_1(t_0)[q(t_0)y_2(t_0)/p(t_0)] - \\ &\qquad [q(t_0)y_1(t_0)/p(t_0)]y_2(t_0) \\ &= 0. \end{aligned}$$

27. Let $-1 < t_0$, $t_1 < 1$ and $t_0 \neq t_1$. If $y_1 = t$ and $y_2 = t^2$ are linearly dependent then $c_1 t_1 + c_2 t_1^2 = 0$ and $c_1 t_0 + c_2 t_0^2 = 0$ have a solution for c_1 and c_2 such that c_1 and c_2 are not both zero. But this system of equations has a non-zero solution only if $t_1 = 0$ or $t_0 = 0$ or $t_1 = t_0$. Hence, the only set c_1 and c_2 that satisfies the system for every choice of t_0 and t_1 in $-1 < t < 1$ is $c_1 = c_2 = 0$. Therefore t and t^2 are linearly independent

on $-1 < t < 1$. Next, $W(t,t^2) = t^2$ clearly vanishes at $t = 0$. Since $W(t,t^2)$ vanishes at $t = 0$, but t and t^2 are linearly independent on $-1 < t < 1$, it follows that t and t^2 cannot be solutions of Eq.(7) on $-1 < t < 1$. To show that the functions $y_1 = t$ and $y_2 = t^2$ are solutions of $t^2 y'' - 2ty' + 2y = 0$, substitute each of them in the equation. Clearly, they are solutions. There is no contradiction to Theorem 3.3.3 since $p(t) = -2/t$ and $q(t) = 2/t^2$ are discontinuous at $t = 0$, and hence the theorem does not apply on the interval $-1 < t < 1$.

28. On $0 < t < 1$, $f(t) = t^3$ and $g(t) = t^3$. Hence there are nonzero constants, $c_1 = 1$ and $c_2 = -1$, such that $c_1 f(t) + c_2 g(t) = 0$ for each t in $(0,1)$. On $-1 < t < 0$, $f(t) = -t^3$ and $g(t) = t^3$; thus $c_1 = c_2 = 1$ defines constants such that $c_1 f(t) + c_2 g(t) = 0$ for each t in $(-1,0)$. Thus f and g are linearly dependent on $0 < t < 1$ and on $-1 < t < 0$. We will show that $f(t)$ and $g(t)$ are linearly independent on $-1 < t < 1$ by demonstrating that it is impossible to find constants c_1 and c_2, not both zero, such that $c_1 f(t) + c_2 g(t) = 0$ for all t in $(-1,1)$. Assume that there are two such nonzero constants and choose two points t_0 and t_1 in $-1 < t < 1$ such that $t_0 < 0$ and $t_1 > 0$. Then $-c_1 t_0^3 + c_2 t_0^3 = 0$ and $c_1 t_1^3 + c_2 t_1^3 = 0$. These equations have a nontrivial solution for c_1 and c_2 only if the determinant of coefficients is zero. But the determinant of coefficients is $-2t_0^3 t_1^3 \neq 0$ for t_0 and t_1 as specified. Hence $f(t)$ and $g(t)$ are linearly independent on $-1 < t < 1$.

<u>Section 3.4, Page 158</u>

1. $\exp(1+2i) = e^{1+2i} = ee^{2i} = e(\cos 2 + i \sin 2)$.

5. Recall that $2^{1-i} = e^{\ln(2^{1-i})} = e^{(1-i)\ln 2}$

7. As in Section 3.1, we seek solutions of the form $y = e^{rt}$. Substituting this into the D.E. yields the characteristic equation $r^2 - 2r + 2 = 0$, which has the roots $r_1 = 1 + i$

and $r_2 = 1 - i$, using the quadratic formula. Thus $\lambda = 1$ and $\mu = 1$ and from Eq.(17) the general solution is $y = c_1 e^t \cos t + c_2 e^t \sin t$.

11. The characteristic equation is $r^2 + 6r + 13 = 0$, which has the roots $r = \dfrac{-6 \pm \sqrt{-16}}{2} = -3 \pm 2i$. Thus $\lambda = -3$ and $\mu = 2$, so Eq.(17) becomes $y = c_1 e^{-3t} \cos 2t + c_2 e^{-3t} \sin 2t$.

14. The characteristic equation is $9r^2 + 9r - 4$, which has the real roots $-4/3$ and $1/3$. Thus the solution has the same form as in Section 3.1, $y(t) = c_1 e^{t/3} + c_2 e^{-4t/3}$.

18. The characteristic equation is $r^2 + 4r + 5 = 0$, which has the roots $r_1, r_2 = -2 \pm i$. Thus $y = c_1 e^{-2t} \cos t + c_2 e^{-2t} \sin t$ and $y' = (-2c_1 + c_2) e^{-2t} \cos t + (-c_1 - 2c_2) e^{-2t} \sin t$, so that $y(0) = c_1 = 1$ and $y'(0) = -2c_1 + c_2 = 0$, or $c_2 = 2$. Hence $y = e^{-2t}(\cos t + 2\sin t)$.

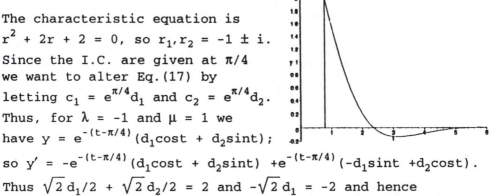

22. The characteristic equation is $r^2 + 2r + 2 = 0$, so $r_1, r_2 = -1 \pm i$. Since the I.C. are given at $\pi/4$ we want to alter Eq.(17) by letting $c_1 = e^{\pi/4} d_1$ and $c_2 = e^{\pi/4} d_2$. Thus, for $\lambda = -1$ and $\mu = 1$ we have $y = e^{-(t-\pi/4)}(d_1 \cos t + d_2 \sin t)$; so $y' = -e^{-(t-\pi/4)}(d_1 \cos t + d_2 \sin t) + e^{-(t-\pi/4)}(-d_1 \sin t + d_2 \cos t)$. Thus $\sqrt{2}\, d_1/2 + \sqrt{2}\, d_2/2 = 2$ and $-\sqrt{2}\, d_1 = -2$ and hence $y = \sqrt{2}\, e^{-(t-\pi/4)}(\cos t + \sin t)$.

23a. The characteristic equation is $3r^2 - r + 2 = 0$, which has the roots $r_1, r_2 = \dfrac{1}{6} \pm \dfrac{\sqrt{23}}{6} i$. Thus $u(t) = e^{t/6}(c_1 \cos\dfrac{\sqrt{23}}{6} t + c_2 \sin\dfrac{\sqrt{23}}{6} t)$ and we obtain $u(0) = c_1 = 2$

and $u'(0) = \dfrac{1}{6}c_1 + \dfrac{\sqrt{23}}{6}c_2 = 0$. Solving for c_2 we find

$$u(t) = e^{t/6}\left(2\cos\frac{\sqrt{23}}{6}t - \frac{2}{\sqrt{23}}\sin\frac{\sqrt{23}}{6}t\right).$$

23b. To estimate the first time that $|u(t)| = 10$ plot the graph of $u(t)$ as found in part (a). Use this estimate in an appropriate computer software program to find $t = 10.7598$.

25a. The characteristic equation is $r^2 + 2r + 6 = 0$, so $r_1, r_2 = -1 \pm \sqrt{5}\,i$ and $y(t) = e^{-t}(c_1\cos\sqrt{5}\,t + c_2\sin\sqrt{5}\,t)$. Thus $y(0) = c_1 = 2$ and $y'(0) = -c_1 + \sqrt{5}\,c_2 = \alpha$ and hence

$$y(t) = e^{-t}\left(2\cos\sqrt{5}\,t + \frac{\alpha+2}{\sqrt{5}}\sin\sqrt{5}\,t\right).$$

25b. $y(1) = e^{-1}\left(2\cos\sqrt{5} + \dfrac{\alpha+2}{\sqrt{5}}\sin\sqrt{5}\right) = 0$ and hence

$$\alpha = -2 - \frac{2\sqrt{5}}{\tan\sqrt{5}} = 1.50878.$$

25c. For $y(t) = 0$ we must have $2\cos\sqrt{5}\,t + \dfrac{\alpha+2}{\sqrt{5}}\sin\sqrt{5}\,t = 0$ or

$$\tan\sqrt{5}\,t = \frac{-2\sqrt{5}}{\alpha+2}.$$ For the given α (actually, for $\alpha > -2$)

this yields $\sqrt{5}\,t = \pi - \arctan\dfrac{2\sqrt{5}}{\alpha+2}$ since $\arctan x < 0$ when $x < 0$.

25d. From part (c) $\arctan\dfrac{2\sqrt{5}}{\alpha+2} \to 0$ as $\alpha \to \infty$, so $t \to \pi/\sqrt{5}$.

31. $\dfrac{d}{dt}[e^{\lambda t}(\cos\mu t + i\sin\mu t)] = \lambda e^{\lambda t}(\cos\mu t + i\sin\mu t)$

$+ e^{\lambda t}(-\mu\sin\mu t + i\mu\cos\mu t) = \lambda e^{\lambda t}(\cos\mu t + i\sin\mu t)$

$+ i\mu e^{\lambda t}(i\sin\mu t + \cos\mu t) = e^{\lambda t}(\lambda+i\mu)(\cos\mu t + i\sin\mu t)$.

Setting $r = \lambda+i\mu$ we then have $\dfrac{d}{dt}e^{rt} = re^{rt}$.

33. Suppose that $t = a$ and $t = b$ ($b>a$) are consecutive zeros of y_1. We must show that y_2 vanishes once and only once in the interval $a < t < b$. Assume that it does not

vanish. Then we can form the quotient y_1/y_2 on the interval $a \le t \le b$. Note $y_2(a) \ne 0$ and $y_2(b) \ne 0$, otherwise y_1 and y_2 would not be linearly independent solutions. Next, y_1/y_2 vanishes at $t = a$ and $t = b$ and has a derivative in $a < t < b$. By Rolles theorem, the derivative must vanish at an interior point. But

$$\left(\frac{y_1}{y_2}\right)' = \frac{y_1'y_2 - y_2'y_1}{y_2^2} = \frac{-W(y_1,y_2)}{y_2^2}, \text{ which cannot be zero}$$

since y_1 and y_2 are linearly independent solutions. Hence we have a contradiction, and we conclude that y_2 must vanish at a point between a and b. Finally, we show that it can vanish at only one point between a and b. Suppose that it vanishes at two points c and d between a and b. By the argument we have just given we can show that y_1 must vanish between c and d. But this contradicts the hypothesis that a and b are consecutive zeros of y_1.

35. We use the result of Problem 34. Note that $q(t) = e^{-t^2} > 0$ for $-\infty < t < \infty$. Next, we find that $(q' + 2pq)/q^{3/2} = 0$. Hence the D.E. can be transformed into an equation with constant coefficients by letting $x = u(t) = \int e^{-t^2/2}dt$. Substituting $x = u(t)$ in the differential equation found in part (b) of Problem 34 we obtain, after dividing by the coefficient of d^2y/dx^2, the D.E. $d^2y/dx^2 - y = 0$. Hence the general solution of the original D.E. is $y = c_1\cos x + c_2\sin x$, $x = \int e^{-t^2/2}dt$.

38. Rewrite the D.E. as $y'' + (\alpha/t)y' + (\beta/t^2)y = 0$ so that $p = \alpha/t$ and $q = \beta/t^2$, which satisfy the conditions of parts (c) and (d) of Problem 34. Thus $x = \int(1/t^2)^{1/2}dt = \ln t$ will transform the D.E. into $dy^2/dx^2 + (\alpha-1)dy/dx + \beta y = 0$. Note that since β is constant, it can be neglected in defining x.

39. By direct substitution, or from Problem 38, $x = \ln t$ will transform the D.E. into $d^2y/dx^2 + y = 0$, since $\alpha = 1$ and $\beta = 1$. Thus $y = c_1\cos x + c_2\sin x$, with $x = \ln t$, $t > 0$.

Section 3.5, Page 166

1. Substituting $y = e^{rt}$ into the D.E., we find that
    $r^2 - 2r + 1 = 0$, which gives $r_1 = 1$ and $r_2 = 1$. Since the
    roots are equal, the second linearly independent solution
    is te^t and thus the general solution is $y = c_1e^t + c_2te^t$.

9. The characteristic equation is $25r^2 - 20r + 4 = 0$, which
    may be written as $(5r-2)^2 = 0$ and hence the roots are
    $r_1, r_2 = 2/5$. Thus $y = c_1e^{2t/5} + c_2\, te^{2t/5}$.

12. The characteristic equation is
    $r^2 - 6r + 9 = (r-3)^2$, which has
    the repeated root $r = 3$. Thus
    $y = c_1e^{3t} + c_2te^{3t}$, which gives
    $y(0) = c_1 = 0$, $y'(t) = c_2(e^{3t}+3te^{3t})$
    and $y'(0) = c_2 = 2$. Hence
    $y(t) = 2te^{3t}$.

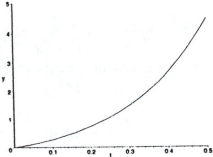

14. The characteristic equation is
    $r^2 + 4r + 4 = (r+2)^2 = 0$, which
    has the repeated root $r = -2$.
    Since the I.C. are given at
    $t = -1$, write the general solution
    as $y = c_1e^{-2(t+1)} + c_2te^{-2(t+1)}$. Then
    $y' = -2c_1e^{-2(t+1)} + c_2e^{-2(t+1)} - 2c_2te^{-2(t+1)}$ and hence
    $c_1-c_2 = 2$ and $-2c_1+3c_2 = 1$ which yield $c_1 = 7$ and $c_2 = 5$.
    Thus $y = 7e^{-2(t+1)} + 5te^{-2(t+1)}$, a decaying exponential as
    shown in the graph.

17a. The characteristic equation is $4r^2 + 4r + 1 = (2r+1)^2 = 0$,
    so we have $y(t) = (c_1+c_2t)e^{-t/2}$. Thus $y(0) = c_1 = 1$ and
    $y'(0) = -c_1/2 + c_2 = 2$ and hence $c_2 = 5/2$ and
    $y(t) = (1 + 5t/2)e^{-t/2}$.

17b. From part (a), $y'(t) = -\dfrac{1}{2}(1 + 5t/2)e^{-t/2} + \dfrac{5}{2}e^{-t/2} = 0$, when
    $-\dfrac{1}{2} - \dfrac{5t}{4} + \dfrac{5}{2} = 0$, or $t_0 = \dfrac{8}{5}$ and $y_0 = 5e^{-4/5}$.

17c. From part (a), $-\dfrac{1}{2} + c_2 = b$ or $c_2 = b + \dfrac{1}{2}$ and

$y(t) = [1 + (b + \dfrac{1}{2})t]e^{-t/2}$.

17d. From part (c), $y'(t) = -\dfrac{1}{2}[1 + (b+\dfrac{1}{2})t]e^{-t/2} + (b+\dfrac{1}{2})e^{-t/2} = 0$

which yields $t_M = \dfrac{4b}{2b+1}$ and

$y_M = (1 + \dfrac{2b+1}{2} \cdot \dfrac{4b}{2b+1})e^{-2b/(2b+1)} = (1 + 2b)e^{-2b/(2b+1)}$.

19. If $r_1 = r_2$ then $y(t) = (c_1 + c_2 t)e^{r_1 t}$. Since the exponential is never zero, $y(t)$ can be zero only if $c_1 + c_2 t = 0$, which yields at most one positive value of t if c_1 and c_2 differ in sign. If $r_2 > r_1$ then

$y(t) = c_1 e^{r_1 t} + c_2 e^{r_2 t} = e^{r_1 t}(c_1 + c_2 e^{(r_2-r_1)t})$. Again, this is zero only if c_1 and c_2 differ in sign, in which case

$t = \dfrac{\ln(-c_1/c_2)}{(r_2-r_1)}$.

21. If $r_2 \neq r_1$ then $\phi(t;r_1,r_2) = (e^{r_2 t} - e^{r_1 t})/(r_2 - r_1)$ is defined for all t. Note that ϕ is a liner combination of two solutions, $e^{r_1 t}$ and $e^{r_2 t}$, of the D.E. Hence, ϕ is a solution of the differential equation. Think of r_1 as fixed and let $r_2 \to r_1$. The limit of ϕ as $r_2 \to r_1$ is indeterminate. If we use L'Hopital's rule, we find

$\lim_{r_2 \to r_1} \dfrac{e^{r_2 t} - e^{r_1 t}}{r_2 - r_1} \cdot = \lim_{r_2 \to r_1} \dfrac{te^{r_2 t}}{1} = te^{r_1 t}$. Hence, the

solution $\phi(t;r_1,r_2) \to te^{r_1 t}$ as $r_2 \to r_1$.

25. Let $y_2 = v/t$. Then $y_2' = v'/t - v/t^2$ and

$y_2'' = v''/t - 2v'/t^2 + 2v/t^3$. Substituting in the D.E. we obtain

$t^2(v''/t - 2v'/t^2 + 2v/t^3) + 3t(v'/t - v/t^2) + v/t = 0$. Simplifying the left side we get $tv'' + v' = 0$, which yields $v' = c_1/t$. Thus $v = c_1\ln t + c_2$. Hence a second solution is $y_2(t) = (c_1\ln t + c_2)/t$. However, we may set $c_2 = 0$ and $c_1 = 1$ without loss of generality and thus we have $y_2(t) = (\ln t)/t$ as a second solution. Note that in the form we actually calculated, $y_2(t)$ is a linear combination of $1/t$ and $\ln t/t$, and hence is the general solution.

27. In this case the calculations are somewhat easier if we do not use the explicit form for $y_1(x) = \sin x^2$ at the beginning but simply set $y_2(x) = y_1 v$. Substituting this form for y_2 in the D.E. gives $x(y_1 v)'' - (y_1 v)' + 4x^3(y_1 v) = 0$. On carrying out the differentiations and making use of the fact that y_1 is a solution, we obtain $xy_1 v'' + (2xy_1' - y_1)v' = 0$. This is a first order linear equation for v', which has the solution $v' = cx/(\sin x^2)^2$. Setting $u = x^2$ allows integration of this to get $v = c_1 \cot x^2 + c_2$. Setting $c_1 = 1$, $c_2 = 0$ and multiplying by $y_1 = \sin x^2$ we obtain $y_2(x) = \cos x^2$ as the second solution of the D.E.

30. Substituting $y_2(x) = y_1(x)v(x)$ in the D.E. gives
$$x^2(y_1 v)'' + x(y_1 v)' + (x^2 - \frac{1}{4})y_1 v = 0.$$ On carrying out the differentiations and making use of the fact that y_1 is a solution, we obtain $x^2 y_1 v'' + (2x^2 y_1' + xy_1)v' = 0$. This is a first order linear equation for v',
$v'' + (2y_1'/y_1 + 1/x)v' = 0$, with solution
$$v'(x) = c \exp[-\int(2\frac{y_1'}{y_1} + \frac{1}{x})dx] = c \exp[-2\ln y_1 - \ln x]$$
$$= c\frac{1}{xy_1{}^2} = \frac{c}{x(x^{-1}\sin^2 x)} = c \csc^2 x,$$
where c is an arbitrary constant, which we will take to be one. Then $v(x) = \int \csc^2 x \, dx = -\cot x + k$ where again k is an arbitrary constant which can be taken equal to zero. Thus $y_2(x) = y_1(x)v(x) = (x^{-1/2}\sin x)(-\cot x) = -x^{-1/2}\cos x$. The second solution is usually taken to be $x^{-1/2}\cos x$. Note that $c = -1$ would have given this solution.

31b. Let $y_2(x) = e^x v(x)$, then $y_2' = e^x v' + e^x v$, and $y_2'' = e^x v'' + 2e^x v' + e^x v$. Substituting in the D.E. we obtain $xe^x v'' + (xe^x - Ne^x)v' = 0$, or $v'' + (1 - N/x)v' = 0$. This is a first order linear D.E. for v' with integrating

factor $\mu(x) = \exp[\int(1-N/x)\,dx] = x^{-N}e^x$. Hence
$(x^{-N}e^xv')' = 0$, and $v' = cx^Ne^{-x}$ which gives
$v(x) = c\int x^Ne^{-x}dx + k$. On taking $k = 0$ we obtain as the
second solution $y_2(x) = ce^x\int x^Ne^{-x}dx$. The integral can be
evaluated by using the method of integration by parts.
At each stage let $u = x^N$ or x^{N-1}, or whatever the power of
x that remains, and let $dv = e^{-x}$. Note that this dv is
not related to the $v(x)$ in $y_2(x)$. For $N = 2$ we have

$$y_2(x) = ce^x\int x^2e^{-x}dx = ce^x[x^2\frac{e^{-x}}{-1} - \int 2x\frac{e^{-x}}{-1}dx]$$

$$= -cx^2 + ce^x[2x\frac{e^{-x}}{-1} - \int 2\frac{e^{-x}}{-1}dx]$$

$$= c(-x^2 - 2x - 2) = -2c(1 + x + x^2/2!).$$

Choosing $c = -1/2!$ gives the desired result. For the
general case $c = -1/N!$.

33. $(y_2/y_1)' = (y_1y_2' - y_1'y_2)/y_1^2 = W(y_1,y_2)/y_1^2$. Abel's identity
is $W(y_1,y_2) = c\,\exp[-\int_{t_0}^t p(r)\,dr]$. Hence

$(y_2/y_1)' = cy_1^{-2}\exp[-\int_{t_0}^t p(r)\,dr]$. Integrating and setting
$c = 1$ (since a solution y_2 can be multiplied by any
constant) and taking the constant of integration to be
zero we obtain

$$y_2(t) = y_1(t)\int_{t_0}^t \frac{\exp[-\int_{s_0}^s p(r)\,dr]}{[y_1(s)]^2}\,ds.$$

35. From Problem 33 and Abel's formula we have
$$(\frac{y_2}{y_1})' = \frac{\exp[\int(1/t)\,dt]}{\sin^2(t^2)} = \frac{e^{\ln t}}{\sin^2(t^2)} = t\csc^2(t^2).$$ Thus
$y_2/y_1 = -(1/2)\cot(t^2)$ and hence we can choose $y_2 = \cos(t^2)$
since $y_1 = \sin^2(t^2)$.

38. The general solution of the D.E. is $y = c_1e^{r_1t} + c_2e^{r_2t}$
where $r_1,r_2 = (-b \pm \sqrt{b^2-4ac})/2a$ provided $b^2 - 4ac \neq 0$.
In this case there are two possibilities. If $b^2 - 4ac > 0$
then $(b^2 - 4ac)^{1/2} < b$ and r_1 and r_2 are real and
<u>negative</u>. Consequently $e^{r_1t} \to 0$ and $e^{r_2t} \to 0$; and hence

$y \to 0$, as $t \to \infty$. If $b^2 - 4ac < 0$ then r_1 and r_2 are complex conjugates with <u>negative</u> real part. Again $e^{r_1 t} \to 0$ and $e^{r_2 t} \to 0$; and hence $y \to 0$, as $t \to \infty$. Finally, if $b^2 - 4ac = 0$, then $y = c_1 e^{r_1 t} + c_2 t e^{r_1 t}$ where $r_1 = -b/2a < 0$. Hence, again $y \to 0$ as $t \to \infty$. This conclusion does not hold if either $b = 0$ (since $y(t) = c_1 \cos \omega t + c_2 \sin \omega t$) or $c = 0$ (since $y_1(t) = c_1$).

42. Substituting $z = \ln t$ into the D.E. gives
$$\frac{d^2 y}{dz^2} + \frac{dy}{dz} + 0.25y = 0,$$ which has the solution
$y(z) = c_1 e^{-z/2} + c_2 z e^{-z/2}$ so that $y(t) = c_1 t^{-1/2} + c_2 t^{-1/2} \ln t$.

Section 3.6, Page 178

1. First we find the solution of the homogeneous D.E., which has the characteristic equation $r^2 - 2r - 3 = (r-3)(r+1) = 0$. Hence $y_c = c_1 e^{3t} + c_2 e^{-t}$ and we can assume $Y = A e^{2t}$ for the particular solution. Thus $Y' = 2A e^{2t}$ and $Y'' = 4A e^{2t}$ and substituting into the D.E. yields
$4A e^{2t} - 2(2A e^{2t}) - 3(A e^{2t}) = 3e^{2t}$. Thus $-3A = 3$ and $A = -1$, yielding $y = c_1 e^{3t} + c_2 e^{-t} - e^{2t}$.

4. Initially we assume $Y = A + B\sin 2t + C\cos 2t$. However, since a constant is a solution of the related homogeneous D.E. we must modify Y by multiplying the constant A by t and thus the correct form is $Y = At + B\sin 2t + C\cos 2t$.

6. Since $y_c = c_1 e^{-t} + c_2 t e^{-t}$ we must assume $Y = At^2 e^{-t}$, so that $Y' = 2At e^{-t} - At^2 e^{-t}$ and $y'' = 2A e^{-t} - 4At e^{-t} + At^2 e^{-t}$. Substituting in the D.E. gives $(At^2 - 4At + 2A)e^{-t} + 2(-At^2 + 2At)e^{-t} + At^2 e^{-t} = 2e^{-t}$. Notice that all terms on the left involving t^2 and t add to zero and we are left with $2A = 2$, or $A = 1$. Hence $y = c_1 e^{-t} + c_2 t e^{-t} + t^2 e^{-t}$.

8. The assumed form is $Y = (At + B)\sin 2t + (Ct + D)\cos 2t$, which is appropriate for both terms appearing on the right side of the D.E. Since none of the terms appearing

in Y are solutions of the homogeneous equation, we do not need to modify Y.

11. First solve the homogeneous D.E. Substituting $y = e^{rt}$ gives $r^2 + r + 4 = 0$. Hence $y_c = e^{-t/2}[c_1\cos(\sqrt{15}\,t/2) + c_2\sin(\sqrt{15}\,t/2)]$. We replace sinht by $(e^t - e^{-t})/2$ and then assume $Y(t) = Ae^t + Be^{-t}$. Since neither e^t nor e^{-t} are solutions of the homogeneous equation, there is no need to modify our assumption for Y. Substituting in the D.E., we obtain $6Ae^t + 4Be^{-t} = e^t - e^{-t}$. Hence, $A = 1/6$ and $B = -1/4$. The general solution is
$$y = e^{-t/2}[c_1\cos(\sqrt{15}\,t/2) + c_2\sin(\sqrt{15}\,t/2)] + e^t/6 - e^{-t}/4.$$
[For this problem we could also have found a particular solution as a linear combination of sinht and cosht: $Y(t) = A\cosh t + B\sinh t$. Substituting this in the D.E. gives $(5A + B)\cosh t + (A + 5B)\sinh t = 2\sinh t$. The solution is $A = -1/12$ and $B = 5/12$. A simple calculation shows that $-(1/12)\cosh t + (5/12)\sinh t = e^t/6 - e^{-t}/4.$]

13. $y_c = c_1e^{-2t} + c_2e^t$ so for the particular solution we assume $Y = At + B$. Since neither At or B are solutions of the homogeneous equation it is not necessary to modify the original assumption. Substituting Y in the D.E. we obtain $0 + A - 2(At+B) = 2t$ or $-2A = 2$ and $A - 2B = 0$. Solving for A and B we obtain $y = c_1e^{-2t} + c_2e^t - t - 1/2$ as the general solution. $y(0) = 0 \rightarrow c_1 + c_2 - 1/2 = 0$ and $y'(0) = 1 \rightarrow -2c_1 + c_2 - 1 = 1$, which yield $c_1 = -1/2$ and $c_2 = 1$. Thus $y = e^t - (1/2)e^{-2t} - t - 1/2$.

16. Since the nonhomogeneous term is the product of a linear polynomial and an exponential, assume Y of the same form: $Y = (At+B)e^{2t}$. Thus $Y' = Ae^{2t} + 2(At+B)e^{2t}$ and $Y'' = 4Ae^{2t} + 4(At+B)e^{2t}$. Substituting into the D.E. we find $-3At = 3t$ and $2A - 3B = 0$, yielding $A = -1$ and $B = -2/3$. Since the characteristic equation is $r^2 - 2r - 3 = 0$, the general solution is
$$y = c_1e^{3t} + c_2e^{-t} - \frac{2}{3}e^{2t} - te^{2t}.$$

19a. The solution of the homogeneous D.E. is $y_c = c_1 e^{-3t} + c_2$.

After inspection of the nonhomogeneous term, for $2t^4$ we must assume a fourth order polynominial, for $t^2 e^{-3t}$ we must assume a quadratic polynomial times the exponential, and for sin3t we must assume Csin3t + Dcos3t. Thus
$$Y(t) = (A_0 t^4 + A_1 t^3 + A_2 t^2 + A_3 t + A_4) + (B_0 t^2 + B_1 t + B_2) e^{-3t} + Csin3t + Dcos3t.$$
However, since e^{-3t} and a constant are solutions of the homogeneous D.E., we must multiply the coefficient of e^{-3t} and the polynomial by t. The correct form is
$$Y(t) = t(A_0 t^4 + A_1 t^3 + A_2 t^2 + A_3 t + A_4) +$$
$$t(B_0 t^2 + B_1 t + B_2) e^{-3t} + Csin3t + Dcos3t.$$

22a. The solution of the homgeneous D.E. is $y_c = e^{-t}[c_1 cost + c_2 sint]$. After inspection of the nonhomogeneous term, we assume $Y(t) = Ae^{-t} + (B_0 t^2 + B_1 t + B_2) e^{-t} cost + (C_0 t^2 + C_1 t + C_2) e^{-t} sint$. Since $e^{-t} cost$ and $e^{-t} sint$ are solutions of the homogeneous D.E., it is necessary to multiply both the last two terms by t. Hence the correct form is
$$Y(t) = Ae^{-t} + t(B_0 t^2 + B_1 t + B_2) e^{-t} cost +$$
$$t(C_0 t^2 + C_1 t + C_2) e^{-t} sint.$$

28. First solve the I.V.P. $y'' + y = t$, $y(0) = 0$, $y'(0) = 1$ for $0 \le t \le \pi$. The solution of the homogeneous D.E. is $y_c(t) = c_1 cost + c_2 sint$. The correct form for Y(t) is $y(t) = A_0 t + A_1$. Substituting in the D.E. we find $A_0 = 1$ and $A_1 = 0$. Hence, $y = c_1 cost + c_2 sint + t$. Applying the I.C., we obtain $y = t$. For $t > \pi$ we have $y'' + y = \pi e^{\pi-t}$ so the form for Y(t) is $Y(t) = Ee^{\pi-t}$. Substituting Y(t) in the D.E., we obtain $Ee^{\pi-t} + Ee^{\pi-t} = \pi e^{\pi-t}$ so $E = \pi/2$. Hence the general solution for $t > \pi$ is $Y = D_1 cost + D_2 sint + (\pi/2)e^{\pi-t}$. If y and y' are to be continuous at $t = \pi$, then the solutions and their derivatives for $t \le \pi$ and $t > \pi$ must have the same value at $t = \pi$. These conditions require $\pi = -D_1 + \pi/2$ and $1 = -D_2 - \pi/2$. Hence $D_1 = -\pi/2$, $D_2 = -(1 + \pi/2)$, and

$$y = \phi(t) = \begin{cases} t, & 0 \leq t \leq \pi \\ -(\pi/2)\cos t - (1 + \pi/2)\sin t + (\pi/2)e^{\pi-t}, & t > \pi. \end{cases}$$

The graphs of the nonhomogeneous term and ϕ follow.

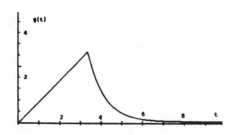   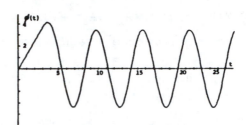

30. According to Theorem 3.6.1, the difference of any two solutions of the linear second order nonhomogeneous D.E. is a solution of the corresponding homogeneous D.E. Hence $Y_1 - Y_2$ is a solution of $ay'' + by' + cy = 0$. In Problem 38 of Section 3.5 we showed that if $a > 0$, $b > 0$, and $c > 0$ then every solution of this D.E. goes to zero as $t \to \infty$. If $b = 0$, then y_c involves only sines and cosines, so $Y_1 - Y_2$ does not approach zero as $t \to \infty$.

33. From Problem 32 we write the D.E. as $(D-4)(D+1)y = 3e^{2t}$. Thus let $(D+1)y = u$ and then $(D-4)u = 3e^{2t}$. This last equation is the same as $du/dt - 4u = 3e^{2t}$, which may be solved by multiplying both sides by e^{-4t} and integrating (see section 2.1). This yields $u = (-3/2)e^{2t} + Ce^{4t}$. Substituting this form of u into $(D+1)y = u$ we obtain $dy/dt + y = (-3/2)e^{2t} + Ce^{4t}$. Again, multiplying by e^t and integrating gives $y = (-1/2)e^{2t} + C_1e^{4t} + C_2e^{-t}$, where $C_1 = C/5$.

Section 3.7, Page 183

2. Two linearly independent solutions of the homogeneous D.E. are $y_1(t) = e^{2t}$ and $y_2(t) = e^{-t}$. Assume $Y = u_1(t)e^{2t} + u_2(t)e^{-t}$, then $Y'(t) = [2u_1(t)e^{2t} - u_2(t)e^{-t}] + [u_1'(t)e^{2t}$

+ $u_2'(t)e^{-t}$]. We set $u_1'(t)e^{2t} + u_2'(t)e^{-t} = 0$.   Then
$Y'' = 4u_1e^{2t} + u_2e^{-t} + 2u_1'e^{2t} - u_2'e^{-t}$ and substituting in
the D.E. gives $2u_1'(t)e^{2t} - u_2'(t)e^{-t} = 2e^{-t}$. Thus we have
two algebraic equations for $u_1'(t)$ and $u_2'(t)$ with the
solution $u_1'(t) = 2e^{-3t}/3$ and $u_2'(t) = -2/3$. Hence
$u_1(t) = -2e^{-3t}/9$ and $u_2(t) = -2t/3$. Substituting in the
formula for $Y(t)$ we obtain $Y(t) = (-2e^{-3t}/9)e^{2t} +$
$(-2t/3)e^{-t} = (-2e^{-t}/9) - (2te^{-t}/3)$. Since e^{-t} is a
solution of the homogeneous D.E., we can choose
$Y(t) = -2te^{-t}/3$.

5.   Since cost and sint are solutions of the homogeneous
     D.E., we assume $Y = u_1(t)\cos t + u_2(t)\sin t$. Thus
     $Y' = -u_1(t)\sin t + u_2(t)\cos t$, after setting
     $u_1'(t)\cos t + u_2'(t)\sin t = 0$. Finding Y'' and substituting
     into the D.E. then yields $-u_1'(t)\sin t + u_2'(t)\cos t = \tan t$.
     The two equations for $u_1'(t)$ and $u_2'(t)$ have the solution:
     $u_1'(t) = -\sin^2 t/\cos t = -\sec t + \cos t$ and
     $u_2'(t) = \sin t$. Thus $u_1(t) = \sin t - \ln(\tan t + \sec t)$ and
     $u_2(t) = -\cos t$, which when substituted into the assumed
     form for Y, simplified, and added to the homogeneous
     solution yields
     $y = c_1\cos t + c_2\sin t - (\cos t)\ln(\tan t + \sec t)$.

11. Two linearly independent solutions of the homogeneous
     D.E. are $y_1(t) = e^{3t}$ and $y_2(t) = e^{2t}$. Applying Theorem
     3.7.1 with $W(y_1, y_2)(t) = -e^{5t}$, we obtain

$$Y(t) = -e^{3t}\int \frac{e^{2s}g(s)}{-e^{5s}}\, ds + e^{2t}\int \frac{e^{3s}g(s)}{-e^{5s}}\, ds$$
$$= \int [e^{3(t-s)} - e^{2(t-s)}]g(s)\, ds.$$

     The complete solution is then obtained by adding
     $c_1e^{3t} + c_2e^{2t}$ to $Y(t)$.

14. That t and te^t are solutions of the homogeneous D.E. can
     be verified by direction substitution. Thus we assume
     $Y = tu_1(t) + te^tu_2(t)$. Following the pattern of earlier
     problems we find $tu_1'(t) + te^tu_2'(t) = 0$ and

$u_1'(t) + (t+1)e^t u_2' = 2t$. [Note that $g(t) = 2t$, since the D.E. must be put into the form of Eq.(16)]. The solution of these equations gives $u_1'(t) = -2$ and $u_2'(t) = 2e^{-t}$.

Hence, $u_1(t) = -2t$ and $u_2(t) = -2e^{-t}$, and

$Y(t) = t(-2t) + te^t(-2e^{-t}) = -2t^2 - 2t$. However, since t is a solution of the homogeneous D.E. we can choose as our particular solution $Y(t) = -2t^2$.

18. For this problem, and for many others, it is probably easier to rederive Eqs.(26) without using the explicit form for $y_1(x)$ and $y_2(x)$ and then to substitute for $y_1(x)$ and $y_2(x)$ in Eqs.(26). In this case if we take

$y_1 = x^{-1/2}\sin x$ and $y_2 = x^{-1/2}\cos x$, then $W(y_1, y_2) = -1/x$. If the D.E. is put in the form of Eq.(16), then $g(x) = 3x^{-1/2}\sin x$ and thus $u_1'(x) = 3\sin x \cos x$ and $u_2'(x) = -3\sin^2 x = 3(-1 + \cos 2x)/2$. Hence $u_1(x) = (3\sin^2 x)/2$ and $u_2(x) = -3x/2 + 3(\sin 2x)/4$, and

$$Y(x) = \frac{3\sin^2 x}{2}\frac{\sin x}{\sqrt{x}} + (-\frac{3x}{2} + \frac{3\sin 2x}{4})\frac{\cos x}{\sqrt{x}}$$

$$= \frac{3\sin^2 x}{2}\frac{\sin x}{\sqrt{x}} + (-\frac{3x}{2} + \frac{3\sin x \cos x}{2})\frac{\cos x}{\sqrt{x}}$$

$$= \frac{3\sin x}{2\sqrt{x}} - \frac{3\sqrt{x}\cos x}{2}.$$

The first term is a multiple of $y_1(x)$ and thus can be neglected for $Y(x)$.

22. Putting limits on the integrals of Eq.(28) and changing the integration variable to s yields

$$Y(t) = -y_1(t)\int_{t_0}^t \frac{y_2(s)g(s)ds}{W(y_1,y_2)(s)} + y_2(t)\int_{t_0}^t \frac{y_1(s)g(s)ds}{W(y_1,y_2)(s)}$$

$$= \int_{t_0}^t \frac{-y_1(t)y_2(s)g(s)ds}{W(y_1,y_2)(s)} + \int_{t_0}^t \frac{y_2(t)y_1(s)g(s)ds}{W(y_1,y_2)(s)}$$

$$= \int_{t_0}^t \frac{[y_1(s)y_2(t) - y_1(t)y_2(s)]g(s)ds}{y_1(s)y_2'(s) - y_1'(s)y_2(s)}.$$ To show that

$Y(t)$ satisfies $L[y] = g(t)$ we must take the derivative of Y using Leibnitz's rule, which says that if

$$Y(t) = \int_{t_0}^t G(t,s)ds, \text{ then } Y'(t) = G(t,t) + \int_{t_0}^t \frac{\partial G}{\partial t}(t,s)ds.$$

Letting $G(t,s)$ be the above integrand, then $G(t,t) = 0$

and $\dfrac{\partial G}{\partial t} = \dfrac{y_1(s)y_2'(t) - y_1'(t)y_2(s)}{W(y_1,y_2)(s)} \, g(s).$ Likewise

$$Y'' = \dfrac{\partial G(t,t)}{\partial t} + \int_{t_0}^t \dfrac{\partial^2 G}{\partial t^2}(t,s)\,ds$$

$$= g(t) + \int_{t_0}^t \dfrac{y_1(s)y_2''(t) - y_1''(t)y_2(s)}{W(y_1,y_2)(s)}\,ds.$$

Since y_1 amd y_2 are solutions of $L[y] = 0$, we have
$L[Y] = g(t)$ since all the terms involving the integral
will add to zero. Clearly $Y(t_0) = 0$ and $Y'(t_0) = 0$.

25. Note that $y_1 = e^{\lambda t}\cos\mu t$ and $y_2 = e^{\lambda t}\sin\mu t$ and thus
$W(y_1,y_2) = \mu e^{2\lambda t}$. From Problem 22 we then have:

$$Y(t) = \int_{t_0}^t \dfrac{e^{\lambda s}\cos\mu s\, e^{\lambda t}\sin\mu t - e^{\lambda t}\cos\mu t\, e^{\lambda s}\sin\mu s}{\mu e^{2\lambda s}}\, g(s)\,ds$$

$$= \mu^{-1}\int_{t_0}^t e^{\lambda(t-s)}[\cos\mu s\,\sin\mu t - \cos\mu t\,\sin\mu s]g(s)\,ds$$

$$= \mu^{-1}\int_{t_0}^t e^{\lambda(t-s)}[\sin\mu(t-s)]g(s)\,ds.$$

29. First, we put the D.E. in standard form by dividing by
t^2: $y'' - 2y'/t + 2y/t^2 = 4$. Assuming that $y = tv(t)$ and
substituting in the D.E. we obtain $tv'' = 4$. Hence
$v'(t) = 4\ln t + c_2$ and $v(t) = 4\int \ln t\,dt + c_2 t = $
$4(t\ln t - t) + c_2 t$. The general solution is
$c_1 y_1(t) + tv(t) = c_1 t + 4(t^2\ln t - t^2) + c_2 t^2$. Since $-4t^2$
is a multiple of $y_2 = c_2 t^2$ we can write
$y = c_1 t + c_2 t^2 + 4t^2\ln t$.

Section 3.8, Page 197

2. From Eq.(15) we have $R\cos\delta = -1$, and $R\sin\delta = \sqrt{13}$. Thus
$R = \sqrt{1+3} = 2$ and $\delta = \tan^{-1}(-\sqrt{3}) + \pi = 2\pi/3 \cong 2.09440$. Note
that we have to "add" π to the inverse tangent value since δ
must be a second quadrant angle. Thus $u = 2\cos(t - 2\pi/3)$.

6. The motion is an undamped free vibration. The units are
in the CGS system. The spring constant
$k = (100\text{ gm})(980\text{cm/sec}^2)/5\text{cm}$. Hence the D.E. for the
motion is $100u'' + [(100 \cdot 980)/5]u = 0$ where u is measured
in cm and time in sec. We obtain $u'' + 196u = 0$ so

u = Acos14t + Bsin14t. The I.C. are u(0) = 0 → A = 0
and u'(0) = 10 cm/sec → B = 10/14 = 5/7. Hence
u(t) = (5/7)sin14t, which first reaches equilibrium when
14t = π, or t = π/14.

8.   We use Eq.(33) without R and E(t) (there is no resistor or
      impressed voltage) and with L = 1 henry and $1/C = 4x10^6$ since
      $C = .25x10^{-6}$ farads. Thus the I.V.P. is $Q'' + 4x10^6 Q = 0$,
      $Q(0) = 10^{-6}$ coulombs and $Q'(0) = 0$.

9.   The spring constant is k = (20)(980)/5 = 3920 dyne/cm.
      The I.V.P. for the motion is 20u'' + 400u' + 3920u = 0 or
      u'' + 20u' + 196u = 0 and u(0) = 2, u'(0) = 0. Here u is
      measured in cm and t in sec. The general solution of the
      D.E. is $u = Ae^{-10t}\cos4\sqrt{6}\,t + Be^{-10t}\sin4\sqrt{6}\,t$. The I.C.
      u(0) = 2 → A = 2 and u'(0) = 0 → $-10A + 4\sqrt{6}\,B = 0$. The
      solution is $u = e^{-10t}[2\cos4\sqrt{6}\,t + 5(\sin4\sqrt{6}\,t)/\sqrt{6}\,]$cm.
      The quasi frequency is $\mu = 4\sqrt{6}$, the quasi period is
      $T_d = 2\pi\mu = \pi/2\sqrt{6}$ and $T_d/T = 7/2\sqrt{6}$ since
      T = 2π/14 = π/7. To find an upper bound for τ, write u
      in the form of Eq.(30): $u(t) = \sqrt{4+25/6}\,e^{-10t}\cos(4\sqrt{6}\,t-\delta)$.
      Now, since $|\cos(4\sqrt{6}\,t-\delta)| \le 1$, we have $|u(t)| < .05 \Rightarrow$
      $\sqrt{4+25/6}\,e^{-10t} < .05$, which yields τ = .4046. A more
      precise answer can be obtained with a computer algebra
      system, which in this case yields τ = .4045. The
      original estimate was unusually close for this problem
      since $\cos(4\sqrt{6}\,t-\delta)$ = -0.9996 for t = .4046.

12.   Substituting the given values for L, C and R in Eq.(33),
      we obtain the D.E. $.2Q'' + 3x10^2 Q' + 10^5 Q = 0$. The I.C.
      are $Q(0) = 10^{-6}$ and Q'(0) = I(0) = 0. Assuming $Q = e^{rt}$,
      we obtain the roots of the characteristic equation as
      $r_1 = -500$ and $r_2 = -1000$. Thus $Q = c_1e^{-500t} + c_2e^{-1000t}$
      and hence $Q(0) = 10^{-6} \rightarrow c_1 + c_2 = 10^{-6}$ and
      $Q'(0) = 0 \rightarrow -500c_1 - 1000c_2 = 0$. Solving for c_1 and c_2
      yields the solution.

17.   The mass is $8/32$ lb-sec^2/ft, and the spring constant is
      8/(1/8) = 64 lb/ft. Hence (1/4)u'' + γu' + 64u = 0 or
      u'' + 4γu' + 256u = 0, where u is measured in ft, t in sec
      and the units of γ are lb-sec/ft. We look for solutions
      of the D.E. of the form $u = e^{rt}$ and find
      $r^2 + 4\gamma r + 256 = 0$, so $r_1,r_2 = [-4\gamma \pm \sqrt{16\gamma^2 - 1024}\,]/2$.
      The system will be overdamped, critically damped or
      underdamped as $(16\gamma^2 - 1024)$ is > 0, =0, or < 0,

respectively. Thus the system is critically damped when γ = 8 lb-sec/ft.

19. The general solution of the D.E. is $u = Ae^{r_1 t} + Be^{r_2 t}$ where $r_1, r_2 = [-\gamma \pm (\gamma^2 - 4km)^{1/2}]/2m$ provided $\gamma^2 - 4km \neq 0$, and where A and B are determined by the I.C. When the motion is overdamped, $\gamma^2 - 4km > 0$ and $r_1 > r_2$. Setting u = 0, we obtain $Ae^{r_1 t} = - Be^{r_2 t}$ or $e^{(r_1 - r_2)t} = - B/A$. Since the exponential function is a monotone function, there is at most one value of t (when B/A < 0) for which this equation can be satisfied. Hence u can vanish at most once. If the system is critically damped, the general solution is $u(t) = (A + Bt)e^{-\gamma t/2m}$. The exponential function is never zero; hence u can vanish only if A + Bt = 0. If B = 0 then u never vanishes; if $B \neq 0$ then u vanishes once at t = - A/B provided A/B < 0.

20. The general solution of Eq.(21) for the case of critical damping is $u = (A + Bt)e^{-\gamma t/2m}$. The I.C. $u(0) = u_0 \rightarrow$ $A = u_0$ and $u'(0) = v_0 \rightarrow A(-\gamma/2m) + B = v_0$. Hence $u = [u_0 + (v_0 + \gamma u_0/2m)t]e^{-\gamma t/2m}$. If $v_0 = 0$, then $u = u_0(1 + \gamma t/2m)e^{-\gamma t/2m}$, which is never zero since γ and m are postive. By L'Hopital's Rule $u \rightarrow 0$ as $t \rightarrow \infty$. Finally for $u_0 > 0$, we want the condition which will insure that v = 0 at least once. Since the exponential function is never zero we require $u_0 + (v_0 + \gamma u_0/2m)t = 0$ at a positive value of t. This requires that $v_0 + \gamma u_0/2m \neq 0$ and that $t = -u_0(v_0 + u_0\gamma/2m)^{-1} > 0$. We know that $u_0 > 0$ so we must have $v_0 + \gamma u_0/2m < 0$ or $v_0 < -\gamma u_0/2m$.

23. From Problem 21: $\Delta = \dfrac{2\pi\gamma}{\mu(2m)} = T_d\gamma/2m$. Substituting the known values we find $\gamma = \dfrac{(1/2)\ (3)}{.3} = 5$ lb sec/ft.

24. From Eq.(13) $\omega_0^2 = \dfrac{2k}{3}$ so $P = 2\pi/\sqrt{2k/3} = \pi \rightarrow k = 6$. Thus $u(t) = c_1\cos 2t + c_2\sin 2t$ and $u(0) = 2 \rightarrow c_1 = 2$ and $u'(0) = v \rightarrow c_2 = v/2$. Hence $u(t) = 2\cos 2t + \dfrac{v}{2}\sin 2t =$

$$\sqrt{4+\frac{v^2}{4}}\cos(2t-\gamma). \quad \text{Thus} \quad \sqrt{4+\frac{v^2}{4}} = 3 \text{ and } v = \pm 2\sqrt{5}.$$

27. First, consider the static case. Let Δl denote the length of the block below the surface of the water. The weight of the block, which is a downward force, is $w = \rho l^3 l^3 g$. This is balanced by an equal and opposite buoyancy force B, which is equal to the weight of the displaced water. Thus $B = (\rho_0 l^2 \Delta l)g = \rho \Delta l^3 g$ so $\rho_0 \Delta l = \rho l$. Now let x be the displacement of the block from its equilibrium position. We take downward as the positive direction. In a displaced position the forces acting on the block are its weight, which acts downward and is unchanged, and the buoyancy force which is now $\rho_0 l^2(\Delta l + x)g$ and acts upward. The resultant force must be equal to the mass of the block times the acceleration, namely $\rho l^3 x''$. Hence $\rho l^3 g - \rho_0 l^2(\Delta l + x)g = \rho l^3 x''$. The D.E. for the motion of the block is $\rho l^3 x'' + \rho_0 l^2 g x = 0$. This gives a simple harmonic motion with frequency $(\rho_0 g/\rho l)^{1/2}$ and natural period $2\pi(\rho l/\rho_0 g)^{1/2}$.

29a. The characteristic equation is $4r^2 + r + 8 = 0$, so $r = (-1 \pm \sqrt{127})/8$ and hence

$$u(t) = e^{-t/8}(c_1 \cos\frac{\sqrt{127}}{8}t + c_2 \sin\frac{\sqrt{127}}{8}t. \quad u(0) = 0 \to c_1 = 0$$

and $u'(0) = 2 \to \dfrac{\sqrt{127}}{8}c_2 = 2$. Thus

$$u(t) = \frac{16}{\sqrt{127}}e^{-t/8}\sin\frac{\sqrt{127}}{8}t.$$

29c. The phase plot is the spiral shown and the direction of motion is clockwise since the graph starts at (0,2) and u increases initially.

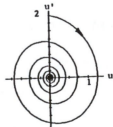

30c. Using u(t) as found in part(b), show that $ku^2/2 + m(u')^2/2 = (ka^2 + mb^2)/2$ for all t.

Section 3.9, Page 205

1. We use the trigonometric identities
    cos(A ± B) = cosA cosB + sinA sinB to obtain
    cos(A + B) - cos(A - B) = -2sinA sinB. If we choose
    A + B = 9t and A - B = 7t, then A = 8t and B = t.
    Substituting in the formula just derived, we obtain
    cos9t - cos7t = -2sin8tsint.

5. The mass m = 4/32 = 1/8 lb-sec^2/ft and the spring
    constant k = 4/(1/8) = 32 lb/ft. Since there is no
    damping, the I.V.P. is (1/8)u″ + 32u = 2cos3t,
    u(0) = 1/6, u′(0) = 0 where u is measured in ft and t in
    sec.

7a. From the solution to Problem 5, we have m = 1/8, F_0 = 2,
    ω_0^2 = 256, and ω^2 = 9, so Eq.(3) becomes
    u = $c_1$cos16t + $c_2$sin16t + $\dfrac{16}{247}$ cos3t. The I.C.
    u(0) = 1/6 → c_1 + 16/247 = 1/6 and u′(0) = 0 → 16c_2 = 0,
    so the solution is u = (151/1482)cos16t + (16/247)cos3t
    ft.

7c. Resonance occurs when the frequency ω of the forcing
    function 4sinωt is the same as the natural frequency ω_0
    of the system. Since ω_0 = 16, the system will resonate
    when ω = 16 rad/sec.

10. The I.V.P. is .25u″ + 16u = 8sin8t, u(0) = 3 and
    u′(0) = 0. Thus, the particular solution has the form
    t(Acos8t + Bsin8t) and resonance occurs.

11a. For this problem the mass m = 8/32 lb-sec^2/ft and the
    spring constant k = 8/(1/2) = 16 lb/ft, so the D.E. is
    0.25u″ + 0.25u′ + 16u = 4cos2t where u is measured in ft
    and t in sec. To determine the steady state response we
    need only compute a particular solution of the
    nonhomogeneous D.E. since the solutions of the
    homogeneous D.E. decay to zero as t → ∞. We assume
    u(t) = Acos2t + Bsin2t, and substitute in the D.E.:
    - Acos2t - Bsin2t + (1/2)(-Asin2t + Bcos2t) + 16(Acos2t +
    Bsin2t) = 4cos2t. Hence 15A + (1/2)B = 4 and
    -(1/2)A + 15B = 0, from which we obtain A = 240/901 and
    B = 8/901. The steady state response is
    u(t) = (240cos2t + 8sin2t)/901.

11b. In order to determine the value of m that maximizes the steady state response, we note that the present problem has exactly the form of the problem considered in the text. Referring to Eqs.(8) and (9), the response is a maximum when Δ is a minimum since F_0 is constant. Δ, as given in Eq.(10), will be a minimum when

$f(m) = m^2(\omega_0{}^2 - \omega^2)^2 + \gamma^2\omega^2$, where $\omega_0{}^2 = k/m$, is a minimum. We calculate df/dm and set this quantity equal to zero to obtain $m = k/\omega^2$. We verify that this value of m gives a minimum of $f(m)$ by the second derivative test. For this problem $k = 16$ lb/ft and $\omega = 2$ rad/sec so the value of m that maximizes the response of the system is $m = 4$ slugs.

15. We must solve the three I.V.P.: $(1) u_1'' + u_1 = F_0 t$,

$0 < t < \pi$, $u_1(0) = u_1'(0) = 0$; (2) $u_2'' + u_2 = F_0(2\pi - t)$,

$\pi < t < 2\pi$, $u_2(\pi) = u_1(\pi)$, $u_2'(\pi) = u_1'(\pi)$; and

(3) $u_3'' + u_3 = 0$, $2\pi < t$, $u_3(2\pi) = u_2(2\pi)$, $u_3'(2\pi) = u_2'(2\pi)$. The conditions at π and 2π insure the continuity of u and u' at those points. The general solutions of the D.E. are $u_1 = b_1\cos t + b_2\sin t + F_0 t$, $u_2 = c_1\cos t + c_2\sin t + F_0(2\pi - t)$, and $u_3 = d_1\cos t + d_2\sin t$. The I.C. and matching conditions, in order, give $b_1 = 0$, $b_2 + F_0 = 0$, $-b_1 + \pi F_0 = -c_1 + \pi F_0$, $-b_2 + F_0 = -c_2 - F_0$, $c_1 = d_1$, and $c_2 - F_0 = d_2$. Solving these equations we obtain

$$u = F_0 \begin{cases} t - \sin t & , \ 0 \le t \le \pi \\ (2\pi - t) - 3\sin t & , \ \pi < t \le 2\pi \\ -4\sin t & , \ 2\pi < t. \end{cases}$$

16. The I.V.P. is $Q'' + 5 \times 10^3 Q' + 4 \times 10^6 Q = 12$, $Q(0) = 0$, and $Q'(0) = 0$. The particular solution is of the form $Q = A$, so that upon substitution into the D.E. we obtain $4 \times 10^6 A = 12$ or $A = 3 \times 10^{-6}$. The general solution of the D.E. is $Q = c_1 e^{r_1 t} + c_2 e^{r_2 t} + 3 \times 10^{-6}$, where r_1 and r_2 satisfy $r^2 + 5 \times 10^3 r + 4 \times 10^6 = 0$ and thus are $r_1 = -1000$ and $r_2 = -4000$. The I.C. yield $c_1 = -4 \times 10^{-6}$ and $c_2 = 10^{-6}$ and thus $Q = 10^{-6}(e^{-4000t} - 4e^{-1000t} + 3)$ coulombs. Substituting $t = .001$ sec we obtain

$Q(.001) = 10^{-6}(e^{-4} - 4e^{-1} + 3) = 1.5468 \times 10^{-6}$ coulombs.

Since exponentials are to a negative power $Q(t) \rightarrow 3 \times 10^{-6}$ coulombs as $t \rightarrow \infty$, which is the steady state charge.

22. The amplitude of the steady state response is seven or
    eight times the amplitude (3) of the forcing term. This
    large an increase is due to the fact that the forcing
    function has the same frequency as the natural frequency,
    ω_0, of the system.

    There also appears to be a phase lag of approximately 1/4
    of a period. That is, the maximum of the response occurs
    1/4 of a period after the maximum of the forcing
    function. Both these results are substantially different
    than those of either Problems 21 or 23.

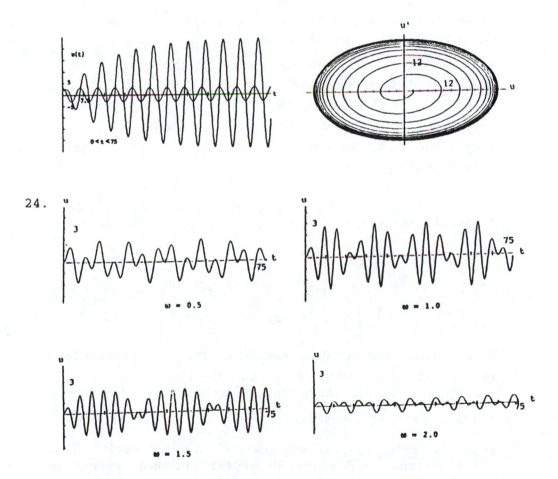

24.

From viewing the above graphs, it appears that the system
exhibits a beat near $\omega = 1.5$, while the pattern for
$\omega = 1.0$ is more irregular. However, the system exhibits
the resonance characteristic of the linear system for ω
near 1, as the amplitude of the response is the largest
here.

CHAPTER 4

2. Writing the equation in standard form, we obtain
    $y''' + [(\sin t)/t]y'' + (3/t)y = \cos t/t$. The functions
    $p_1(t) = \sin t/t$, $p_3(t) = 3/t$ and $g(t) = \cos t/t$ have
    discontinuities at $t = 0$. Hence Theorem 4.1.1 guarantees
    that a solution exists for $t < 0$ and for $t > 0$.

8. We have $W(f_1, f_2, f_3) = \begin{vmatrix} 2t-3 & 2t^2+1 & 3t^2+t \\ 2 & 4t & 6t+1 \\ 0 & 4 & 6 \end{vmatrix} = 0$ for all t.

    Thus by the extension of Theorem 3.3.1 the given
    functions are linearly dependent. Thus
    $c_1(2t-3) + c_2(2t^2+1) + c_3(3t^2+t) =$

    $(2c_2+3c_3)t^2 + (2c_1+c_3)t + (-3c_1+c_2) = 0$ when

    $(2c_2 + 3c_3) = 0$, $2c_1 + c_3 = 0$ and $-3c_1 + c_2 = 0$. Thus $c_1 = 1$
    $c_2 = 3$ and $c_3 = -2$.

13. That e^t, e^{-t}, and e^{-2t} are solutions can be verified by
    direct substitution. Computing the Wronskian we obtain,

    $$W(e^t, e^{-t}, e^{-2t}) = \begin{vmatrix} e^t & e^{-t} & e^{-2t} \\ e^t & -e^{-t} & -2e^{-2t} \\ e^t & e^{-t} & 4e^{-2t} \end{vmatrix} = e^{-2t} \begin{vmatrix} 1 & 1 & 1 \\ 1 & -1 & -2 \\ 1 & 1 & 4 \end{vmatrix} = -6e^{-2t}$$

17. To show that the given Wronskian is zero, it is helpful,
    in evaluating the Wronskian, to note that $(\sin^2 t)' =$
    $2\sin t\cos t = \sin 2t$. This result can be obtained directly
    since $\sin^2 t = (1 - \cos 2t)/2 = \dfrac{1}{10}(5) + (-1/2)\cos 2t$ and

    hence $\sin^2 t$ is a linear combination of 5 and $\cos 2t$. Thus
    the functions are linearly dependent and their Wronskian
    is zero.

19c. If we let $L[y] = y^{iv} - 5y'' + 4y$ and if we use the result
    of Problem 19b, we have $L[e^{rt}] = (r^4 - 5r^2 + 4)e^{rt}$. Thus
    e^{rt} will be a solution of the D.E. provided
    $(r^2-4)(r^2-1) = 0$. Solving for r, we obtain the four
    solutions e^t, e^{-t}, e^{2t} and e^{-2t}. Since

$W(e^t, e^{-t}, e^{2t}, e^{-2t}) \neq 0$, the four functions form a fundamental set of solutions.

21. Comparing this D.E. to that of Problem 20 we see that $p_1(t) = 2$ and thus from the results of Problem 20 we have
$$W = ce^{-\int 2dt} = ce^{-2t}.$$

27. As in Problem 26, let $y = v(t)e^t$. Differentiating three times and substituting into the D.E. yields $(2-t)e^t v''' + (3-t)e^t v'' = 0$. Dividing by $(2-t)e^t$ and letting $w = v''$ we obtain the first order separable

equation $w' = -\dfrac{t-3}{t-2}w = (-1 + \dfrac{1}{t-2})w$. Separating t and w,

integrating, and then solving for w yields $w = v'' = c_1(t-2)e^{-t}$. Integrating this twice then gives $v = c_1 te^{-t} + c_2 t + c_3$ so that $y = ve^t = c_1 t + c_2 te^t + c_3 e^t$, which is the complete solution, since it contains the given $y_1(t)$ and three constants.

Section 4.2, Page 219

2. If $-1 + i\sqrt{3} = Re^{i\theta}$, then $R = [(-1)^2 + (\sqrt{3})^2]^{1/2} = 2$. The angle θ is given by $R\cos\theta = 2\cos\theta = -1$ and $R\sin\theta = 2\sin\theta = \sqrt{3}$. Hence $\cos\theta = -1/2$ and $\sin\theta = \sqrt{3}/2$ which has the solution $\theta = 2\pi/3$. The angle θ is only determined up to an additive integer multiple of $\pm 2\pi$.

8. Writing $(1-i)$ in the form $Re^{i\theta}$, we obtain $(1-i) = \sqrt{2}\, e^{i(-\pi/4+2m\pi)}$ where m is any integer. Hence, $(1-i)^{1/2} = [2^{1/2}e^{i(-\pi/4+2m\pi)}]^{1/2} = 2^{1/4}e^{i(-\pi/8+m\pi)}$. We obtain the two square roots by setting m = 0,1. They are $2^{1/4}e^{-i\pi/8}$ and $2^{1/4}e^{i7\pi/8}$. Note that any other integer value of m gives one of these two values. Also note that 1-i could be written as $1-i = \sqrt{2}\, e^{i(7\pi/4 + 2m\pi)}$.

12. We look for solutions of the form $y = e^{rt}$. Substituting in the D.E., we obtain the characteristic equation $r^3 - 3r^2 + 3r - 1 = 0$ which has roots r = 1,1,1. Since the roots are repeated, the general solution is $y = c_1 e^t + c_2 te^t + c_3 t^2 e^t$.

15. We look for solutions of the form $y = e^{rt}$. Substituting in the D.E. we obtain the characteristic equation $r^6 + 1 = 0$. The six roots of -1 are obtained by setting $m = 0,1,2,3,4,5$ in $(-1)^{1/6} = e^{i(\pi+2m\pi)/6}$. They are $e^{i\pi/6} = (\sqrt{3} + i)/2$, $e^{i\pi/2} = i$, $e^{i5\pi/6} = (-\sqrt{3} + i)/2$, $e^{i7\pi/6} = (-\sqrt{3} - i)/2$, $e^{i3\pi/2} = -i$, and $e^{i11\pi/6} = (\sqrt{3} - i)/2$. Note that there are three pairs of conjugate roots. The general solution is

$$y = e^{\sqrt{3}\,t/2}[c_1\cos(t/2) + c_2\sin(t/2)]$$
$$+ e^{-\sqrt{3}\,t/2}[c_3\cos(t/2) + c_4\sin(t/2)] + c_5\cos t + c_6\sin t.$$

23. The characteristic equation is $r^3 - 5r^2 + 3r + 1 = 0$. Using the procedure suggested following Eq. (12) we try $r = 1$ as a root and find that indeed it is. Factoring out $(r-1)$ we are then left with $r^2 - 4r - 1 = 0$, which has the roots $2 \pm \sqrt{5}$.

27. The characteristic equation in this case is $12r^4 + 31r^3 + 75r^2 + 37r + 5 = 0$. Using an equation solver we find $r = -\dfrac{1}{4}, -\dfrac{1}{3}, -1 \pm 2i$. Thus $y = c_1 e^{-t/4} + c_2 e^{-t/3} + e^{-t}(c_3\cos 2t + c_4\sin 2t)$. As in Problem 23, it is possible to find the first two of these roots without using an equation solver.

29. The characteristic equation is $r^3 + r = 0$ and hence $r = 0, +i, -i$ are the roots and the general solution is $y(t) = c_1 + c_2\cos t + c_3\sin t$. $y(0) = 0$ implies $c_1 + c_2 = 0$, $y'(0) = 1$ implies $c_3 = 1$ and $y''(0) = 2$ implies $-c_2 = 2$. Use this last equation in the first to find $c_1 = 2$ and thus $y(t) = 2 - 2\cos t + \sin t$, which continues to oscillate as $t \to \infty$.

30. The general solution is given by Eq. (21).

31. The general solution would normally be written $y(t) = c_1 + c_2 t + c_3 e^{2t} + c_4 t e^{2t}$. However, in order to evaluate the c's when the initial conditions are given at $t = 1$, it is advantageous to rewrite $y(t)$ as $y(t) = c_1 + c_2 t + c_5 e^{2(t-1)} + c_6(t-1)e^{2(t-1)}$.

34. The characteristic equation is $4r^3 + r + 5 = 0$, which has roots -1, $\frac{1}{2} \pm i$. Thus

$$y(t) = c_1 e^{-t} + e^{t/2}(c_2\cos t + c_3\sin t),$$

$$y'(t) = -c_1 e^{-t} + e^{t/2}[(c_2/2 + c_3)\cos t + (-c_2 + c_3/2)\sin t]$$

and

$$y''(t) = c_1 e^{-t} + e^{t/2}[(-3c_2/4 + c_3)\cos t + (-c_2 - 3c_3/4)\sin t].$$

The I.C. then yield $c_1 + c_2 = 2$, $-c_1 + c_2/2 + c_3 = 1$ and $c_1 - 3c_2/4 + c_3 = -1$. Solving these last three equations give $c_1 = 2/13$, $c_2 = 24/13$ and $c_3 = 3/13$.

37. The approach developed in this section for solving the D.E. would normally yield $y(t) = c_1\cos t + c_2\sin t + c_5 e^t + c_6 e^{-t}$ as the solution. Now use the definition of $\cosh x$ and $\sinh t$ to yield the desired result. It is convenient to use $\cosh t$ and $\sinh t$ rather than e^t and e^{-t} because the I.C. are given at $t = 0$. Since $\cosh t$ and $\sinh t$ and all of their derivatives are either 0 or 1 at $t = 0$, the algebra in satisfying the I.C. is simplified.

38a. Since $p_1(t) = 0$, $W = ce^{-\int 0\,dt} = c$.

39a. As in Section 3.8, the force that the spring designated by k_1 exerts on mass m_1 is $-3u_1$. By an analysis similar to that shown in Section 3.8, the middle spring exerts a force of $-2(u_1-u_2)$ on mass m_1 and a force of $-2(u_2-u_1)$ on mass m_2. In all cases the positive direction is taken in the direction shown in Figure 4.2.4.

39c. From Eq.(i) we have $u_1''(0) = 2u_2(0) - 5u_1(0) = -1$ and $u_1'''(0) = 2u_2'(0) - 5u_1'(0) = 0$. From Prob.39b we have $u_1 = c_1\cos t + c_2\sin t + c_3\cos\sqrt{6}\,t + c_4\sin\sqrt{6}\,t$. Thus $c_1+c_3 = 1$, $c_2+\sqrt{6}\,c_4 = 0$, $-c_1-6c_3 = -1$ and $-c_2-6\sqrt{6}\,c_4 = 0$, which yield $c_1 = 1$ and $c_2 = c_3 = c_4 = 0$, so that $u_1 = \cos t$. The first of Eqs.(i) then gives u_2.

Section 4.3, Page 224

1. First solve the homogeneous D.E. The characteristic

equation is $r^3 - r^2 - r + 1 = 0$, and the roots are $r = -1$, 1, 1; hence $y_c(t) = c_1 e^{-t} + c_2 e^t + c_3 t e^t$. Using the superposition principle, we can write a particular solution as the sum of particular solutions corresponding to the D.E. $y''' - y'' - y' + y = 2e^{-t}$ and $y''' - y'' - y' + y = 3$. Our initial choice for a particular solution, Y_1, of the first equation is Ae^{-t}; but e^{-t} is a solution of the homogeneous equation so we multiply by t. Thus, $Y_1(t) = Ate^{-t}$. For the second equation we choose $Y_2(t) = B$, and there is no need to modify this choice. The constants are determined by substituting into the individual equations. We obtain $A = 1/2$, $B = 3$. Thus, the general solution is

$$y = c_1 e^{-t} + c_2 e^t + c_3 t e^t + 3 + (te^{-t})/2.$$

5.   The characteristic equation is $r^4 - 4r^2 = r^2(r^2 - 4) = 0$, so $y_c(t) = c_1 + c_2 t + c_3 e^{-2t} + c_4 e^{2t}$. For the particular solution correspnding to t^2 we assume $Y_1 = t^2(At^2 + Bt + C)$ and for the particular solution corresponding to e^t we assume $Y_2 = De^t$. Substituting Y_1, in the D.E. yields $-48A = 1$, $B = 0$ and $24A - 8C = 0$ and substituting Y_2 yields $-3D = 1$. Solving for A, B, C and D gives the desired solution.

9.   The characteristic equation for the related homogeneous D.E. is $r^3 + 4r = 0$ with roots $r = 0$, $+2i$, $-2i$. Hence $y_c(t) = c_1 + c_2 \cos 2t + c_3 \sin 2t$. The initial choice for $Y(t)$ is $At + B$, but since B is a solution of the homogeneous equation we must multiply by t and assume $Y(t) = t(At+B)$. A and B are found by substituting in the D.E., which gives $A = 1/8$, $B = 0$, and thus the general solution is $y(t) = c_1 + c_2 \cos 2t + c_3 \sin 2t + (1/8)t^2$. Applying the I.C. we have $y(0) = 0 \rightarrow c_1 + c_2 = 0$, $y'(0) = 0 \rightarrow 2c_3 = 0$, and $y''(0) = 1 \rightarrow -4c_2 + 1/4 = 1$, which have the solution $c_1 = 3/16$, $c_2 = -3/16$, $c_3 = 0$. For small t the graph will approximate $3(1-\cos 2t)/16$ and for large t it will be approximated by $t^2/8$.

13. The characteristic equation for the homogeneous D.E. is $r^3 - 2r^2 + r = 0$ with roots $r = 0,1,1$. Hence the complementary solution is $y_c(t) = c_1 + c_2 e^t + c_3 t e^t$. We

consider the differential equations $y''' - 2y'' + y' = t^3$
and $y''' - 2y'' + y' = 2e^t$ separately. Our initial choice
for a particular solution, Y_1, of the first equation is
$A_0 t^3 + A_1 t^2 + A_2 t + A_3$; but since a constant is a solution
of the homogeneous equation we must multiply by t. Thus
$Y_1(t) = t(A_0 t^3 + A_1 t^2 + A_2 t + A_3)$. For the second equation
we first choose $Y_2(t) = Be^t$, but since both e^t and te^t
are solutions of the homogeneous equation, we multiply by
t^2 to obtain $Y_2(t) = Bt^2 e^t$. Then $Y(t) = Y_1(t) + Y_2(t)$ by
the superposition principle and $y(t) = y_c(t) + Y(t)$.

17. The complementary solution is $y_c(t) = c_1 + c_2 e^{-t} + c_3 e^t +$
$c_4 te^t$. The superposition principle allows us to consider
separately the D.E. $y^{iv} - y''' - y'' + y' = t^2 + 4$ and
$y^{iv} - y''' - y'' + y' = t \sin t$. For the first equation our
initial choice is $Y_1(t) = A_0 t^2 + A_1 t + A_2$; but this must
be multiplied by t since a constant is a solution of the
homogeneous D.E. Hence $Y_1(t) = t(A_0 t^2 + A_1 t + A_2)$. For
the second equation our initial choice that
$Y_2 = (B_0 t + B_1) \cos t + + (C_0 t + C_1) \sin t$ does not need to be
modified. Hence
$Y(t) = t(A_0 t^2 + A_1 t + A_2) + (B_0 t + B_1) \cos t + (C_0 t + C_1) \sin t$.

20. $(D-a)(D-b)f = (D-a)(Df-bf) = D^2 f - (a+b)Df + abf$ and
$(D-b)(D-a)f = (D-b)(Df-af) = D^2 f - (b+a)Df + baf$. Since
$a+b = b+a$ and $ab = ba$, we find the given equation holds
for any function f.

22a. The D.E. of Problem 13 can be written as
$D(D-1)^2 y = t^3 + 2e^t$. Since D^4 annihilates t^3 and $(D-1)$
annihilates $2e^t$, we have $D^5 (D-1)^3 y = 0$, which corresponds
to Eq.(ii) of Problem 21. The solution of this equation
is $y(x) = A_1 t^4 + A_2 t^3 + A_3 t^2 + A_4 t + A_5 +$
$(B_1 t^2 + B_2 t + B_3)e^{-t}$. Since $A_5 + (B_2 t + B_3)e^{-t}$ are
solutions of the homogeneous equation related to the
original D.E., they may be deleted and thus
$Y(t) = A_1 t^4 + A_2 t^3 + A_3 t^2 + A_4 t + B_1 t^2 e^{-t}$.

22b. $(D+1)^2(D^2+1)$ annihilates the right side of the D.E. of
     Problem 14.

22e. $D^3(D^2+1)^2$ annihilates the right side of the D.E. of
     Problem 17.

Section 4.4, Page 229

1.   The complementary solution is $y_c = c_1 + c_2\cos t + c_3\sin t$
     and thus we assume a particular solution of the form
     $Y = u_1(t) + u_2(t)\cos t + u_3(t)\sin t$.   Differentiating and
     assuming Eq.(5), we obtain $Y' = -u_2\sin t + u_3\cos t$ and
     $$u_1' + u_2'\cos t + u_3'\sin t = 0 \qquad (a).$$
     Continuing this process we obtain $Y'' = -u_2\cos t - u_3\sin t$,
     $Y''' = u_2\sin t - u_3\cos t - u_2'\cos t - u_3'\sin t$ and
     $$-u_2'\sin t + u_3'\cos t = 0 \qquad (b).$$
     Substituting Y and its derivatives, as given above, into
     the D.E. we obtain the third equation:
     $$-u_2'\cos t - u_3'\sin t = \tan t \qquad (c).$$
     Equations (a), (b) and (c) constitute Eqs.(10) of the
     text for this problem and may be solved to give
     $u_1' = \tan t$, $u_2' = -\sin t$, and $u_3' = -\sin^2 t/\cos t$.   Thus
     $u_1 = -\ln\cos t$, $u_2 = \cos t$ and $u_3 = \sin t - \ln(\sec t + \tan t)$
     and $Y = -\ln\cos t + 1 - (\sin t)\ln(\sec t + \tan t)$.   Note that
     the constant 1 can be absorbed in c_1.

4.   Replace tant in Eq. (c) of Prob. 1 by sect and use
     Eqs. (a) and (b) as in Prob. 1 to obtain $u_1' = \sec t$,
     $u_2' = -1$ and $u_3' = -\sin t/\cos t$.

5.   Replace sect in Problem 7 with $e^{-t}\sin t$.

7.   Since e^t, cost and sint are solutions of the related
     homogenous equation we have
     $Y(t) = u_1 e^t + u_2\cos t + u_3\sin t$.   Eqs. (10) then are
     $$u_1'e^t + u_2'\cos t + u_3'\sin t = 0$$
     $$u_1'e^t - u_2'\sin t + u_3'\cos t = 0$$
     $$u_1'e^t - u_2'\cos t - u_3'\sin t = \sec t.$$
     Using Abel's identity, $W(t) = c\exp(-\int p_1(t)dt) = ce^t$.

Using the above equations, $W(0) = 2$, so $c = 2$ and

$W(t) = 2e^t$. From Eq.(11), we have $u'_1(t) = \dfrac{sect\ W_1(t)}{2e^t}$,

where $W_1 = \begin{vmatrix} 0 & cost & sint \\ 0 & -sint & cost \\ 1 & -cost & -sint \end{vmatrix} = 1$ and thus

$u'_1(t) = \dfrac{1}{2}e^{-t}/cost$. Likewise

$u'_2 = \dfrac{sect\ W_2(t)}{2e^t} = -\dfrac{1}{2}sect(cost - sint)$ and

$u'_3 = \dfrac{sect\ W_3(t)}{2e^t} = -\dfrac{1}{2}sect(sint + cost)$. Thus

$u_1 = \dfrac{1}{2}\displaystyle\int_{t_0}^{t}\dfrac{e^{-s}ds}{coss}$, $u_2 = -\dfrac{1}{2}t - \dfrac{1}{2}\ln(cost)$ and $u_3 = -\dfrac{1}{2}t + \dfrac{1}{2}\ln(cost)$

which, when substituted into the assumed form for Y, yields
the desired solution.

11. Since the D.E. is the same as in Problem 7, we may use
the complete solution from that, with $t_0 = 0$. Thus

$y(0) = c_1 + c_2 = 2$, $y'(0) = c_1 + c_3 - \dfrac{1}{2} + \dfrac{1}{2} = -1$ and

$y''(0) = c_1 - c_2 + \dfrac{1}{2} - 1 + \dfrac{1}{2} = 1$. Again, a computer

algebra system may be used to yield the respective
derivatives.

14. Since a fundamental set of solutions of the homogeneous
D.E. is $y_1 = e^t$, $y_2 = cost$, $y_3 = sint$, a particular
solution is of the form $Y(t) = e^t u_1(t) + (cost)u_2(t) + (sint)u_3(t)$. Differentiating and making the same
assumptions that lead to Eqs.(10), we obtain

$$u'_1e^t + u'_2cost + u'_3sint = 0$$
$$u'_1e^t - u'_2sint + u'_3cost = 0$$
$$u'_1e^t - u'_2cost - u'_3sint = g(t)$$

Solving these equations using either determinants or by
elimination, we obtain $u'_1 = (1/2)e^{-t}g(t)$,

$u'_2 = (1/2)(sint - cost)g(t)$, $u'_3 = -(1/2)(sint + cost)g(t)$.
Integrating these and substituting into Y yields

$$Y(t) = \frac{1}{2}\{e^t\int_{t_0}^t e^{-s}g(s)ds + \cos t\int_{t_0}^t(\sin s - \cos s)g(s)ds$$

$$-\sin t\int_{t_0}^t(\sin s + \cos s)g(s)ds\}.$$

This can be written in the form

$$Y(t) = (1/2)\int_{t_0}^t(e^{t-s} + \cos t\sin s - \cos t\cos s$$

$$-\sin t\sin s - \sin t\cos s)g(s)ds.$$

If we use the trigonometric identities $\sin(A-B) = \sin A\cos B - \cos A\sin B$ and $\cos(A-B) = \cos A\cos B + \sin A\sin B$, we obtain the desired result. Note: Eqs.(11) and (12) of this section give the same result, but it is not recommended to memorize these equations.

16. The particular solution has the form $Y = e^t u_1(t) + te^t u_2(t) + t^2 e^t u_3(t)$. Differentiating, making the same assumptions as in the earlier problems, and solving the three linear equations for u_1', u_2', and u_3' yields

$u_1' = (1/2)t^2 e^{-t}g(t)$, $u_2' = -te^{-t}g(t)$ and $u_3' = (1/2)e^{-t}g(t)$.

Integrating and substituting into Y yields the desired solution. For instance

$$te^t u_2 = -te^t\int_{t_0}^t se^{-s}g(s)ds = -\frac{1}{2}\int_{t_0}^t 2tse^{(t-s)}g(s)ds, \text{ and}$$

likewise for u_1 and u_3. If $g(t) = t^{-2}e^t$ then $g(s) = e^s/s^2$ and the integration is accomplished using the power rule. Note that terms involving t_0 become part of the complimentary solution.

CHAPTER 5

2. Use the ratio test:
$$\lim_{n \to \infty} \frac{|(n+1)x^{n+1}/2^{n+1}|}{|nx^n/2^n|} = \lim_{n \to \infty} \frac{n+1}{n} \frac{1}{2} |x| = \frac{|x|}{2}.$$
Therefore the series converges absolutely for $|x| < 2$. For $x = 2$ and $x = -2$ the nth term does not approach zero as $n \to \infty$ so the series diverge. Hence the radius of convergence is $\rho = 2$.

5. Use the ratio test:
$$\lim_{n \to \infty} \frac{|(2x+1)^{n+1}/(n+1)^2|}{|(2x+1)^n/n^2|} = \lim_{n \to \infty} \frac{n^2}{(n+1)^2} |2x+1| = |2x+1|.$$
Therefore the series converges absolutely for $|2x+1| < 1$, or $|x+1/2| < 1/2$. At $x = 0$ and $x = -1$ the series also converge absolutely. However, for $|x+1/2| > 1/2$ the series diverges by the ratio test. The radius of convergence is $\rho = 1/2$.

9. For this problem $f(x) = \sin x$. Hence $f'(x) = \cos x$, $f''(x) = -\sin x$, $f'''(x) = -\cos x, \ldots$. Then $f(0) = 0$, $f'(0) = 1$, $f''(0) = 0$, $f'''(0) = -1, \ldots$. The even terms in the series will vanish and the odd terms will alternate in sign. We obtain $\sin x = \sum_{n=0}^{\infty} (-1)^n x^{2n+1}/(2n+1)!$. From the ratio test it follows that $\rho = \infty$.

12. For this problem $f(x) = x^2$. Hence $f'(x) = 2x$, $f''(x) = 2$, and $f^{(n)}(x) = 0$ for $n > 2$. Then $f(-1) = 1$, $f'(-1) = -2$, $f''(-1) = 2$ and $x^2 = 1 - 2(x+1) + 2(x+1)^2/2! = 1 - 2(x+1) + (x+1)^2$. Since the series terminates after a finite number of terms, it converges for all x. Thus $\rho = \infty$.

13. For this problem $f(x) = \ln x$. Hence $f'(x) = 1/x$, $f''(x) = -1/x^2$, $f'''(x) = 1 \cdot 2/x^3, \ldots$, and $f^{(n)}(x) = (-1)^{n+1}(n-1)!/x^n$. Then $f(1) = 0$, $f'(1) = 1$, $f''(1) = -1$, $f'''(1) = 1 \cdot 2, \ldots$, $f^{(n)}(1) = (-1)^{n+1}(n-1)!$ The Taylor series is $\ln x = (x-1) - (x-1)^2/2 + (x-1)^3/3 - \ldots = \sum_{n=1}^{\infty} (-1)^{n+1}(x-1)^n/n$. It follows from the ratio test that

the series converges absolutely for $|x-1| < 1$. However,
the series diverges at $x = 0$ so $\rho = 1$.

18. Writing the individual terms of y, we have
$$y = a_0 + a_1 x + a_2 x^2 + \ldots + a_a x^n + \ldots, \text{ so}$$
$$y' = a_1 + 2a_2 x + 3a_3 x^2 + \ldots + (n+1)a_{n+1} x^n + \ldots, \text{ and}$$
$$y'' = 2a_2 + 3 \cdot 2a_3 x + 4 \cdot 3a_4 x^2 + \ldots + (n+2)(n+1)a_{n+2} x^n + \ldots \, .$$
If $y'' = y$, we then equate coefficients of like powers of x to
obtain $2a_2 = a_0$, $3 \cdot 2a_3 = a_1$, $4 \cdot 3a_4 = a_2$, $\ldots$ $(n+2)(n+1)a_{n+2} = a_n$,
which yields the desired result for $n = 0,1,2,3 \ldots$.

19. Set $m = n-1$ on the right hand side of the equation. Then
$n = m+1$ and when $n = 1$, $m = 0$. Thus the right hand side

becomes $\displaystyle\sum_{m=0}^{\infty} a_m (x-1)^{m+1}$, which is the same as the left hand

side when m is replaced by n.

23. Multiplying each term of the first series by x yields

$$x\sum_{n=1}^{\infty} na_n x^{n-1} = \sum_{n=1}^{\infty} na_n x^n = \sum_{n=0}^{\infty} na_n x^n, \text{ where the last}$$

equality can be verified by writing out the first few
terms. Changing the index from k to n ($n=k$) in the
second series then yields

$$\sum_{n=0}^{\infty} na_n x^n + \sum_{n=0}^{\infty} a_n x^n = \sum_{n=0}^{\infty} (n+1)a_n x^n.$$

25. $\displaystyle\sum_{m=2}^{\infty} m(m-1)a_m x^{m-2} + x\sum_{k=1}^{\infty} ka_k x^{k-1} =$

$\displaystyle\sum_{n=0}^{\infty} (n+2)(n+1)a_{n+2} x^n + \sum_{k=1}^{\infty} ka_k x^k =$

$\displaystyle\sum_{n=0}^{\infty} [(n+2)(n+1)a_{n+2} + na_n]x^n$. In the first case we have

let $n = m - 2$ in the first summation and multiplied each
term of the second summation by x. In the second case we
have let $n = k$ and noted that for $n = 0$, $na_n = 0$.

28. If we shift the index of summation in the first sum by
letting $m = n-1$, we have

$$\sum_{n=1}^{\infty} na_n x^{n-1} = \sum_{m=0}^{\infty} (m+1)a_{m+1}x^m.$$ Substituting this into the given equation and letting $m = n$ again, we obtain:

$$\sum_{n=0}^{\infty} (n+1)a_{n+1} x^n + 2 \sum_{n=0}^{\infty} a_n x^n = 0, \text{ or}$$

$$\sum_{n=0}^{\infty} [(n+1)a_{n+1} + 2a_n]x^n = 0.$$

Hence $a_{n+1} = -2a_n/(n+1)$ for $n = 0,1,2,3,\dots$. Thus $a_1 = -2a_0$, $a_2 = -2a_1/2 = 2^2a_0/2$, $a_3 = -2a_2/3 = -2^3a_0/2\cdot 3 = -2^3a_0/3!\dots$ and $a_n = (-1)^n 2^n a_0/n!$. Notice that for $n = 0$ this formula reduces to a_0 so we can write

$$\sum_{n=0}^{\infty} a_n x^n = \sum_{n=0}^{\infty} (-1)^n 2^n a_0 x^n/n! = a_0 \sum_{n=0}^{\infty} (-2x)^n/n! = a_0 e^{-2x}.$$

Section 5.2, Page 247

2. $y = \sum_{n=0}^{\infty} a_n x^n$; $y' = \sum_{n=1}^{\infty} na_n x^{n-1}$ and since we must multiply y' by x in the D.E. we do not shift the index; and

$$y'' = \sum_{n=2}^{\infty} n(n-1)a_n x^{n-2} = \sum_{n=0}^{\infty} (n+2)(n+1)a_{n+2}x^n.$$ Substituting in the D.E., we obtain

$$\sum_{n=0}^{\infty} (n+2)(n+1)a_{n+2}x^n - \sum_{n=1}^{\infty} na_n x^n - \sum_{n=0}^{\infty} a_n x^n = 0.$$ In order to have the starting point the same in all three summations, we let $n = 0$ in the first and third terms to obtain the following

$$(2\cdot 1\ a_2 - a_0)x^0 + \sum_{n=1}^{\infty} [(n+2)(n+1)a_{n+2} - (n+1)a_n]x^n = 0.$$

Thus $a_{n+2} = a_n/(n+2)$ for $n = 1,2,3,\dots$. Note that the recurrence relation is also correct for $n = 0$. We show how to calculate the odd a's:
$a_3 = a_1/3$, $a_5 = a_3/5 = a_1/5\cdot 3$, $a_7 = a_5/7 = a_1/7\cdot 5\cdot 3, \dots$.

Now notice that $a_3 = 2a_1/(2 \cdot 3) = 2a_1/3!$, that
$a_5 = 2 \cdot 4a_1/(2 \cdot 3 \cdot 4 \cdot 5) = 2^2 2a_1/5!$, and that
$a_7 = 2 \cdot 4 \cdot 6a_1/(2 \cdot 3 \cdot 4 \cdot 5 \cdot 6 \cdot 7) = 2^3 3! \, a_1/7!$. Likewise
$a_9 = a_7/9 = 2^3 3! \, a_1/(7!)9 = 2^3 3! \, 8a_1/9! = 2^4 4! \, a_1/9!$.
Continuing we have $a_{2m+1} = 2^m m! \, a_1/(2m+1)!$. In the same
way we find that the even a's are given by $a_{2m} = a_0/2^m \, m!$.
Thus

$$y = a_0 \sum_{m=0}^{\infty} \frac{x^{2m}}{2^m m!} + a_1 \sum_{m=0}^{\infty} \frac{2^m m! \, x^{2m+1}}{(2m+1)!}.$$

3. $y = \sum_{n=0}^{\infty} a_n(x-1)^n; \quad y' = \sum_{n=1}^{\infty} na_n(x-1)^{n-1} = \sum_{n=0}^{\infty} (n+1)a_{n+1}(x-1)^n,$

and

$$y'' = \sum_{n=2}^{\infty} n(n-1)a_n(x-1)^{n-2} = \sum_{n=0}^{\infty} (n+2)(n+1)a_{n+2}(x-1)^n.$$

Substituting in the D.E. and setting $x = 1 + (x-1)$ we
obtain

$$\sum_{n=0}^{\infty} (n+2)(n+1)a_{n+2}(x-1)^n - \sum_{n=0}^{\infty} (n+1)a_{n+1}(x-1)^n - \sum_{n=1}^{\infty} na_n(x-1)^n$$

$$- \sum_{n=0}^{\infty} a_n(x-1)^n = 0,$$

where the third term comes from:

$$-(x-1)y' = \sum_{n=1}^{\infty} (n+1)a_{n+1}(x-1)^{n+1} = -\sum_{n=1}^{\infty} na_n(x-1)^n.$$

Letting $n = 0$ in the first, second, and the fourth sums,
we obtain

$$(2 \cdot 1 \cdot a_2 - 1 \cdot a_1 - a_0)(x-1)^0 + \sum_{n=1}^{\infty} [(n+2)(n+1)a_{n+2}$$

$$- (n+1)a_{n+1} - (n+1)a_n](x-1)^n = 0.$$

Thus $(n+2)a_{n+2} - a_{n+1} - a_n = 0$ for $n = 0,1,2,\ldots$. This
recurrence relation can be used to solve for a_2 in terms
of a_0 and a_1, then for a_3 in terms of a_0 and a_1, etc. In
many cases it is easier to first take $a_0 = 0$ and generate
one solution and then take $a_1 = 0$ and generate the second

linearly independent solution. Thus, choosing $a_0 = 0$ we
find that $a_2 = a_1/2$, $a_3 = (a_2+a_1)/3 = a_1/2$, $a_4 = (a_3+a_2)/4 =$
$a_1/4$, $a_5 = (a_4+a_3)/5 = 3a_1/20,\ldots$. This yields the
solution $y_2(x) = a_1[(x-1) + (x-1)^2/2 + (x-1)^3/2 +$
$(x-1)^4/4 + 3(x-1)^5/20 + \ldots]$. The second independent
solution may be obtained by choosing $a_1 = 0$. Then
$a_2 = a_0/2$, $a_3 = (a_2+a_1)/3 = a_0/6$, $a_4 = (a_3+a_2)/4 = a_0/6$,
$a_5 = (a_4+a_3)/5 = a_0/15,\ldots$. This yields the solution
$y_1(x) = a_0[1+(x-1)^2/2+(x-1)^3/6+(x-1)^4/6+(x-1)^5/15+\ldots]$.

5. $y = \sum_{n=0}^{\infty} a_n x^n$; $y' = \sum_{n=1}^{\infty} n a_n x^{n-1}$; and $y'' = \sum_{n=2}^{\infty} n(n-1) a_n x^{n-2}$.

Substituting in the D.E. and shifting the index in both
summations for y'' gives

$$\sum_{n=0}^{\infty} (n+2)(n+1) a_{n+2} x^n - \sum_{n=1}^{\infty} (n+1)n\, a_{n+1} x^n + \sum_{n=0}^{\infty} a_n x^n =$$

$$(2\cdot1\cdot a_2 + a_0) x^0 + \sum_{n=1}^{\infty} [(n+2)(n+1) a_{n+2} - (n+1)n a_{n+1} + a_n] x^n = 0.$$

Thus $a_2 = -a_0/2$ and $a_{n+2} = n a_{n+1}/(n+2) - a_n/(n+2)(n+1)$,
$n = 1,2,\ldots$. Choosing $a_0 = 0$ yields $a_2 = 0$, $a_3 = -a_1/6$,
$a_4 = 2a_3/4 = -a_1/12,\ldots$ which gives one solution as
$y_2(x) = a_1(x - x^3/6 - x^4/12 + \ldots)$. A second linearly
independent solution is obtained by choosing $a_1 = 0$. Then
$a_2 = -a_0/2$, $a_3 = a_2/3 = -a_0/6$, $a_4 = 2a_3/4 - a_2/12 = -a_0/24,\ldots$
which gives $y_1(x) = a_0(1 - x^2/2 - x^3/6 - x^4/24 + \ldots)$.

8. If $y = \sum_{n=1}^{\infty} a_n(x-1)^n$ then

$$xy = [1+(x-1)]y = \sum_{n=1}^{\infty} a_n(x-1)^n + \sum_{n=1}^{\infty} a_n(x-1)^{n+1},$$

$$y' = \sum_{n=1}^{\infty} n a_n(x-1)^{n-1}, \text{ and}$$

$$xy'' = [1+(x-1)]y''$$

$$= \sum_{n=1}^{\infty} n(n-1) a_n(x-1)^{n-2} + \sum_{n=1}^{\infty} n(n-1) a_n(x-1)^{n-1}.$$

14. You will need to rewrite x+1 as 3 + (x-2) in order to multiply x+1 times y' as a power series about $x_0 = 2$.

16a. From Problem 6 we have

$$y(x) = c_1(1 - x^2 + \frac{1}{6}x^4 + \ldots) + c_2(x - \frac{1}{4}x^3 + \frac{7}{160}x^5 + \ldots).$$

Now $y(0) = c_1 = -1$ and $y'(0) = c_2 = 3$ and thus

$$y(x) = -1+x^2-\frac{1}{6}x^4+3x-\frac{3}{4}x^3 = -1+3x+x^2-\frac{3}{4}x^3-\frac{1}{6}x^4 + \ldots .$$

16c. By plotting $f = -1 + 3x + x^2 - 3x^3/4$ and $g = f - x^4/6$ between -1 and 1 it appears that f is a reasonable approximation for $|x| < 0.7$.

19. The D.E. transforms into $u''(t) + t^2 u'(t) + (t^2+2t)u(t) = 0$.

Assuming that $u(t) = \sum\limits_{n=0}^{\infty} a_n t^n$, we have $u'(t) = \sum\limits_{n=1}^{\infty} n a_n t^{n-1}$ and

$u''(t) = \sum\limits_{n=2}^{\infty} n(n-1)a_n t^{n-2}$. Substituting in the D.E. and shifting indices yields

$$\sum_{n=0}^{\infty} (n+2)(n+1)a_{n+2}t^n + \sum_{n=2}^{\infty} (n-1)a_{n-1}t^n + \sum_{n=2}^{\infty} a_{n-2}t^n$$

$$+ \sum_{n=1}^{\infty} 2a_{n-1}t^n = 0,$$

$$2 \cdot 1 \cdot a_2 t^0 + (3 \cdot 2 \cdot a_3 + 2 \cdot a_0)t^1 + \sum_{n=2}^{\infty} [(n+2)(n+1)a_{n+2}$$

$$+ (n+1)a_{n-1} + a_{n-2}]t^n = 0.$$

It follows that $a_2 = 0$, $a_3 = -a_0/3$ and $a_{n+2} = -a_{n-1}/(n+2) - a_{n-2}/[(n+2)(n+1)]$, $n = 2,3,4\ldots$. We obtain one solution by choosing $a_1 = 0$. Then $a_4 = -a_0/12$, $a_5 = -a_2/5 - a_1/20 = 0$, $a_6 = -a_3/6 - a_2/30 = a_0/18,\ldots$. Thus one solution is $u_1(t) = a_0(1 - t^3/3 - t^4/12 + t^6/18 + \ldots)$ so $y_1(x) = u_1(x-1) = a_0[1 - (x-1)^3/3 - (x-1)^4/12 + (x-1)^6/18 + \ldots]$. We obtain a second solution by choosing $a_0 = 0$. Then $a_4 = -a_1/4$, $a_5 = -a_2/5 - a_1/20 = -a_1/20$, $a_6 = -a_3/6 - a_2/30 = 0$, $a_7 = -a_4/7 - a_3/42 = a_1/28,\ldots$.

Thus a second linearly independent solution is

$u_2(t) = a_1[t - t^4/4 - t^5/20 + t^7/28 + \ldots]$ or

$y_2(x) = u_2(x-1)$

$\qquad = a_1[(x-1) - (x-1)^4/4 - (x-1)^5/20 + (x-1)^7/28 + \ldots].$

The Taylor series for $x^2 - 1$ about $x = 1$ may be obtained by writing $x = (x-1) + 1$ so $x^2 = (x-1)^2 + 2(x-1) + 1$ and $x^2 - 1 = (x-1)^2 + 2(x-1)$. The D.E. now appears as $y'' + (x-1)^2y' + [(x-1)^2 + 2(x-1)]y = 0$ which is identical to the transformed equation with $t = x - 1$.

22b. $y = a_0 + a_1x + a_2x^2 + \ldots$, $y^2 = a_0^2 + 2a_0a_1x + (2a_0a_2 + a_1^2)x^2$

$\quad + \ldots$, $y' = a_1 + 2a_2x + 3a_3x^2 + \ldots$, and

$(y')^2 = a_1^2 + 4a_1a_2x + (6a_1a_3 + 4a_2^2)x^2 + \ldots$. Substituting

these into $(y')^2 = 1 - y^2$ and collecting coefficients of

like powers of x yields $(a_1^2 + a_0^2 - 1) + (4a_1a_2 + 2a_0a_1)x +$

$(6a_1a_3 + 4a_2^2 + 2a_0a_2 + a_1^2)x^2 + \ldots = 0$. As in the earlier

problems, each coefficient must be zero. The I.C. $y(0) = 0$

requires that $a_0 = 0$, and thus $a_1^2 + a_0^2 - 1 = 0$ gives $a_1^2 = 1$.

However, the D.E. indicates that y' is always positive, so

$y'(0) = a_1 > 0$ implies $a_1 = 1$. Then $4a_1a_2 + 2a_0a_1 = 0$ implies

that $a_2 = 0$; and $6a_1a_3 + 4a_2^2 + 2a_0a_2 + a_1^2 = 6a_1a_3 + a_1^2 = 0$

implies that $a_3 = -1/6$. Thus $y = x - x^3/3! + \ldots$.

23.   26.

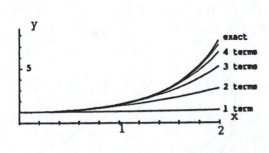
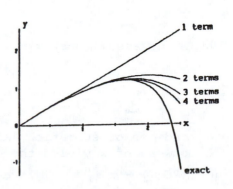

26. We have $y(x) = a_0y_1 + a_1y_2$, where y_1 and y_2 are found in
     Problem 10. Now $y(0) = a_0 = 0$ and $y'(0) = a_1 = 1$. Thus

$$y(x) = x - \frac{x^3}{12} - \frac{x^5}{240} - \frac{x^7}{2240}.$$

Section 5.3, Page 253

1. The D.E. can be solved for y'' to yield $y'' = -xy' - y$. If
    $y = \phi(x)$ is a solution, then $\phi''(x) = -x\phi'(x) - \phi(x)$ and
    thus setting $x = 0$ we obtain $\phi''(0) = -0 - 1 = -1$.
    Differentiating the equation for y'' yields
    $y''' = -xy'' - 2y'$ and hence setting $y = \phi(x)$ again yields
    $\phi'''(0) = -0 - 0 = 0$. In a similar fashion
    $y^{iv} = -xy''' - 3y''$ and thus $\phi^{iv}(0) = -0 - 3(-1) = 3$. The
    process can be continued to calculate higher derivatives
    of $\phi(x)$.

6. The zeros of $P(x) = x^2 - 2x - 3$ are $x = -1$ and $x = 3$. For
    $x_0 = 4$, $x_0 = -4$, and $x_0 = 0$ the distance to the nearest
    zero of $P(x)$ is 1,3, and 1, respectively. Thus a lower
    bound for the radius of convergence for series solutions
    in powers of $(x-4)$, $(x+4)$, and x is $\rho = 1$, $\rho = 3$, and
    $\rho = 1$, respectively.

9a. Since $P(x) = 1$ has no zeros, the radius of convergence
    for $x_0 = 0$ is $\rho = \infty$.

9f. Since $P(x) = x^2 + 2$ has zeros at $x = \pm\sqrt{2}\ i$, the lower
    bound for the radius of convergence of the series
    solution about $x_0 = 0$ is $\rho = \sqrt{2}$.

9h. Since $x_0 = 1$ and $P(x) = x$ has a zero at $x = 0$, $\rho = 1$.

10a. If we assume that $y = \displaystyle\sum_{n=2}^{\infty} a_n x^n$, then $y' = \displaystyle\sum_{n=2}^{\infty} n a_n x^{n-1}$ and

$y'' = \displaystyle\sum_{n=2}^{\infty} n(n-1) a_n x^{n-2}$. Substituting in the D.E., shifting
indices of summation, and collecting coefficients of like
powers of x yields the equation
$$(2{\cdot}1{\cdot}a_2 + \alpha^2 a_0)x^0 + [3{\cdot}2{\cdot}a_3 + (\alpha^2-1)a_1]x^1$$
$$+ \sum_{n=2}^{\infty}[(n+2)(n+1)a_{n+2} + (\alpha^2-n^2)a_n]x^n = 0.$$
Hence the recurrence relation is
$a_{n+2} = (n^2-\alpha^2)a_n/(n+2)(n+1)$, $n = 0, 1, 2, \ldots$. For the
first solution we choose $a_1 = 0$. We find that

$a_2 = -\alpha^2 a_0/2 \cdot 1$, $a_3 = 0$, $a_4 = (2^2-\alpha^2)a_2/4 \cdot 3 = -(2^2-\alpha^2)\alpha^2 a_0/4!$

$\ldots$, $a_{2m} = -[(2m-2)^2 - \alpha^2] \ldots (2^2-\alpha^2)\alpha^2 a_0/(2m)!$,

and $a_{2m+1} = 0$, so $y_1(x) = 1 - \dfrac{\alpha^2}{2!}x^2 - \dfrac{(2^2-\alpha^2)\alpha^2}{4!}x^4 - \ldots$

$$- \frac{[(2m-2)^2-\alpha^2] \ldots (2^2-\alpha^2)\alpha^2}{(2m)!}x^{2m} - \ldots,$$

where we have set $a_0 = 1$. For the second solution we take $a_0 = 0$ and $a_1 = 1$ in the recurrence relation to obtain the desired solution.

10b. If α is an even integer $2k$ then $(2m-2)^2 - \alpha^2 = (2m-2)^2 - 4k^2 = 0$. Thus when $m = k+1$ all terms in the series for $y_1(x)$ are zero after the x^{2k} term. A similar argument shows that if $\alpha = 2k+1$ then all terms in $y_2(x)$ are zero after the x^{2k+1}.

11. The Taylor series about $x = 0$ for $\sin x$ is

$\sin x = x - x^3/3! + x^5/5! - \ldots$. Assuming that

$y = \displaystyle\sum_{n=2}^{\infty} a_n x^n$ we find $y'' + (\sin x)y = 2a_2 + 6a_3 x + 12a_4 x^2$

$+ 20a_5 x^3 + 30a_6 x^4 + 42a_7 x^5 + \ldots$

$+ (x-x^3/3!+x^5/5!-\ldots)(a_0+a_1 x+a_2 x^2+a_3 x^3+a_4 x^4+\ldots)$

$= 2a_2 + (6a_3+a_0)x + (12a_4+a_1)x^2 + (20a_5+a_2-a_0/6)x^3 +$

$(30a_6+ a_3-a_1/6)x^4 + (42a_7+a_4+a_0/5!)x^5 + \ldots = 0$. Hence

$a_2 = 0$, $a_3 = -a_0/6$, $a_4 = -a_1/12$, $a_5 = a_0/120$,

$a_6 = (a_1+a_0)/180$, $a_7 = -a_0/7! + a_1/504$, $\ldots$. We set

$a_0 = 1$ and $a_1 = 0$ and obtain

$y_1(x) = (1 - x^3/6 + x^5/120 + x^6/180 + \ldots)$. Next we set

$a_0 = 0$ and $a_1 = 1$ and obtain

$y_2(x) = (x - x^4/12 + x^6/180 + x^7/504 +\ldots)$. Since

$p(x) = 1$ and $q(x) = \sin x$ both have $\rho = \infty$, the solution in this case converges for all x, that is, $\rho = \infty$

18. We know that $e^x = 1 + x + x^2/2! + x^3/3! + \ldots$, and

therefore $e^{x^2} = 1 + x^2 + x^4/2! + x^6/3! + \ldots$. Hence, if

$y = \Sigma a_n x^n$, we have

$a_1 + 2a_2x + 3a_3x^2 + \ldots = (1+x^2 + x^4/2+\ldots)(a_0+a_1x+a_2x^2+\ldots)$

$$= a_0 + a_1x + (a_0+a_2)x^2 + \ldots:$$

Thus, $a_1 = a_0$ $2a_2 = a_1$ and $3a_3 = a_0 + a_2$, which yield the desired solution.

20. Substituting $y = \sum\limits_{n=2}^{\infty} a_nx^n$ into the D.E. we obtain

$\sum\limits_{n=2}^{\infty} na_nx^{n-1} - \sum\limits_{n=2}^{\infty} a_nx^n = x^2$. Shifting indices in the summation

yields $\sum\limits_{n=2}^{\infty} [(n+1)a_{n+1} - a_n]x^n = x^2$. Equating coefficients of

both sides then gives: $a_1 - a_0 = 0$, $2a_2 - a_1 = 0$, $3a_3 - a_2 = 1$ and $(n+1)a_{n+1} = a_n$ for $n = 3,4,\ldots$. Thus $a_1 = a_0$, $a_2 = a_1/2 = a_0/2$, $a_3 = 1/3 + a_2/3 = 1/3 + a_0/2\cdot3$, $a_4 = a_3/4 = 1/3\cdot4 + a_0/2\cdot3\cdot4$, $\ldots$, $a_n = a_{n-1}/n = 2/n! + a_0/n!$ and hence

$$y(x) = a_0(1 + x + \frac{x^2}{2!}+\ldots+\frac{x^n}{n!}\ldots) + 2(\frac{x^3}{3!}+\frac{x^4}{4!}+\ldots+\frac{x^n}{n!}+\ldots).$$

Using the power series for e^x, the first and second sums can be rewritten as $a_0e^x + 2(e^x - 1 - x - x^2/2)$.

22. Substituting $y = \sum\limits_{n=2}^{\infty} a_nx^n$ into the Legendre equation, shifting indices, and collecting coefficients of like powers of x yields

$[2\cdot1\cdot a_2 + \alpha(\alpha+1)a_0]x^0 + \{3\cdot2\cdot a_3 - [2\cdot1 - \alpha(\alpha+1)]a_1\}x^1 +$

$\sum\limits_{n=2}^{\infty} \{(n+2)(n+1)a_{n+2} - [n(n+1) - \alpha(\alpha+1)]a_n\}x^n = 0$. Thus

$a_2 = -\alpha(\alpha+1)a_0/2!$, $a_3 = [2\cdot1 - \alpha(\alpha+1)]a_1/3! =$ $-(\alpha-1)(\alpha+2)a_1/3!$ and the recurrence relation is $(n+2)(n+1)a_{n+2} = -[\alpha(\alpha+1) - n(n+1)]a_n = -(\alpha-n)(\alpha+n+1)a_n$, $n = 2,3,\ldots$. Setting $a_1 = 0$, $a_0 = 1$ yields a solution with $a_3 = a_5 = a_7 = \ldots = 0$ and

$a_4 = \alpha(\alpha-2)(\alpha+1)(\alpha+3)/4!,\ldots$, $a_{2m} = (-1)^m\alpha(\alpha-2)(\alpha-4) \ldots$ $(\alpha-2m+2)(\alpha+1)(\alpha+3) \ldots (\alpha+2m-1)/(2m)!,\ldots$. The second linearly independent solution is obtained by setting

$a_0 = 0$ and $a_1 = 1$. The coefficients are $a_2 = a_4 = a_6 = \ldots = 0$ and $a_3 = -(\alpha-1)(\alpha+2)/3!$, $a_5 = -(\alpha-3)(\alpha+4)a_3/5\cdot4 = (\alpha-1)(\alpha-3)(\alpha+2)(\alpha+4)/5!, \ldots$.

26. Using the chain rule we have:

$$\frac{dF(\phi)}{d\phi} = \frac{dF[\phi(x)]}{dx}\frac{dx}{d\phi} = -f'(x)\sin\phi(x) = -f'(x)\sqrt{1-x^2},$$

$$\frac{d^2F(\phi)}{d\phi^2} = \frac{d}{dx}[-f'(x)\sqrt{1-x^2}]\frac{dx}{d\phi} = (1-x^2)f''(x) - xf'(x),$$

which when substituted into the D.E. yields the desired result.

28. Since $[(1-x^2)y']' = (1-x^2)y'' - 2xy'$, the Legendre Equation, from Problem 22, can be written as shown. Thus, carrying out the steps indicated yields the two equations:

$$P_m[(1-x^2)P_n']' = -n(n+1)P_nP_m$$

$$P_n[(1-x^2)P_m']' = -m(m+1)P_nP_m.$$

As long as $n \neq m$ the second equation can be subtracted from the first and the result integrated from -1 to 1 to obtain

$$\int_{-1}^{1}\{P_m[(1-x^2)P_n']'-P_n[(1-x^2)P_m']'\}dx = [m(m+1)-n(n+1)]\int_{-1}^{1}P_nP_mdx$$

The left side may be integrated by parts to yield

$$[P_m(1-x^2)P_n' - P_n(1-x^2)P_m']_{-1}^{1} + \int_{-1}^{1}[P_m'(1-x^2)P_n' - P_n'(1-x^2)P_m']dx,$$

which is zero. Thus $\int_{-1}^{1}P_n(x)P_m(x)dx = 0$ for $n \neq m$.

Section 5.4, Page 259

1. Since the coefficients of y, y' and y'' have no common factors and since $P(x)$ vanishes only at $x = 0$ we conclude that $x = 0$ is a singular point. Writing the D.E. in the form $y'' + p(x)y' + q(x)y = 0$, we obtain $p(x) = (1-x)/x$ and $q(x) = 1$. Thus for the singular point we have

$$\lim_{x\to0} x\, p(x) = \lim_{x\to0} 1-x = 1, \quad \lim_{x\to0} x^2 q(x) = 0 \text{ and thus } x = 0$$

is a regular singular point.

5. Writing the D.E. in the form $y'' + p(x)y' + q(x)y = 0$, we find $p(x) = x/(1-x)(1+x)^2$ and $q(x) = 1/(1-x^2)(1+x)$. Therefore $x = \pm1$ are singular points. Since

$\lim\limits_{x\to 1} (x-1)p(x)$ and $\lim\limits_{x\to 1} (x-1)^2 q(x)$ both exist, we conclude
$x = 1$ is a regular singular point. Finally, since
$\lim\limits_{x\to -1} (x+1)p(x)$ does not exist, we find that $x = -1$ is an
irregular singular point.

12. Writing the D.E. in the form $y + p(x)y' + q(x)y = 0$, we
see that $p(x) = e^x/x$ and $q(x) = (3\cos x)/x$. Thus $x = 0$ is
a singular point. Since $xp(x) = e^x$ is analytic at $x = 0$
and $x^2 q(x) = 3x\cos x$ is analytic at $x = 0$ the point $x = 0$
is a regular singular point.

17. Writing the D.E. in the form $y'' + p(x)y' + q(x)y = 0$, we
see that $p(x) = \dfrac{x}{\sin x}$ and $q(x) = \dfrac{4}{\sin x}$. Since $\lim\limits_{x\to 0} q(x)$
does not exist, the point $x_0 = 0$ is a singular point and
since neither $\lim\limits_{x\to \pm n\pi} p(x)$ nor $\lim\limits_{x\to \pm n\pi} q(x)$ exist either the
points $x_0 = \pm n\pi$ are also singular points. To determine
whether the singular points are regular or irregular we
must use Eq.(8) and the result #7 of multiplication and
division of power series from Section 5.1. For $x_0 = 0$,
we have

$$xp(x) = \dfrac{x^2}{\sin x} = \dfrac{x^2}{x - \dfrac{x^3}{6} + \ldots} = x[1 + \dfrac{x^2}{6} + \ldots]$$

$$= x + \dfrac{x^3}{6} + \ldots,$$

which converges about $x_0 = 0$ and thus $xp(x)$ is analytic
at $x_0 = 0$. $x^2 q(x)$, by similar steps, is also analytic at
$x_0 = 0$ and thus $x_0 = 0$ is a regular singular point. For
$x_0 = n\pi$, we have

$$(x-n\pi)p(x) = \dfrac{(x-n\pi)x}{\sin x} = \dfrac{(x-n\pi)[(x-n\pi) + n\pi]}{\pm(x-n\pi) + \dfrac{-(x-n\pi)^3}{6} \pm \ldots}$$

$$= [(x-n\pi)+n\pi][\pm 1 \pm \dfrac{(x-n\pi)^2}{6} \pm \ldots], \text{ which}$$

converges about $x_0 = n\pi$ and thus $(x-n\pi)p(x)$ is analytic
at $x = n\pi$. Similarly $(x+n\pi)p(x)$ and $(x\pm n\pi)^2 q(x)$ are
analytic and thus $x_0 = \pm n\pi$ are regular singular points.

19. Substituting $y = \sum_{n=0}^{\infty} a_n x^n$ into the D.E. yields

$$2 \sum_{n=2}^{\infty} n(n-1) a_n x^{n-1} + 3 \sum_{n=1}^{\infty} n a_n x^{n-1} + \sum_{n=0}^{\infty} a_n x^{n+1} = 0.$$ The last sum

becomes $\sum_{n=2}^{\infty} a_{n-2} x^{n-1}$ by replacing n+1 by n-1, the first term
of the middle sum is $3a_1$, and thus we have

$$3a_1 + \sum_{n=2}^{\infty} \{[2n(n-1)+3n] a_n + a_{n-2}\} x^{n-1} = 0.$$ Hence $a_1 = 0$ and

$a_n = \dfrac{-a_{n-2}}{n(2n+1)}$, which is the desired recurrance relation.

Thus all even coefficients are found in terms of a_0 and
all odd coefficients are zero, thereby yielding only one
solution of the desired form.

21. If $\xi = 1/x$ then
$$\frac{dy}{dx} = \frac{dy}{d\xi} \frac{d\xi}{dx} = -\frac{1}{x^2} \frac{dy}{d\xi} = -\xi^2 \frac{dy}{d\xi},$$

$$\frac{d^2 y}{dx^2} = \frac{d}{d\xi} (-\xi^2 \frac{dy}{d\xi}) \frac{d\xi}{dx} = (-2\xi \frac{dy}{d\xi} - \xi^2 \frac{d^2 y}{d\xi^2}) (-\frac{1}{x^2})$$

$$= \xi^4 \frac{d^2 y}{d\xi^2} + 2\xi^3 \frac{dy}{d\xi}.$$

Substituting in the D.E. we have

$$P(1/\xi) [\xi^4 \frac{d^2 y}{d\xi^2} + 2\xi^3 \frac{dy}{d\xi}] + Q(1/\xi) [-\xi^2 \frac{dy}{d\xi}] + R(1/\xi) y = 0,$$

$$\xi^4 P(1/\xi) \frac{d^2 y}{d\xi^2} + [2\xi^3 P(1/\xi) - \xi^2 Q(1/\xi)] \frac{dy}{d\xi} + R(1/\xi) y = 0.$$

The result then follows from the theory of singular
points at $\xi = 0$.

23. Since $P(x) = x^2$, $Q(x) = x$ and $R(x) = -4$ we have
$f(\xi) = [2P(1/\xi)/\xi - Q(1/\xi)/\xi^2]/P(1/\xi) = 2/\xi - 1/\xi = 1/\xi$
and $g(\xi) = R(1/\xi)/\xi^4 P(1/\xi) = -4/\xi^2$. Thus the point at
infinity is a singular point. Since both $\xi f(\xi)$ and
$\xi^2 g(\xi)$ are analytic at $\xi = 0$, the point at infinity is a
regular singular point.

25. Since $P(x) = x^2$, $Q(x) = x$, and $R(x) = x^2 - v^2$,
     $f(\xi) = [2P(1/\xi)/\xi - Q(1/\xi)/\xi^2]/P(1/\xi) = 2/\xi - 1/\xi = 1/\xi$
     and $g(\xi) = R(1/\xi)/\xi^4 P(1/\xi) = (1/\xi^2 - v^2)/\xi^2 = 1/\xi^4 - v^2/\xi^2$.
     Thus the point at infinity is a singular point. Although
     $\xi f(\xi) = 1$ is analytic at $\xi = 0$, $\xi^2 g(\xi) = 1/\xi^2 - v^2$ is not,
     so the point at infinity is an irregular singular point.

Section 5.5, Page 265

2.   Comparing the D.E. to Eq.(27), we seek solutions of the
     form $y = (x+1)^r$ for $x + 1 > 0$. Substitution of y into
     the D.E. yields $[r(r-1) + 3r + 3/4](x+1)^r = 0$. Thus
     $r^2 + 2r + 3/4 = 0$, which yields $r = -3/2, -1/2$. The
     general solution of the D.E. is then
     $y = c_1|x+1|^{-1/2} + c_2|x+1|^{-3/2}$, $x \neq -1$.

4.   If $y = x^r$ then $r(r-1) + 3r + 5 = 0$. So $r^2 + 2r + 5 = 0$
     and $r = (-2 \pm \sqrt{4-20})/2 = -1 \pm 2i$. Thus the general
     solution of the D.E. is
     $y = c_1 x^{-1}\cos(2\ln|x|) + c_2 x^{-1}\sin(2\ln|x|)$, $x \neq 0$.

9.   Again let $y = x^r$ to obtain $r(r-1) - 5r + 9 = 0$, or
     $(r-3)^2 = 0$. Thus the roots are $x = 3,3$ and
     $y = c_1 x^3 + c_2 x^3 \ln|x|$, $x \neq 0$, is the solution of the D.E.

13. If $y = x^r$, then $F(r) = 2r(r-1) + r - 3 = 2r^2 - r - 3 =$
     $(2r-3)(r+1) = 0$, so $y = c_1 x^{3/2} + c_2 x^{-1}$ and
     $y' = \dfrac{3}{2}c_1 x^{1/2} - c_2 x^{-2}$. Setting $x = 1$ in y and y' we obtain
     $c_1 + c_2 = 1$ and $\dfrac{3}{2}c_1 - c_2 = 4$, which yield $c_1 = 2$ and
     $c_2 = -1$. Hence $y = 2x^{3/2} - x^{-1}$. As $x \to 0^+$ we have
     $y \to -\infty$ due to the second term.

16. We have $F(r) = r(r-1) + 3r + 5 = r^2 + 2r + 5 = 0$. Thus
     $r_1, r_2 = -1 \pm 2i$ and $y = x^{-1}[c_1\cos(2\ln x) + c_2\sin(2\ln x)]$.
     Then $y(1) = c_1 = 1$ and $y' = -x^{-2}[\cos(2\ln x) + c_2\sin(2\ln x)]$
     $+ x^{-1}[-\sin(2\ln x)2/x + c_2\cos(2\ln x)2/x]$ so that
     $y'(1) = -1-2c_2 = -1$, or $c_2 = 0$.

17. Substituting $y = x^r$, we find that $r(r-1) + \alpha r + 5/2 = 0$ or $r^2 + (\alpha-1)r + 5/2 = 0$. Thus $r_1, r_2 = [-(\alpha-1) \pm \sqrt{(\alpha-1)^2-10}]/2$. In order for solutions to approach zero as $x \to 0$ it is necessary that the real parts of r_1 and r_2 be positive. Suppose that $\alpha > 1$, then $\sqrt{(\alpha-1)^2-10}$ is either imaginary or real and less than $\alpha - 1$; hence the real parts of r_1 and r_2 will be negative. Suppose that $\alpha = 1$, then $r_1, r_2 = \pm i\sqrt{10}$ and the solutions are oscillatory. Suppose that $\alpha < 1$, then $\sqrt{(\alpha-1)^2-10}$ is either imaginary or real and less than $|\alpha-1| = 1 - \alpha$; hence the real parts of r_1 and r_2 will be positive. Thus if $\alpha < 1$ the solutions of the D.E. will approach zero as $x \to 0$.

21. In all cases the roots of $F(r) = 0$ are given by Eq.(5) and the forms of the solution are given in Theorem 5.5.1.

21a. The real part of the root must be positive so, from Eq.(5), $\alpha < 0$. Also $\beta > 0$, since the $\sqrt{(\alpha-1)^2-4\beta}$ term must be less than $|\alpha-1|$.

22. Assume that $y = v(x)x^{r_1}$. Then $y' = v(x)r_1 x^{r_1-1} + v'(x)x^{r_1}$ and $y'' = v(x)r_1(r_1-1)x^{r_1-2} + 2v'(x)r_1 x^{r_1-1} + v''(x)x^{r_1}$. Substituting in the D.E. and collecting terms yields $x^{r_1+2}v'' + (\alpha + 2r_1)x^{r_1+1}v' + [r_1(r_1-1) + \alpha r_1 + \beta]x^{r_1}v = 0$. Now we make use of the fact that r_1 is a double root of $f(r) = r(r-1) + \alpha r + \beta$. This means that $f(r_1) = 0$ and $f'(r_1) = 2r_1 - 1 + \alpha = 0$. Hence the D.E. for v reduces to $x^{r_1+2}v'' + x^{r_1+1}v'$. Since $x > 0$ we may divide by x^{r_1+1} to obtain $xv'' + v' = 0$. Thus $v(x) = \ln x$ and a second solution is $y = x^{r_1}\ln x$.

25. The change of variable $x = e^z$ transforms the D.E. into $u'' - 4u' + 4u = z$, which has the solution $u(z) = c_1 e^{2z} + c_2 z e^{2z} + (1/4)z + 1/4$. Hence $y(x) = c_1 x^2 + c_2 x^2 \ln x + (1/4)\ln x + 1/4$.

31. If $x > 0$, then $|x| = x$ and $|x|^{r_1} = x^{r_1}$ so we can choose $c_1 = k_1$. If $x < 0$, then $|x| = -x$ and $|x|^{r_1} = (-x)^{r_1} = (-1)^{r_1}x^{r_1}$ and we can choose $c_1 = (-1)^{r_1}k_1$,

or $k_1 = (-1)^{-r_1}c_1 = (-1)^{r_1}c_1$. In both cases we have
$c_2 = k_2$.

Section 5.6, Page 271

2. If the D.E. is put in the standard form $y'' + p(x)y + q(x)y = 0$, then $p(x) = x^{-1}$ and $q(x) = 1 - 1/9x^2$. Thus $x = 0$ is a singular point. Since $xp(x) \to 1$ and $x^2 q(x) \to -1/9$ as $x \to 0$ it follows that $x = 0$ is a regular singular point. In determining a series solution of the D.E. it is more convenient to leave the equation in the form given rather than divide by the x^2, the

coefficient of y''. If we substitute $y = \sum\limits_{n=0}^{\infty} a_n x^{n+r}$, we have

$$\sum_{n=0}^{\infty} (n+r)(n+r-1)a_n x^{n+r} + \sum_{n=0}^{\infty} (n+r)a_n x^{n+r} + (x^2 - \frac{1}{9})\sum_{n=0}^{\infty} a_n x^{n+r} = 0.$$

Note that $x^2 \sum\limits_{n=0}^{\infty} a_n x^{n+r} = \sum\limits_{n=0}^{\infty} a_n x^{n+r+2} = \sum\limits_{n=2}^{\infty} a_{n-2} x^{n+r}$. Thus we

have $[r(r-1) + r - \frac{1}{9}]a_0 x^r + [(r+1)r + (r+1) - \frac{1}{9}]a_1 x^{r+1} +$

$$\sum_{n=2}^{\infty} \{[(n+r)(n+r-1) + (n+r) - \frac{1}{9}]a_n + a_{n-2}\} x^{n+r} = 0.$$ From

the first term, the indicial equation is $r^2 - 1/9 = 0$ with roots $r_1 = 1/3$ and $r_2 = -1/3$. For either value of r it is necessary to take $a_1 = 0$ in order that the coefficient of x^{r+1} be zero. The recurrence relation is $[(n+r)^2 - 1/9]a_n = -a_{n-2}$. For $r = 1/3$ we have

$$a_n = \frac{-a_{n-2}}{(n + \frac{1}{3})^2 - (\frac{1}{3})^2} = -\frac{a_{n-2}}{(n + \frac{2}{3})n}, \quad n = 2,3,4,\ldots .$$

Since $a_1 = 0$ it follows from the recurrence relation that $a_3 = a_5 = a_7 = \ldots = 0$. For the even coefficients it is convenient to let $n = 2m$, $m = 1,2,3,\ldots .$ Then $a_{2m} = -a_{2m-2}/2^2 m(m + \frac{1}{3})$. The first few coefficients are given by

$$a_2 = \frac{(-1)a_0}{2^2(1 + \frac{1}{3})1}, \quad a_4 = \frac{(-1)a_2}{2^2(2 + \frac{1}{3})2} = \frac{a_0}{2^4(1 + \frac{1}{3})(2 + \frac{1}{3})2!}$$

$$a_6 = \frac{(-1)a_4}{2^2(3 + \frac{1}{3})3} = \frac{(-1)a_0}{2^6(1 + \frac{1}{3})(2 + \frac{1}{3})(3 + \frac{1}{3})3!}, \quad \text{and the}$$

coefficent of x^{2m} for m = 1, 2,... is

$$a_{2m} = \frac{(-1)^m a_0}{2^{2m}m!(1 + \frac{1}{3})(2 + \frac{1}{3}) \ldots (m + \frac{1}{3})}. \quad \text{Thus one}$$

solution (on setting $a_0 = 1$) is

$$y_1(x) = x^{1/3}[1 + \sum_{m=1}^{\infty} \frac{(-1)^m}{m!(1 + \frac{1}{3})(2 + \frac{1}{3})\ldots(m + \frac{1}{3})} (\frac{x}{2})^{2m}].$$

Since $r_2 = -1/3 \neq r_1$ and $r_1 - r_2 = 2/3$ is not an integer, we can calculate a second series solution corresponding to $r = -1/3$. The recurrence relation is $n(n-2/3)a_n = -a_{n-2}$, which yields the desired solution following the steps just outlined. Note that $a_1 = 0$, as in the first solution, and thus all the odd coefficients are zero.

4. Putting the D.E. in standard form $y''+p(x)y'+q(x)y = 0$, we see that $p(x) = 1/x$ and $q(x) = -1/x$. Thus $x = 0$ is a singular point, and since $xp(x) \to 1$ and $x^2q(x) \to 0$, as $x \to 0$, $x = 0$ is a regular singular point. Substituting

$$y = \sum_{n=0}^{\infty} a_n x^{n+r} \text{ in } xy'' + y' - y = 0 \text{ and shifting indices we}$$

obtain

$$\sum_{n=-1}^{\infty} a_{n+1}(r+n+1)(r+n)x^{n+r} + \sum_{n=-1}^{\infty} a_{n+1}(r+n+1)x^{n+r} - \sum_{n=0}^{\infty} a_n x^{n+r} = 0,$$

$$[r(r-1) + r]a_0 x^{-1+r} + \sum_{n=0}^{\infty} [(r+n+1)^2 a_{n+1} - a_n]x^{n+r} = 0. \quad \text{The}$$

indicial equation is $r^2 = 0$ so $r = 0$ is a double root. Thus we will obtain only one series of the form

$$y = x^r \sum_{n=0}^{\infty} a_n x^n. \quad \text{The recurrence relation is}$$

$(n+1)^2 a_{n+1} = a_n$, $n = 0,1,2,\ldots$. The coefficients are
$a_1 = a_0$, $a_2 = a_1/2^2 = a_0/2^2$, $a_3 = a_2/3^2 = a_0/3^2 \cdot 2^2$,
$a_4 = a_3/4^2 = a_0/4^2 \cdot 3^2 \cdot 2^2$, $\ldots$ and $a_n = a_0/(n!)^2$. Thus one

solution (on setting $a_0 = 1$) is $y = \displaystyle\sum_{n=0}^{\infty} x^n/(n!)^2$.

11. If we make the change of variable $t = x-1$ and let
$y = u(t)$, then the Legendre equation transforms to
$(t^2 + 2t)u''(t) + 2(t+1)u'(t) - \alpha(\alpha+1)u(t) = 0$. Since
$x = 1$ is a regular singular point of the original
equation, we know that $t = 0$ is a regular singular point

of the transformed equation. Substituting $u = \displaystyle\sum_{n=0}^{\infty} a_n t^{n+r}$

in the transformed equation and shifting indices, we
obtain

$$\sum_{n=0}^{\infty} (n+r)(n+r-1)a_n t^{n+r} + 2\sum_{n=-1}^{\infty} (n+r+1)(n+r)a_{n+1}t^{n+r}$$

$$+ 2\sum_{n=0}^{\infty} (n+r)a_n t^{n+r} + 2\sum_{n=-1}^{\infty} (n+r+1)a_{n+1}t^{n+r}$$

$$- \alpha(\alpha+1)\sum_{n=0}^{\infty} a_n t^{n+r} = 0, \quad \text{or}$$

$$[2r(r-1) + 2r]a_0 t^{r-1} + \sum_{n=0}^{\infty} \{2(n+r+1)^2 a_{n+1}$$

$$+ [(n+r)(n+r+1) - \alpha(\alpha+1)]a_n\}t^{n+r} = 0.$$

The indicial equation is $2r^2 = 0$ so $r = 0$ is a double
root. Thus there will be only one series solution of the

form $y = \displaystyle\sum_{n=0}^{\infty} a_n t^{n+r}$. The recurrence relation is

$2(n+1)^2 a_{n+1} = [\alpha(\alpha+1) - n(n+1)]a_n, n = 0,1,2,\ldots$. We have
$a_1 = [\alpha(\alpha+1)]a_0/2 \cdot 1^2$, $a_2 = [\alpha(\alpha+1)][\alpha(\alpha+1) - 1 \cdot 2]a_0/2^2 \cdot 2^2 \cdot 1^2$,
$a_3 = [\alpha(\alpha+1)][\alpha(\alpha+1) - 1 \cdot 2][\alpha(\alpha+1) - 2 \cdot 3]a_0/2^3 \cdot 3^2 \cdot 2^2 \cdot 1^2, \ldots$,
and $a_n = [\alpha(\alpha+1)][\alpha(\alpha+1)-1 \cdot 2]..[\alpha(\alpha+1)-(n-1)n]a_0/2^n(n!)^2$.

Reverting to the variable x it follows that one solution
of the Legendre equation in powers of x-1 is

$$y_1(x) = \sum_{n=0}^{\infty} [\alpha(\alpha+1)][\alpha(\alpha+1) - 1\cdot 2] \ldots$$

$[\alpha(\alpha+1) - (n-1)n](x-1)^n/2^n(n!)^2$ where we have set $a_0 = 1$,
which is equivalent to the answer in the text if a (-1)
is taken out of each square bracket.

14. The standard form is $y'' + p(x)y' + q(x)y = 0$, with
$p(x) = 1/x$ and $q(x) = 1$. Thus x = 0 is a singular point;
and since $xp(x) \to 1$ and $x^2q(x) \to 0$ as $x \to 0$, x = 0 is a

regular singular point. Substituting $y = \sum_{n=0}^{\infty} a_n x^{n+r}$ into

$x^2y'' + xy' + x^2y = 0$ and shifting indices appropriately,
we obtain

$$\sum_{n=0}^{\infty} (n+r)(n+r-1)a_n x^{n+r} + \sum_{n=0}^{\infty} (n+r)a_n x^{n+r} + \sum_{n=2}^{\infty} a_{n-2}x^{n+r} = 0,$$

or

$[r(r-1)+r]a_0 x^r + [(1+r)r+1+r]a_1 x^{r+1}$

$$+\sum_{n=2}^{\infty} [(n+r)^2 a_n + a_{n-2}]x^{n+r} = 0.$$ The indicial equation

is $r^2 = 0$ so r = 0 is a double root. It is necessary to
take $a_1 = 0$ in order that the coefficient of x^{r+1} be zero.
The recurrence relation in $n^2 a_n = -a_{n-2}$, n = 2,3,... .
Since $a_1 = 0$ it follows that $a_3 = a_5 = a_7 = \ldots = 0$. For
the even coefficients we let n = 2m, m = 1,2,... . Then
$a_{2m} = -a_{2m-2}/2^2 m^2$ so $a_2 = -a_0/2^2\cdot 1^2$, $a_4 = a_0/2^2\cdot 2^2\cdot 1^2\cdot 2^2$,... ,
and $a_{2m} = (-1)^m a_0/2^{2m}(m!)^2$. Thus one solution of the Bessel

equation of order zero is $J_0(x) = 1 + \sum_{m=1}^{\infty} (-1)^m x^{2m}/2^{2m}(m!)^2$

where we have set $a_0 = 1$. Using the ratio test it can be
shown that the series converges for all x. Also note
that $J_0(x) \to 1$ as $x \to 0$.

15. In order to determine the form of the integral for x near
zero we must study the integrand for x small. Using the

above series for J_0, we have

$$\frac{1}{x[J_0(x)]^2} = \frac{1}{x[1 - x^2/2 + \ldots]^2} = \frac{1}{x[1 - x^2 + \ldots]} =$$

$\frac{1}{x}[1 + x^2 + \ldots]$ for x small. Thus

$$y_2(x) = J_0(x)\int\frac{dx}{x[J_0(x)]^2} = J_0(x)\int[\frac{1}{x} + x + \ldots]dx$$

$$= J_0(x)[\ln x + \frac{x^2}{x} + \ldots], \text{ and it is clear that } y_2(x)$$

will contain a logarithmic term.

16a. Putting the D.E. in the standard form
$y'' + p(x)y' + q(x)y = 0$ we see that $p(x) = 1/x$ and
$q(x) = (x^2-1)/x^2$. Thus $x = 0$ is a singular point and
since $xp(x) \to 1$ and $x^2q(x) \to -1$ as $x \to 0$, $x = 0$ is a

regular singular point. Substituting $y = \sum\limits_{n=0}^{\infty} a_n x^{n+r}$ into

$x^2y'' + xy' + (x^2-1)y = 0$, shifting indices appropriately,
and collecting coefficients of common powers of x we
obtain $[r(r-1) + r - 1]a_0x^r + [(1+r)r + 1 + r - 1]a_1x^{r+1}$

$$+ \sum\limits_{n=2}^{\infty} \{[(n+r)^2 - 1]a_n + a_{n-2}\}x^{n+r} = 0.$$

The indicial equation is $r^2-1 = 0$ so the roots are $r_1 = 1$
and $r_2 = -1$. For either value of r it is necessary to
take $a_1 = 0$ in order that the coefficient of x^{r+1} be zero.
The recurrence relation is $[(n+r)^2 - 1]a_n = -a_{n-2}$,
$n = 2,3,4\ldots$. For $r = 1$ we have $a_n = -a_{n-2}/[n(n+2)]$,
$n = 2,3,4,\ldots$. Since $a_1 = 0$ it follows that $a_3 = a_5 = a_7$
$= \ldots = 0$. Let $n = 2m$. Then $a_{2m} = -a_{2m-2}/2^2m(m+1)$, $m =$
$1,2,\ldots$, so $a_2 = -a_0/2^2 \cdot 1 \cdot 2$, $a_4 = -a_2/2^2 \cdot 1 \cdot 2 \cdot 3 =$
$a_0/2^2 \cdot 2^2 \cdot 1 \cdot 2 \cdot 2 \cdot 3,\ldots$, and $a_{2m} = (-1)^m a_0/2^{2m}m!(m+1)!$. Thus one
solution (set $a_0 = 1/2$) of the Bessel equation of order

one is $J_1(x) = (x/2)\sum\limits_{n=0}^{\infty} (-1)^n x^{2n}/(n+1)!n!2^{2n}$. The ratio

test shows that the series converges for all x. Also
note that $J_1(x) \to 0$ as $x \to 0$.

16b. For r = -1 the recurrence relation is
$[(n-1)^2 - 1]a_n = -a_{n-2}$, n = 2,3,... . Substituting n = 2
into the relation yields $[(2-1)^2 - 1]a_2 = 0$ $a_2 = -a_0$.
Hence it is impossible to determine a_2 and consequently
impossible to find a series solution of the form
$$x^{-1} \sum_{n=0}^{\infty} b_n x^n.$$

Secton 5.7, Page 278

1.   The D.E. has the form $P(x)y'' + Q(x)y' + R(x)y = 0$ with
P(x) = x, Q(x) = 2x, and R(x) = $6e^x$. From this we find
p(x) = Q(x)/P(x) = 2 and q(x) = R(x)/P(x) = $6e^x/x$ and
thus x = 0 is a singular point. Since xp(x) = 2x and
$x^2 q(x) = 6xe^x$ are analytic at x = 0 we conclude that
x = 0 is a regular singular point. Next, we have
$xp(x) \to 0 = p_0$ and $x^2 q(x) \to 0 = q_0$ as $x \to 0$ and thus the
indicial equation is $r(r-1) + 0 \cdot r + 0 = r^2 - r = 0$, which
has the roots $r_1 = 1$ and $r_2 = 0$.

3.   The equation has the form $P(x)y'' + Q(x)y' + R(x)y = 0$
with P(x) = x(x-1), Q(x) = $6x^2$ and R(x) = 3. Since P(x),
Q(x), and R(x) are polynomials with no common factors and
P(0) = 0 and P(1) = 0, we conclude that x = 0 and x = 1
are singular points. The first point, x = 0, can be shown
to be a regular singular point using steps similar to
those to shown in Problem 1. For x = 1, we must put the
D.E. in a form similar to Eq.(1) for this case. To do
this, divide the D.E. by x and multiply by (x-1) to
obtain $(x-1)^2 y'' + 6x(x-1)y + \dfrac{3}{x}(x-1)y = 0$. Comparing this
to Eq.(1) we find that (x-1)p(x) = 6x and
$(x-1)^2 q(x) = 3(x-1)/x$ which are both analytic at
x = 1 and hence x = 1 is a regular singular point. These
last two expressions approach $p_0 = 6$ and $q_0 = 0$
respectively as $x \to 1$, and thus the indicial equation is
$r(r-1) + 6r + 0 = r(r+5) = 0$.

9.   For this D.E., $p(x) = \dfrac{-(1+x)}{x^2(1-x)}$ and $q(x) = \dfrac{2}{x(1-x)}$ and thus
x = 0, -1 are singular points. Since xp(x) is not
analytic at x = 0, x = 0 is not a regular singular point.

Looking at $(x-1)p(x) = \dfrac{1+x}{x^2}$ and $(x-1)^2 q(x) = \dfrac{2(1-x)}{x}$ we see that $x = 1$ is a regular singular point and that $p_0 = 2$ and $q_0 = 0$.

17a. We have $p(x) = \dfrac{\sin x}{x^2}$ and $q(x) = -\dfrac{\cos x}{x^2}$, so that $x = 0$ is a singular point. Note that $xp(x) = (\sin x)/x \to 1 = p_0$ as $x \to 0$ and $x^2 q(x) = -\cos x \to -1 = q_0$ as $x \to 0$. In order to assert that $x = 0$ is a regular singular point we must demonstrate that $xp(x)$ and $x^2 q(x)$, with $xp(x) = 1$ at $x = 0$ and $x^2 q(x) = -1$ at $x = 0$, have convergent power series (are analytic) about $x = 0$. We know that $\cos x$ is analytic so we need only consider $(\sin x)/x$. Now

$$\sin x = \sum_{n=0}^{\infty} (-1)^n x^{2n+1} / (2n+1)! \quad \text{for} \; -\infty < x < \infty \quad \text{so}$$

$$(\sin x)/x = \sum_{n=0}^{\infty} (-1)^n x^{2n} / (2n+1)! \quad \text{and hence is analytic.}$$

Thus we may conclude that $x = 0$ is a regular singular point.

17b. From part a) it follows that the indicial equation is $r(r-1) + r - 1 = r^2 - 1 = 0$ and the roots are $r_1 = 1$, $r_2 = -1$.

17c. To find the first few terms of the solution corresponding to $r_1 = 1$, assume that

$y = x(a_0 + a_1 x + a_2 x^2 + \ldots) = a_0 x + a_1 x^2 + a_2 x^3 + \ldots$.
Substituting this series for y in the D.E. and expanding $\sin x$ and $\cos x$ about $x = 0$ yields
$x^2(2a_1 + 6a_2 x + 12a_3 x^2 + 20a_4 x^3 + \ldots) +$
$(x - x^3/3! + x^5/5! - \ldots)(a_0 + 2a_1 x + 3a_2 x^2 + 4a_3 x^3 + 5a_4 x^4 +$
$\ldots) - (1 - x^2/2! + x^4/4! - \ldots)(a_0 x + a_1 x^2 + a_2 x^3 + a_3 x^4 +$
$a_4 x^5 + \ldots) = 0$. Collecting terms, $(2a_1 + 2a_1 - a_1)x^2 +$
$(6a_2 + 3a_2 - a_0/6 - a_2 + a_0/2)x^3 + (12a_3 + 4a_3 - 2a_1/6 - a_3 +$
$a_1/2)x^4 + (20a_4 + 5a_4 - 3a_2/6 + a_0/120 - a_4 + a_2/2 -$
$a_0/24)x^5 + \ldots = 0$. Simplifying, $3a_1 x^2 + (8a_2 + a_0/3)x^3 +$
$(15a_3 + a_1/6)x^4 + (24a_4 - a_0/30)x^5 + \ldots = 0$. Thus, $a_1 = 0$,

$a_2 = -a_0/4!$, $a_3 = 0$, $a_4 = a_0/6!$, ... Hence
$y_1(x) = x - x^3/4! + x^5/6! + \ldots$ where we have set $a_0 = 1$.
From Eq. (24) the second solution has the form

$$y_2(x) = ay_1(x)\ln x + x^{-1}(1 + \sum_{n=1}^{\infty} c_n x^n)$$

$$= ay_1(x)\ln x + \frac{1}{x} + c_1 + c_2 x + c_3 x^2 + c_4 x^3 + \ldots, \text{ so}$$

$y_2' = ay_1'\ln x + ay_1 x^{-1} - x^{-2} + c_2 + 2c_3 x + 3c_4 x^2 + \ldots, \text{ and}$
$y_2'' = ay_1''\ln x + 2ay_1'x^{-1} - ay_1 x^{-2} + 2x^{-3} + 2c_3 + 3c_4 x + \ldots.$
When these are substituted in the given D.E. the terms
including $\ln x$ will appear as
$a[x^2 y_1'' + (\sin x)y_1' - (\cos x)y_1]$, which is zero since y_1 is
a solution. For the remainder of the terms, use
$y_1 = x - x^3/24 + x^5/720$ and the $\cos x$ and $\sin x$ series as
shown earlier to obtain
$-c_1 + (2/3+2a)x + (3c_3+c_1/2)x^2 + (4/45+c_2/3+8c_4)x^3 + \ldots = 0.$
These yield $c_1 = 0$, $a = -1/3$, $c_3 = 0$, and
$c_4 = -c_2/24 - 1/90$. We may take $c_2 = 0$, since this term
will simply generate $y_1(x)$ over again. Thus
$y_2(x) = -\frac{1}{3}y_1(x)\ln x + x^{-1} - \frac{1}{90}x^3$. If a computer algebra
system is used, then additional terms in each series may
be obtained without much additional effort. The next
terms, in each case, are shown here:
$y_1(x) = x - \dfrac{x^3}{24} + \dfrac{x^5}{720} - \dfrac{43x^7}{1451520} + \ldots$ and

$y_2(x) = -\dfrac{1}{3}y_1(x)\ln x + \dfrac{1}{x}[1 - \dfrac{x^4}{90} + \dfrac{41x^6}{120960} - \ldots].$

18. We first write the D.E. in the standard form as given for
Theorem 5.7.1 except that we are expanding in powers of
$(x-1)$ rather than powers of x:
$(x-1)^2 y'' + (x-1)[(x-1)/2\ln x]y' + [(x-1)^2/\ln x]y = 0.$ since
$\ln 1 = 0$, $x = 1$ is a singular point. To show it is a
regular singular point of this D.E. we must show that
$(x-1)/\ln x$ is analytic at $x = 1$; it will then follow that
$(x-1)^2/\ln x = (x-1)[(x-1)/\ln x]$ is also analytic at
$x = 1$. If we expand $\ln x$ in a Taylor series about $x = 1$
we find that $\ln x = (x-1) - \dfrac{1}{2}(x-1)^2 + \dfrac{1}{3}(x-1)^3 - \ldots.$

Thus

$(x-1)/\ln x = [1 - \frac{1}{2}(x-1) + \frac{1}{3}(x-1)^2 - \ldots]^{-1} = 1 + \frac{1}{2}(x-1) + \ldots$

has a power series expansion about $x = 1$, and hence is analytic. We can use the above result to obtain the indicial equation at $x = 1$. We have

$(x-1)^2 y'' + (x-1)[\frac{1}{2} + \frac{1}{4}(x-1) + \ldots]y' + [(x-1) +$

$\frac{1}{2}(x-1)^2 + \ldots]y = 0$. Thus $p_0 = 1/2$, $q_0 = 0$ and the

indicial equation is $r(r-1) + r/2 = 0$. Hence $r = 1/2$ and $r = 0$. In order to find the first three non-zero terms in a series solution corresponding to $r = 1/2$, it is better to keep the differential equation in its original form and to substitute the above power series for $\ln x$:

$[(x-1) - \frac{1}{2}(x-1)^2 + \frac{1}{3}(x-1)^3 - \frac{1}{4}(x-1)^4 + \ldots]y'' + \frac{1}{2}y' + y = 0.$

Next we substitute $y = a_0(x-1)^{1/2} + a_1(x-1)^{3/2} + a_2(x-1)^{5/2}$ + ... and collect coefficients of like powers of $(x-1)$ which are then set equal to zero. This requires some algebra before we find that $6a_1/4 + 9a_0/8 = 0$ and $5a_2 + 5a_1/8 - a_0/12 = 0$. These equations yield $a_1 = -3a_0/4$ and $a_2 = 53a_0/480$. With $a_0 = 1$ we obtain the solution

$y_1(x) = (x-1)^{1/2} - \frac{3}{4}(x-1)^{3/2} + \frac{53}{480}(x-1)^{5/2} + \ldots$. Since

the radius of convergence of the series for $\ln x$ is 1, we would expect $\rho = 1$.

20a. If we write the D.E. in the standard form as given in Theorem 5.7.1 we obtain $x^2 y'' + x[\alpha/x]y' + [\beta/x]y = 0$ where $xp(x) = \alpha/x$ and $x^2 q(x) = \beta/x$. Neither of these terms are analytic at $x = 0$ so $x = 0$ is an irregular singular point.

20b. Substituting $y = x^r \sum_{n=0}^{\infty} a_n x^n$ in $x^3 y'' + \alpha x y' + \beta y = 0$ gives

$$\sum_{n=0}^{\infty}(n+r)(n+r-1)a_n x^{n+r+1} + \alpha \sum_{n=0}^{\infty}(n+r)a_n x^{n+r} + \beta \sum_{n=0}^{\infty} a_n x^{n+r} = 0.$$

Shifting the index in the first series and collecting coefficients of common powers of x we obtain $(\alpha r + \beta)a_0 x^r$

$$+ \sum_{n=1}^{\infty} (n+r-1)(n+r-2)a_{n-1} + [\alpha(n+r) + \beta]a_n x^{n+r} = 0. \quad \text{Thus}$$

the indicial equation is $\alpha r + \beta = 0$ with the single root
$r = -\beta/\alpha$.

20c. From part b, the recurrence relation is

$$a_n = \frac{(n+r-1)(n+r-2)a_{n-1}}{\alpha(n+r) + \beta}, \quad n = 1, 2, \ldots$$

$$= \frac{(n - \frac{\beta}{\alpha} - 1)(n - \frac{\beta}{\alpha} - 2)a_{n-1}}{\alpha n}, \quad \text{for } r = -\beta/\alpha.$$

For $\dfrac{\beta}{\alpha} = -1$, then, $a_n = \dfrac{n(n-1)a_{n-1}}{\alpha n}$, which is zero for
$n = 1$ and thus $y(x) = x$ is the solution. Similarly for
$\dfrac{\beta}{\alpha} = 0$, $a_n = \dfrac{(n-1)(n-2)}{\alpha n}$ and again for $n = 1$ $a_1 = 0$ and
$y(x) = 1$ is the solution. Continuing in this fashion, we
see that the series solution will terminate for β/α any
positive integer as well as 0 and -1. For other values
of β/α, we have $\dfrac{a_n}{a_{n-1}} = \dfrac{(n-\frac{\beta}{2}-1)(n-\frac{\beta}{\alpha}-2)}{\alpha n}$, which approaches
∞ as $n \to \infty$ and thus the ratio test yields a zero radius
of convergence.

21b. Substituting $y = \sum_{n=0}^{\infty} a_n x^{n+r}$ in the D.E. in standard form

gives

$$\sum_{n=0}^{\infty} (n+r)(n+r-1)a_n x^{n+r} + \alpha \sum_{n=0}^{\infty} (n+r)a_n x^{n+r+1-s}$$

$$+ \beta \sum_{n=0}^{\infty} a_n x^{n+r+2-t} = 0.$$

If $s = 2$ and $t = 2$ the first term in each of the three
series is $r(r-1)a_0 x^r$, $\alpha r a_0 x^{r-1}$, and $\beta a_0 x^r$, respectively.
Thus we must have $\alpha r a_0 = 0$ which requires $r = 0$. Hence
there is at most one solution of the assumed form.

21d. In order for the indicial equation to be quadratic in r
it is necessary that the first term in the first series

contribute to the indicial equation. This means that the
first term in the second and the third series cannot
appear before the first term of the first series. The
first terms are $r(r-1)a_0x^r$, αra_0x^{r+1-s}, and βa_0x^{r+2-t},
respectively. Thus if $s \leq 1$ and $t \leq 2$ the quadratic term
will appear in the indicial equation.

Section 5.8, Page 289

1. It is clear that $x = 0$ is a singular point. The D.E. is
    in the standard form given in Theorem 5.7.1 with
    $xp(x) = 2$ and $x^2q(x) = x$. Both are analytic at $x = 0$, so
    $x = 0$ is a regular singular point. Substituting

    $$y = \sum_{n=0}^{\infty} a_nx^{n+r} \text{ in the D.E., shifting indices}$$

    appropriately, and collecting coefficients of like powers
    of x yields

    $$[r(r-1) + 2r]a_0x^r + \sum_{n=1}^{\infty} [(r+n)(r+n+1)a_n + a_{n-1}]x^{r+n} = 0.$$

    The indicial equation is $F(r) = r(r+1) = 0$ with roots
    $r_1 = 0$, $r_2 = -1$. Treating a_n as a function of r, we see
    that $a_n(r) = -a_{n-1}(r)/F(r+n)$, $n = 1,2,...$ if $F(r+n) \neq 0$.
    Thus $a_1(r) = -a_0/F(r+1)$, $a_2(r) = a_0/F(r+1)F(r+2),...,$ and
    $a_n(r) = (-1)^na_0/F(r+1)F(r+2)...F(r+n)$, provided $F(r+n) \neq$
    0 for $n = 1,2,...$. For the case $r_1 = 0$, we have
    $a_n(0) = (-1)^na_0/F(1)F(2) \ ... \ F(n) = (-1)^na_0/n!(n+1)!$ so

    $$\text{one solution is } y_1(x) = \sum_{n=0}^{\infty} (-1)^nx^n/n!(n+1)! \text{ where we have}$$

    set $a_0 = 1$.

    If we try to use the above recurrence relation for
    the case $r_2 = -1$ we find that $a_n(-1) = -a_{n-1}/n(n-1)$,
    which is undefined for $n = 1$. Thus we must follow the
    procedure described at the end of Section 5.7 to
    calculate a second solution of the form given in Eq.(24).
    Specifically, we use Eqs.(19) and (20) of that section to
    calculate a and $c_n(r_2)$, where $r_2 = -1$. Since
    $r_1 - r_2 = 1 = N$, we have $a_N(r) = a_1(r) = -1/F(r+1)$, with
    $a_0 = 1$. Hence

$a = \lim_{r \to -1} [(r+1)(-1)/F(r+1)] = \lim_{r \to -1} [-(r+1)/(r+1)(r+2)] = -1.$

Next

$$c_n(-1) = \frac{d}{dr}[(r+1)a_n(r)]\Big|_{r=-1} = (-1)^n\frac{d}{dr}[\frac{(r+1)}{F(r+1) \ldots F(r+n)}]\Big|_{r=-1},$$

where we again have set $a_0 = 1$. Observe that

$(r+1)/F(r+1) \ldots F(r+n)=1/[(r+2)^2(r+3)^2 \ldots (r+n)^2(r+n+1)]=1/G_n(r)$.

Hence $c_n(-1) = (-1)^{n+1}G_n'(-1)/G_n^2(-1)$. Notice that

$G_n(-1) = 1^2 \cdot 2^2 \cdot 3^2 \ldots (n-1)^2 n = (n-1)!n!$ and

$G_n'(-1)/G_n(-1) = 2[1/1 + 1/2 + 1/3 + \ldots + 1/(n-1)] + 1/n =$

$H_n + H_{n-1}$. Thus $c_n(-1) = (-1)^{n+1}(H_n + H_{n-1})/(n-1)!n!$.

From Eq.(24) of Section 5.7 we obtain the second solution

$$y_2(x) = -y_1(x)\ln x + x^{-1}[1 - \sum_{n=1}^{\infty}(-1)^n(H_n + H_{n-1})x^n/n!(n-1)!].$$

2.   It is clear that $x = 0$ is a singular point. The D.E. is
     in the standard form given in Theorem 5.7.1 with
     $xp(x) = 3$ and $x^2q(x) = 1+x$. Both are analytic at $x = 0$,
     so $x = 0$ is a regular singular point. Substituting

$$y = \sum_{n=0}^{\infty} a_n x^{n+r} \text{ in the D.E., shifting indices}$$

appropriately, and collecting coefficients of like powers
of x yields

$$[r(r-1) + 3r + 1]a_0 x^r + \sum_{n=1}^{\infty} \{[(r+n)(r+n+2) + 1]a_n$$

$$+ a_{n-1}\} x^{n+r} = 0.$$

The indicial equation is $F(r) = r^2 + 2r + 1 = (r+1)^2 = 0$
with the double root $r_1 = r_2 = -1$. Treating a_n as a
function of r, we see that $a_n(r) = -a_{n-1}(r)/F(r+n)$,
$n = 1,2,\ldots$. Thus $a_1(r) = -a_0/F(r+1)$,
$a_2(r) = a_0/F(r+1)F(r+2),\ldots$, and

$a_n(r) = (-1)^n a_0/F(r+1)F(r+2) \ldots F(r+n)$. Setting $r = -1$ we
find that $a_n(-1)=(-1)^n a_0/(n!)^2$, $n = 1,2,\ldots$. Hence one

solution is $y_1(x) = x^{-1}\sum_{n=0}^{\infty}(-1)^n x^n/(n!)^2$ where we have set

$a_0 = 1$. To find a second solution we follow the
procedure described in Section 5.7 for the case when the

roots of the indicial equation are equal. Specifically, the second solution will have the form given in Eq.(17) of that section. We must calculate $a_n'(-1)$. If we let

$$G_n(r) = F(r+1)...F(r+n) = (r+2)^2(r+3)^2...(r+n+1)^2 \text{ and}$$

take $a_0 = 1$, then $a_n'(-1) = (-1)^n[1/G_n(r)]'$ evaluated

$r = -1$. Hence $a_n'(-1) = (-1)^{n+1}G_n'(-1)/G_n^2(-1)$. But

$G_n(-1) = (n!)^2$ and $G_n'(-1)/G_n(-1) = 2[1/1 + 1/2 + 1/3 + ... + 1/n] = 2H_n$. Thus a second solution is

$$y_2(x) = y_1(x)\ln x - 2x^{-1}\sum_{n=1}^{\infty}(-1)^n H_n x^n/(n!)^2.$$

3.   The roots of the indicial equation are r_1 and $r_2 = 0$ and thus the analysis is similar to that for Problem 2.

4.   The roots of the indicial equation are $r_1 = -1$ and $r_2 = -2$ and thus the analysis is similar to that for Problem 1.

5.   Since $x = 0$ is a regular singular point, substitute

$$y = \sum_{n=0}^{\infty}a_n x^{n+r} \text{ in the D.E., shift indices appropriately,}$$

and collect coefficients of like powers of x to obtain
$[r^2 - 9/4]a_0 x^r + [(r+1)^2 - 9/4]a_1 x^{r+1}$

$$+ \sum_{n=2}^{\infty}\{[(r+n)^2 - 9/4]a_n + a_{n-2}\} x^{n+r} = 0.$$

The indicial equation is $F(r) = r^2 - 9/4 = 0$ with roots $r_1 = 3/2$, $r_2 = -3/2$. Treating a_n as a function of r we see that $a_n(r) = -a_{n-2}(r)/F(r+n)$, $n = 2,3,..$ if $F(r+n) \neq 0$. For the case $r_1 = 3/2$, $F(r_1+1)$, which is the

coefficient of x^{r_1+1} is $\neq 0$ so we must set $a_1 = 0$. It follows that $a_3 = a_5 = ... = 0$. For the even coefficients, set $n = 2m$ so
$a_{2m}(3/2) = -a_{2m-2}(3/2)/F(3/2 + 2m) = -a_{2m-2}/2^2 m(m+3/2)$,

$m = 1,2...$. Thus $a_2(3/2) = -a_0/2^2 \cdot 1(1 + 3/2)$,

$a_4(3/2) = a_0/2^4 \cdot 2!(1 + 3/2)(2 + 3/2),...$, and

$a_{2m}(3/2) = (-1)^m/2^{2m}m! \cdot (1 + 3/2)...(m + 3/2)$.   Hence one solution is

$$y_1(x) = x^{3/2}[1 + \sum_{m=1}^{\infty} \frac{(-1)^m}{m!(1 + 3/2)(2 + 3/2)\ldots(m + 3/2)}(\frac{x}{2})^{2m}],$$

where we have set $a_0 = 1$. For this problem, the roots r_1 and r_2 of the indicial equation differ by an integer: $r_1 - r_2 = 3/2 - (-3/2) = 3$. Hence we can anticipate that there may be difficulty in calculating a second solution corresponding to $r = r_2$. This difficulty will occur in calculating $a_3(r) = -a_1(r)/F(r+3)$ since when $r = r_2 = -3/2$ we have $F(r_2+3) = F(r_1) = 0$. However, in this problem we are fortunate because $a_1 = 0$ and it will not be necessary to use the theory described at the end of Section 5.7. Notice for $r = r_2 = -3/2$ that the coefficient of x^{r_2+1} is $[(r_2+1)^2 - 9/4]a_1$, which does not vanish unless $a_1 = 0$. Thus the recurrence relation for the odd coefficients yields $a_5 = -a_3/F(7/2)$, $a_7 = -a_5/F(11/2) = a_3/F(11/2)F(7/2)$ and so forth. Substituting these terms into the assumed form we see that a multiple of $y_1(x)$ has been obtained and thus we may take $a_3 = 0$ without loss of generality. Hence $a_3 = a_5 = a_7 = \ldots = 0$. The even coefficients are given by $a_{2m}(-3/2) = -a_{2m-2}(-3/2)/F(2m - 3/2)$, $m = 1,2\ldots$.

Thus $a_2(-3/2) = -a_0/2^2 \cdot 1 \cdot (1 - 3/2)$,

$a_4(-3/2) = a_0/2^4 \cdot 2!(1 - 3/2)(2 - 3/2),\ldots$, and

$a_{2m}(-3/2) = (-1)^m a_0/2^{2m} m!(1 - 3/2)(2 - 3/2) \ldots (m - 3/2)$.

Thus a second solution is

$$y_2(x) = x^{-3/2}[1 + \sum_{m=1}^{\infty} \frac{(-1)^m}{m!(1 - 3/2)(2 - 3/2) \ldots (m - 3/2)}(\frac{x}{2})^{2m}].$$

7. Apply the ratio test:

$$\lim_{m \to \infty} \frac{|(-1)^{m+1} x^{2m+2}/2^{2m+2}[(m+1)!]^2|}{|(-1)^m x^{2m}/2^{2m}(m!)^2|} = |x^2| \lim_{m \to \infty} \frac{1}{2^2(m+1)^2} = 0$$

for every x. Thus the series for $J_0(x)$ converges absolutely for all x.

12. If $\xi = \alpha x^\beta$, then $dy/dx = \frac{1}{2}x^{-1/2}f + x^{1/2}f'\alpha\beta x^{\beta-1}$ where f' denotes $df/d\xi$. Find d^2y/dx^2 in a similar fashion and use

algebra to show that f satisfies the D.E.
$$\xi^2 f'' + \xi f' + [\xi^2 - v^2] f = 0.$$

13. To compare $y'' - xy = 0$ with the D.E. of Problem 12, we must multiply by x^2 to get $x^2 y'' - x^3 y = 0$. Thus $2\beta = 3$, $\alpha^2 \beta^2 = -1$ and $1/4 - v^2 \beta^2 = 0$. Hence $\beta = 3/2$, $\alpha = 2i/3$ and $v^2 = 1/9$ which yields the desired result.

14. First we verify that $J_0(\lambda_j x)$ satisfies the D.E. We know that $J_0(t)$ is a solution of the Bessel equation of order zero:
$$t^2 J_0''(t) + t J_0'(t) + t^2 J_0(t) = 0 \text{ or}$$
$$J_0''(t) + t^{-1} J_0'(t) + J_0(t) = 0.$$
Let $t = \lambda_j x$. Then
$$\frac{d}{dx} J_0(\lambda_j x) = \frac{d}{dt} J_0(t) \frac{dt}{dx} = \lambda_j J_0'(t)$$
$$\frac{d^2}{dx^2} J_0(\lambda_j x) = \lambda_j \frac{d}{dt} [J_0'(t)] \frac{dt}{dx} = \lambda_j^2 J_0''(t).$$
Substituting $y = J_0(\lambda_j x)$ in the given D.E. and making use of these results, we have
$$\lambda_j^2 J_0''(t) + (\lambda_j / t) \lambda_j J_0'(t) + \lambda_j^2 J_0(t) =$$
$$\lambda_j^2 [J_0''(t) + t^{-1} J_0'(t) + J_0(t)] = 0.$$
Thus $y = J_0(\lambda_j x)$ is a solution of the given D.E. For the second part of the problem we follow the hint. First, rewrite the D.E. by multiplying by x to yield
$xy'' + y' + \lambda_j^2 xy = 0$, which can be written as
$(xy')' = -\lambda_j^2 xy$. Now let $y_i(x) = J_0(\lambda_i x)$ and $y_j(x) = J_0(\lambda_j x)$ and we have, respectively: $(xy_i')' = -\lambda_i^2 xy_i$
$$(xy_j')' = -\lambda_j^2 xy_j.$$
Now multiply the first equation by y_j, the second by y_i, integrate each from 0 to 1, and subtract the second from the first:
$$\int_0^1 [y_j(xy_i')' - y_i(xy_j')'] dx = -(\lambda_i^2 - \lambda_j^2) \int_0^1 xy_i y_j dx.$$
If we integrate each term on the left side once by parts and note that $y_i = y_j = 0$ at $x = 0$ and $x = 1$, we find that the left side of this equation is identically zero. Hence the right side is identically zero and for $\lambda_i \neq \lambda_j$ this gives the desired result.

CHAPTER 6

Section 6.1, Page 298

1. The graph of $f(t)$ is shown.
Since the function is
continuous on each interval,
but has a jump discontinuity
at $t = 1$, $f(t)$ is piecewise
continuous.

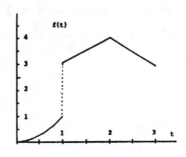

2. Note that $\lim\limits_{t \to 1^+} (t-1)^{-1} = \infty$.

5b. Since t^2 is continuous for $0 \le t \le A$ for any positive A
and since $t^2 \le e^{at}$ for any $a > 0$ and for t sufficiently
large, it follows from Theorem 6.1.2 that $\pounds\{t^2\}$ exists
for $s > 0$. $\pounds\{t^2\} = \int_0^\infty e^{-st}t^2 dt = \lim\limits_{M \to \infty} \int_0^M e^{-st}t^2 dt$

$$= \lim\limits_{M \to \infty} \left[\frac{-t^2}{s} e^{-st}\Big|_0^M + \frac{2}{s} \int_0^M e^{-st}t dt \right]$$

$$= \frac{2}{s} \lim\limits_{M \to \infty} \left[-\frac{1}{s}te^{-st}\Big|_0^M + \frac{1}{s} \int_0^M e^{-st} dt \right]$$

$$= \frac{2}{s^2} \lim\limits_{M \to \infty} - \frac{1}{s}e^{-st}\Big|_0^M = \frac{2}{s^3}.$$

6. That $f(t) = \cos at$ satisfies the hypotheses of Theorem
6.1.2 can be verified by recalling that $|\cos at| \le 1$ for
all t. To determine $\pounds\{\cos at\} = \int_0^\infty e^{-st} \cos at\, dt$ we

must integrate by parts twice to get $\int_0^\infty e^{-st} \cos at\, dt =$

$\lim\limits_{M \to \infty} [(-s^{-1}e^{-st} \cos at + as^{-2} e^{-st} \sin at)\Big|_0^M$

$- (a^2/s^2) \int_0^M e^{-st} \cos at\, dt]$. Evaluating the first two

terms, letting $M \to \infty$, and adding the third term to both

sides, we obtain $[1 + a^2/s^2]\int_0^\infty e^{-st} \cos at\, dt = 1/s$, $s > 0$.

Division by $[1 + a^2/s^2]$ and simplification yields the
desired solution.

9. From the definition for $\cosh bt$ we have
$\pounds\{e^{at}\cosh bt\} = \pounds\{\frac{1}{2}[e^{(a+b)t} + e^{(a-b)t}]\}$. Using the linearity

property of $\mathcal{L}$, Eq.(5), the right side becomes

$\dfrac{1}{2}\mathcal{L}\{e^{(a+b)t}\} + \dfrac{1}{2}\mathcal{L}\{e^{(a-b)t}\}$ which can be evaluated using the result of Example 5 and thus

$$\mathcal{L}\{e^{at}\cosh bt\} = \frac{1/2}{s-(a+b)} + \frac{1/2}{s-(a-b)}$$

$$= \frac{s-a}{(s-a)^2-b^2}, \quad \text{for } s-a > |b|.$$

13. We write $\sin at = (e^{iat} - e^{-iat})/2i$, then the linearity of the Laplace transform operator allows us to write
$\mathcal{L}\{e^{at}\sin bt\} = (1/2i)\mathcal{L}\{e^{(a+ib)t}\} - (1/2i)\mathcal{L}\{e^{(a-ib)t}\}$. Each of these two terms can be evaluated by using the result of Example 5, where we now have to require s to be greater than the real part of the complex numbers $a \pm ib$ in order for the integrals to converge. Complex algebra then gives the desired result. An alternate method of evaluation would be to use integration on the integral appearing in the definition of $\mathcal{L}\{e^{at}\sin bt\}$, but that method requires integration by parts twice.

16. As in Problem 13,
$\mathcal{L}\{t\sin at\} = (1/2i)\mathcal{L}\{te^{iat}\} - (1/2i)\mathcal{L}\{te^{-iat}\}$. Using the result of Problem 15 we obtain
$\mathcal{L}\{t\sin at\} = (1/2i)[(s-b)^{-2} - (s+b)^{-2}]$ where $b = ia$ and $s > 0$. Hence $\mathcal{L}\{t\sin at\} = 2as/(s^2+a^2)^2$, $s > 0$.

19. Use the approach shown in Problem 16 with the result of Problem 18, for $n = 2$. A computer algebra system may also be used.

21. The integral $\displaystyle\int_0^A (t^2 + 1)^{-1}dt$ can be evaluated in terms of the arctan function and then Eq. (3) can be used. To illustrate Theorem 6.1.1, however, consider that

$\dfrac{1}{t^2+1} < \dfrac{1}{t^2}$ for $t \geq 1$ and, from Example 3, $\displaystyle\int_1^\infty t^{-2}dt$

converges and hence $\displaystyle\int_1^\infty (t^2 + 1)^{-1}dt$ also converges.

$\displaystyle\int_0^1 (t^2 + 1)^{-1}dt$ is finite and hence does not affect the

convergence of $\displaystyle\int_0^\infty (t^2 + 1)^{-1}dt$ at infinity.

25. If we let $u = f$ and $dv = e^{-st}dt$ then $F(s) = \int_0^\infty e^{-st}f(t)\,dt$

$$= \lim_{M \to \infty} - \frac{1}{s}e^{-st}f(t)\,|_0^M + \frac{1}{s}\int_0^\infty e^{-st}f'(t)\,dt.$$ Now use an

argument similar to that given to establish Theorem 6.1.2.

27a. Make a transformation of variables with $x = st$ and
     $dx = s\,dt$. Then use the definition of $\Gamma(P+1)$ from
     Problem 26.

27b. From part a, $\pounds\{t^n\} = \dfrac{1}{s^{n+1}}\int_0^\infty e^{-x}x^n dx = \dfrac{n}{s^{n+1}}\int_0^\infty e^{-x}x^{n-1}dx$

$$= \frac{n!}{s^{n+1}}\int_0^\infty e^{-x}dx,$$ using integration by

parts successively. Evaluation of the last integral
yields the desired answer.

27c. From part a, $\pounds\{t^{-1/2}\} = \dfrac{1}{\sqrt{s}}\int_0^\infty e^{-x}x^{-1/2}dx.$ Let $x = y^2$, then

$2dy = x^{-1/2}dx$ and thus $\pounds\{t^{-1/2}\} = \dfrac{2}{\sqrt{s}}\int_0^\infty e^{-y^2}dy.$

27d. Use the definition of $\pounds\{t^{1/2}\}$ and integrate by parts once
     to get $\pounds\{t^{1/2}\} = (1/2s)\pounds\{t^{-1/2}\}$. The result follows from
     part c.

Section 6.2, Page 307

Problems 1 through 10 are solved by using partial fractions
and algebra to manipulate the given function into a form
matching one of the functions appearing in the middle column
of Table 6.2.1.

2.   We have $\dfrac{4}{(s-1)^3} = 2\dfrac{2!}{(s-1)^{2+1}}$ and thus the inverse Laplace

transform is $2t^2e^t$, using line 11.

4.   We have $\dfrac{3s}{s^2-s-6} = \dfrac{3s}{(s-3)(s+2)} = \dfrac{9/5}{s-3} + \dfrac{6/5}{s+2}$ using partial

fractions. Thus $(9/5)e^{3t} + (6/5)e^{-2t}$ is the inverse
transform, from line 2.

7.   We have $\dfrac{2s+1}{s^2-2s+2} = \dfrac{2s+1}{(s-1)^2+1} = \dfrac{2(s-1)}{(s-1)^2+1} + \dfrac{3}{(s-1)^2+1}$, where

we first used the concept of completing the square (in the denominator) and then added and subtracted appropriately to put the numerator in the desired form. Lines 9 and 10 may now be used to find the desired result.

In each of the Problems 11 through 23 it is assumed that the I.V.P. has a solution y = ϕ(t) which, with its first two derivatives, satisfies the conditions of the Corollary to Theorem 6.2.1.

11. Take the Laplace transform of the D.E., using Eq.(1) and Eq.(2), to get
$s^2 Y(s) - sy(0) - y'(0) - [sY(s) - y(0)] - 6Y(s) = 0.$
Using the I.C. and solving for Y(s) we obtain
$Y(s) = \dfrac{s-2}{s^2-s-6}$. Following the pattern of Eq.(12) we have
$\dfrac{s-2}{s^2-s-6} = \dfrac{a}{s+2} + \dfrac{b}{s-3} = \dfrac{a(s-3)+b(s+2)}{(s+2)(s-3)}$. Equating like
powers in the numerators we find a+b = 1 and
-3a + 2b = -2. Thus a = 4/5 and b = 1/5 and
$Y(s) = \dfrac{4+5}{s+2} + \dfrac{1/5}{s-3}$, which yields the desired solution
using Table 6.2.1.

14. Taking the Laplace transform we have $s^2 Y(s) - sy(0) - y'(0) - 4[sY(s)-y(0)] + 4Y(s) = 0.$ Using the I.C. and solving for
Y(s) we find $Y(s) = \dfrac{s-3}{s^2-4s+4}$. Since the denominator is a
perfect square, the partial fraction form is $\dfrac{s-3}{s^2-4s+4} =$
$\dfrac{a}{(s-2)^2} + \dfrac{b}{s-2}$. Solving for a and b, as shown in examples of
this section or in Problem 11, we find a = -1 and b = 1.
Thus $Y(s) = \dfrac{1}{s-2} - \dfrac{1}{(s-2)^2}$, from which we find
$y(t) = e^{2t} - te^{2t}$ (lines 2 and 11 in Table 6.2.1).

15. Note that $Y(s) = \dfrac{2s-4}{s^2-2s-2} = \dfrac{2s-4}{(s-1)^2-3} = \dfrac{2(s-1)}{(s-1)^2-3} - \dfrac{2}{(s-1)^2-3}$.
Three formulas in Table 6.2.1 are now needed: F(s-c) in line 14 in conjunction with the ones for coshat and sinhat, lines 7 and 8.

17. The Laplace transform of the D.E. is
$s^4Y(s) - s^3y(0) - s^2y'(0) - sy''(0) - y'''(0) - 4[s^3Y(s)-s^2y(0)$
$-sy'(0) - y''(0)] + 6[s^2Y(s) - sy(0) - y'(0)] - 4[sY(s) - y(0)]$
$+Y(s) = 0.$ Using the I.C. and solving for Y(s) we find

$$Y(s) = \frac{s^2 - 4s + 7}{s^4-4s^3+6s^2-4s+1}.$$ The correct partial fraction

form for this is $\dfrac{a}{(s-1)^4} + \dfrac{b}{(s-1)^3} + \dfrac{c}{(s-1)^2} + \dfrac{d}{s-1}.$
Setting this equal to Y(s) above and equating the
numerators we have $s^2-4s+7 = a + b(s-1) + c(s-1)^2 +$
$d(s-1)^3$. Solving for a,b,c, and d and use of Table 6.2.1
yields the desired solution.

20. The Laplace transform of the D.E. is
$s^2Y(s) - sy(0) - y'(0) + \omega^2Y(s) = s/(s^2+4).$ Applying the
I.C. and solving for Y(s) we get $Y(s) = s/[(s^2+4)(s^2+\omega^2)]$
$+ s/(s^2+\omega^2).$ Decomposing the first term by partial
fractions we have

$$Y(s) = \frac{s}{(\omega^2-4)(s^2+4)} - \frac{s}{(\omega^2-4)(s^2+\omega^2)} + \frac{s}{s^2+\omega^2}$$

$$= (\omega^2-4)^{-1}[\frac{(\omega^2-5)s}{s^2+\omega^2} + \frac{s}{s^2+4}].$$

Then, using Table 6.1.2, we have
$y = (\omega^2-4)^{-1}[(\omega^2-5)\cos\omega t + \cos 2t].$

22. Solving for Y(s) we find $Y(s) = 1/[(s-1)^2 + 1] +$
$1/(s+1)[(s-1)^2 + 1].$ Using partial fractions on the
second term we obtain
$Y(s) = 1/[(s-1)^2 + 1] + \{1/(s+1) - (s-3)/[(s-1)^2 + 1]\}/5$
$= (1/5)\{(s+1)^{-1}-(s-1)[(s-1)^2 + 1]^{-1} + 7[(s-1)^2 + 1]^{-1}\}.$
Hence, $y = (1/5)(e^{-t} - e^t\cos t + 7e^t\sin t).$

24. Under the standard assumptions, the Lapace transform of
the left side of the D.E. is $s^2Y(s) - sy(0) - y'(0) +$
$4Y(s).$ To transform the right side we must revert to the
definition of the Laplace trasnform to determine
$\int_0^\infty e^{-st}f(t)dt.$ Since f(t) is piecewise continuous we are
able to calculate $£\{f(t)\}$ by

$$\int_0^\infty e^{-st}f(t)\,dt = \int_0^\pi e^{-st}\,dt + \lim_{M \to \infty}\int_\pi^M (e^{-st})(0)\,dt$$

$$= \int_0^\pi e^{-st}\,dt = (1 - e^{-\pi s})/s.$$

Hence, the Laplace transform $Y(s)$ of the solution is given by $Y(s) = s/(s^2+4) + (1 - e^{-\pi s})/s(s^2+4)$.

27b. The Taylor series for f about $t = 0$ is

$$f(t) = \sum_{n=0}^{\infty}(-1)^n t^{2n}/(2n+1)!,$$ which is obtained from

part(a) by dividing each term of the sine series by t. Also, f is continuous for $t > 0$ since $\lim_{t \to 0+}(\sin t)/t = 1$.

Assuming that we can compute the Laplace transform of f

term by term, we obtain $\mathcal{L}\{f(t)\} = \mathcal{L}\{\sum_{n=0}^{\infty}(-1)^n t^{2n}/(2n+1)!\}$

$$= \sum_{n=0}^{\infty}[(-1)^n/(2n+1)!\,\mathcal{L}\{t^{2n}\}$$

$$= \sum_{n=0}^{\infty}[(-1)^n(2n)!/(2n+1)!]s^{-(2n+1)}$$

$$= \sum_{n=0}^{\infty}[(-1)^n/(2n+1)]s^{-(2n+1)},$$ which converges for $s > 1$.

The Taylor series for $\arctan x$ is given by

$$\sum_{n=0}^{\infty}(-1)^n x^{2n+1}/(2n+1),$$ for $|x| < 1$. Comparing $\mathcal{L}\{f(t)\}$ with

the Taylor series for $\arctan x$, we conclude that $\mathcal{L}\{f(t)\} = \arctan(1/s)$, $s > 1$.

30. Setting $n = 2$ in Problem 28b, we have

$$\mathcal{L}\{t^2\sin bt\} = \frac{d^2}{ds^2}[b/(s^2+b^2)] = \frac{d}{ds}[-2bs/(s^2+b^2)^2] =$$

$$-2b/(s^2+b^2)^2 + 8bs^2/(s^2+b^2)^3 = 2b(3s^2-b^2)/(s^2+b^2)^3.$$

32. Using the result of Problem 28a. we have

$$\mathcal{L}\{te^{at}\} = -\frac{d}{ds}(s-a)^{-1} = (s-a)^{-2}$$

$$\mathcal{L}\{t^2e^{at}\} = -\frac{d}{ds}(s-a)^{-2} = 2(s-a)^{-3}$$

$$\mathcal{L}\{t^3e^{at}\} = -\frac{d}{ds}2(s-a)^{-3} = 3!(s-a)^{-4}.$$ Continuing in this

fashion, or using induction, we obtain the desired result.

36a. Taking the Laplace transform of the D.E. we obtain
$$£\{y''\} - £\{ty\} = £\{y''\} + £\{-ty\}$$
$$= s^2Y(s) - sy(0) - y'(0) + Y'(s) = 0.$$
Hence, Y satisfies $Y' + s^2Y = s$.

38a. From Eq(i) we have $A_k = \lim_{s \to r_k} (s-r_k) \dfrac{P(r_k)}{Q(r_k)}$, since Q has

distinct zeros. Thus $A_k = P(r_k) \lim_{s \to r_k} \dfrac{s-r_k}{Q(r_k)} = \dfrac{P(r_k)}{Q'(r_k)}$, by

L'Hopital's Rule.

38b. Since $£^{-1}\left\{\dfrac{1}{s-r_k}\right\} = e^{r_k t}$, the result follows.

Section 6.3, Page 314

2. From the definition of $u_c(t)$
    we have:
    $$g(t) = (t-3)u_2(t) - (t-2)u_3(t)$$
    $$= \begin{cases} 0 - 0 = 0, & 0 \leq t < 2 \\ (t-3) - 0 = t-3, & 2 \leq t < 3. \\ (t-3) - (t-2) = -1, & 3 \leq t \end{cases}$$

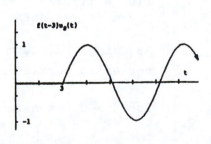

4. As indicated in the discussion
    following Example 1, the unit
    step function can be used to
    translate a given function f,
    with domain $t \geq 0$, a distance c
    to the right by the
    multiplication $u_c(t)f(t-c)$.
    Hence the required graph of
    $y = u_3(t)f(t-3)$ for $f(t) = \sin t$ is shown.

8. In order to use Theorem 6.3.1 we must write $f(t)$ in terms
    of $u_c(t)$. Since $t^2 - 2t + 2 = (t-1)^2 + 1$ (by completing
    the square), we can thus write $f(t) = u_1(t)g(t-1)$, where
    $g(t) = t^2 + 1$. Now applying Theorem 6.3.1 we have
    $$£\{f(t)\} = £\{u_1(t)g(t-1)\} = e^{-s}£\{g(t)\} = e^{-s}(2/s^3 + 1/s).$$

14. Use partial fractions to write
$F(s) = e^{-2s}[(s-1)^{-1} - (s+2)^{-1}]/3$. For ease in calculations
let us define $G(s) = (s-1)^{-1}$ and $H(s) = (s+2)^{-1}$. Then
$F(s) = [e^{-2s} G(s) - e^{-2s} H(s)]/3$. Using the fact that
$£\{e^{at}\} = (s-a)^{-1}$ and applying Theorem 6.3.1, we have
$F(s) = [e^{-2s} £\{e^t\} - e^{-2s} £\{e^{-2t}\}]/3$. Thus
$F(s) = [£\{u_2(t)e^{(t-2)}\} - £\{u_2(t)e^{-2(t-2)}\}]/3$. Using the
linearity of the Laplace transform, we have
$£\{f(t)\} = £\{u_2(t)[e^{t-2} - e^{-2(t-2)}]/3\}$. Hence,
$f(t) = [u_2(t)(e^{t-2} - e^{-2(t-2)})]/3$. An alternate method is
to complete the square in the denominator:
$$F(s) = \frac{e^{-2s}}{(s+1/2)^2 - 9/4}.$$ This gives
$f(t) = (2/3)u_2(t)e^{-(t-2)/2} \sinh\frac{3}{2}(t-2)$, which can be shown
to be the same as that found above.

21. By completing the square in the denominator of F we can
write $F(s) = (2s+1)/[(2s+1)^2 + 4]$. This has the form
$G(2s+1)$ where $G(u) = u/(u^2+4)$. We must find
$£^{-1}\{G(2s+1)\}$. Applying the results of Problem 19(c), we
have $£^{-1}\{F(s)\} = \frac{1}{2}e^{-t/2} \cos(\frac{2t}{2})$.

22. If the approach of Problem 21 is used we find
$f(t) = (1/3)e^{2t/3} \sinh(t/3)$, which is equivalent to the
given answer using the definition of sinh t.

27. Assuming that term-by-term integration of the infinite
series is permissible and recalling that $£\{u_c(t)\} = e^{-cs}/s$

for $s > 0$, we have $£\{f(t)\} = (1/s) + \sum_{k=1}^{\infty}(-1)^k £\{u_k(t)\}$

$= (1/s) + \sum_{k=1}^{\infty}(-1)^k e^{-ks}/s = [\sum_{k=0}^{\infty}(-e^{-s})^k]/s$. We recognize

the last infinite series as the geometric series, $\sum_{k=0}^{\infty}ar^k$,

with $a = 1$ and $r = -e^{-s}$. This series converges to
$[1/(1+e^{-s})]$ if $|r| < 1$. Hence,
$£\{f(t)\} = (1/s)[1/(1+e^{-s})]$, $s > 0$.

28. Using the definition of the Laplace transform we have
$F(s) = \mathcal{L}\{f(t)\} = \int_0^\infty e^{-st} f(t) dt.$ Since f is periodic with period T, we have $f(t+T) = f(t)$. This suggests that we rewrite the improper integral as $\int_0^\infty e^{-st} f(t) dt =$

$\sum_{n=0}^\infty \int_{nT}^{(n+1)T} e^{-st} f(t) dt.$ The periodicity of f also suggests that we make the change of variable $t = r + nT$. Hence,

$F(s) = \sum_{n=0}^\infty \int_0^T e^{-s(r+nT)} f(r+nT) dr = \sum_{n=0}^\infty (e^{-sT})^n \int_0^T e^{-rs} f(r) dr,$

where we have used the fact that
$f(r+nT) = f(r+(n-1)T) = \ldots = f(r+T) = f(r)$ from the definition that f is periodic. We recognize this last

series as the geometric series, $\sum_{n=0}^\infty a u^n$, with

$a = \int_0^T e^{-rs} f(r) dr$ and $u = e^{-sT}$. The geometric series converges to $a/(1-u)$ for $|u| < 1$ and consequently we obtain

$F(s) = (1 - e^{-sT})^{-1} \int_0^T e^{-rs} f(r) dr, \quad s > 0.$

30. The function f is periodic with period 2. The result of Problem 28 gives us $\mathcal{L}\{f(t)\} = \int_0^2 e^{-st} f(t) dt / (1-e^{-2s})$. Calculating the integral we have
$\int_0^2 e^{-st} f(t) dt = \int_0^1 e^{-st} dt - \int_1^2 e^{-st} dt$

$= (1-e^{-s})/s + (e^{-2s}-e^{-s})/s$
$= (e^{-2s}-2e^{-s}+1)/s$
$= (1-e^{-s})^2/s.$ Since the denominator of
$\mathcal{L}\{f(t)\}, 1 - e^{-2s},$ may be written as $(1-e^{-s})(1+e^{-s})$ we obtain the desired answer.

Section 6.4, Page 321

1. f(t) can be written in the form $f(t) = 1 - u_{\pi/2}(t)$ and thus the Laplace transforms of the D.E. is
$(s^2+1)Y(s) - sy(0) - y'(0) = (1/s) - e^{-\pi s/2}/s.$ Introducing the I.C. and solving for Y(s), we obtain
$Y(s) = (s^2+1)^{-1} + [s(s^2+1)]^{-1} - e^{-\pi s/2}/s(s^2+1).$ Using partial fractions on the second and third terms we find

$Y(s) = (1/s) + (s^2+1)^{-1} - s/(s^2+1) - e^{-\pi s/2}/s + e^{-\pi s/2}s/(s^2+1)$.
The inverse transform of the first three terms can be
obtained directly from Table 6.2.1. Using Theorem 6.3.1
to find the inverse transform of the last two terms we
have $\mathcal{L}^{-1}\{e^{-\pi s/2}/s\} = u_{\pi/2}(t)g(t - \pi/2)$ where

$g(t) = \mathcal{L}^{-1}\{1/s\} = 1$ and
$\mathcal{L}^{-1}\{e^{-\pi s/2}s/(s^2+1)\} = u_{\pi/2}(t)h(t - \pi/2)$ where

$h(t) = \mathcal{L}^{-1}\{s/(s^2+1)\} = \cos t$. Hence,
$y = 1 + \sin t - \cos t + u_{\pi/2}(t)[\cos(t - \pi/2) - 1]$
 $= 1 + \sin t - \cos t - u_{\pi/2}(t)[1 - \sin t]$. The graph of the
forcing function is a unit pulse for $0 \le t < \pi/2$ and 0
thereafter. The graph of the solution will be composed
of two segments. The first, for $0 \le t < \pi/2$, is a
sinusoid oscillating about 1, which represents the system
response to a unit forcing function and the given initial
conditions. For $t \ge \pi/2$, the forcing function, $f(t)$, is
zero and the "initial" conditions are
$y(\pi/2) = \lim_{t \to \pi^-/2} 1 + \sin t - \cos t = 2$ and

$y'(\pi/2) = \lim_{t \to \pi^-/2} \cos t + \sin t = 1$. In this case the system

response is $y(t) = 2\sin t - \cos t$, which is a sinusoid
oscillating about zero.

3. According to Theorem 6.3.1,
   $\mathcal{L}\{u_{2\pi}(t)\sin(t-2\pi)\} = e^{-2\pi s}\mathcal{L}\{\sin t\} = e^{-2\pi s}/(s^2+1)$.
   Transforming the D.E., we have
   $(s^2+4)Y(s) - sy(0) - y'(0) = 1/(s^2+1) - e^{-2\pi s}/(s^2+1)$.
   Introducing the I.C. and solving for $Y(s)$, we obtain
   $Y(s) = (1-e^{-2\pi s})/(s^2+1)(s^2+4)$. We apply partial fractions
   to write
   $Y(s) = [(s^2+1)^{-1} - (s^2+4)^{-1} - e^{-2\pi s}(s^2+1)^{-1} + e^{-2\pi s}(s^2+4)^{-1}]/3$.
   We compute the inverse transform of the first two terms
   directly from Table 6.2.1 after noting that
   $(s^2+4)^{-1} = (1/2)[2/(s^2+4)]$. We apply Theorem 6.3.1 to the
   last two terms to obtain the solution,
$y = (1/3)\{\sin t - (1/2)\sin 2t - u_{2\pi}(t)[\sin(t-2\pi) - (1/2)\sin 2(t-2\pi)]\}$.
   This may be simplified, using trigonometric identities,
   to $y = [(2\sin t - \sin 2t)(1-u_{2\pi}(t))]/6$. Note that the
   forcing function is $\sin t - \sin(t-2\pi) = 0$ for $t \ge 2\pi$. The
   solution is $y(t) = 2\sin t - \sin 2t$ for $0 \le t < 2\pi$. Thus
   $y(2\pi^-) = 0$ and $y'(2\pi^-) = 2\cos 2\pi - 2\cos 4\pi = 0$. Hence the
   "initial" value problem for $t \ge 2\pi$ is $y'' + 4y = 0$,
   $y(2\pi) = 0$, $y'(2\pi) = 0$, which has the trivial solution
   $y \equiv 0$.

8. Taking the Laplace transform, applying the I.C. and using
    Theorem 6.3.1 we have $(s^2+s+5/4)Y(s) = (1-e^{-\pi s/2})/s^2$. Thus

$$Y(s) = \frac{1-e^{-s/2}}{s^2(s^2+s+5/4)}$$

$$= (1-e^{-\pi s/2})\left\{\frac{4/5}{s^2} - \frac{16/25}{s} + \frac{(16/25)s-4/25}{(s+1/2)^2+1}\right\}$$

$= (1-e^{-\pi s/2})H(s)$, where we have used partial
fractions and completed the square in the denominator of
the last term. Since the numerator of the last term of H
can be written as $\dfrac{16}{25}[(s+1/2) - 3/4]$, we see that

$\mathcal{L}^{-1}\{H(s)\} = (4/25)(5t - 4 + 4e^{-t/2}\cos t - 3e^{-t/2}\sin t)$,
which yields the desired solution. The graph of the
forcing function is a ramp $(f(t) = t)$ for $0 \le t < \pi/2$ and
a constant $(f(t) = \pi/2)$ for $t \ge \pi/2$. The solution will
be a damped sinusoid oscillating about the "ramp"
$(20t-16)/25$ for $0 \le t < \pi/2$ and oscillating about $2\pi/5$
for $t \ge \pi/2$.

10. Note that $g(t) = \sin t - u_\pi(t)\sin t = \sin t + u_\pi(t)\sin(t-\pi)$.
    Proceeding as in Problem 8 we find

$$Y(s) = (1+e^{-\pi s})\frac{1}{(s^2+1)(s^2+s+5/4)}.$$ The correct partial

fraction expansion of the quotient is $\dfrac{as+b}{s^2+1} + \dfrac{cs+d}{s^2+s+5/4}$,

where
$a+c = 0$, $a+b+d = 0$, $(5/4)a+b+c = 0$ and $(5/4)b+d = 1$ by
equating coefficients. Solving for the constants yields
the desired solution.

16b. Taking the Laplace transform of the D.E. we obtain
    $U(s^2+s/4+1) = k(e^{-3s/2}-e^{-5s/2})/s$, since the I.C. are zero.
    Solving for U and using partial fractions yields

$$U(s) = k(e^{-3s/2}-e^{-5s/2})(\frac{1}{s}-\frac{s+1/4}{s^2+s/4+1}).$$ Thus, if

$$H(s) = (\frac{1}{s}-\frac{s+1/4}{s^2+s/4+1}),$$ then

$$h(t) = 1 - e^{-t/8}(\cos\frac{3\sqrt{7}}{8}t+\frac{\sqrt{7}}{21}\sin\frac{3\sqrt{7}}{8}t)$$ and

$u(t) = ku_{3/2}(t)h(t-3/2) - ku_{5/2}(t)h(t-5/2)$.

16c. In all cases the plot will be zero for $0 \le t < 3/2$. For
    $3/2 \le t < 5/2$ the plot will be the system response

(damped sinusoid) to a step input of magnitude k. For
$t \geq 5/2$, the plot will be the system response to the
I.C. $u(5^-/2)$, $u'(5^-/2)$ with
no forcing function. The
graph shown is for $k = 2$.
Varying k will just affect
the amplitude. Note that
the amplitude never
reaches 2, which would be
the steady state response
for the step input $2u_{3/2}(t)$.

Note also that the solution
and its derivative are
continuous at $t = 5/2$.

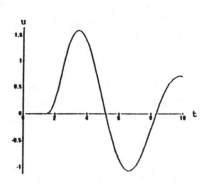

19a. The graph on $0 \leq t < 6\pi$ will depend on how large n is.
For instance, if $n = 2$ then
$$f(t) = \begin{cases} 1 & 0 \leq t < \pi, \ 2\pi \leq t < 6\pi \\ -1 & -\pi \leq t < 2\pi \end{cases}. \quad \text{For}$$

$$n \geq 6, \ f(t) = \begin{cases} 1 & 0 \leq t < \pi, \ 2\pi \leq t < 3\pi, \ 4\pi \leq t < 5\pi \\ -1 & \pi \leq t < 2\pi, \ 3\pi \leq t < 4\pi, \ 5\pi \ t < 6\pi \end{cases}.$$

19b. Taking the Laplace transform of the D.E. and using the I.C. we
have $Y(s) = \dfrac{1}{s(s^2+1)}[1 + 2\sum\limits_{k=1}^{\infty}(-1)^k e^{-\pi ks}]$, since

$\mathcal{L}\{u_{\pi k}(t)\} = \dfrac{e^{-\pi ks}}{s}$. Since $\dfrac{1}{s(s^2+1)} = \dfrac{1}{s} - \dfrac{s}{s^2+1}$, we then obtain

$y(t) = 1 - \cos t + 2\sum\limits_{k=1}^{\infty}(-1)^{-k}u_{\pi k}(t)[1 - \cos(t-\pi k)]$, using line

13 in Table 6.2.1.

19d. Since $\cos(t-\pi k) = (-1)^k \cos t$, the solution in part b can
be written as

$$y(t) = 1 - \cos t + 2\sum\limits_{k=1}^{\infty}(-1)^k u_{\pi k}(t) - 2\sum\limits_{k=1}^{\infty}(-1)^{2k}\cos t$$

$$= 1 - \cos t - 2n\cos t + 2\sum\limits_{k=1}^{\infty}(-1)^k u_{\pi k}(t) \quad \text{which diverges for } n\to\infty.$$

20. In this case

$$Y(s) = \frac{1}{s(s^2+.1s+1)} [1 + 2\sum_{n=1}^{\infty}(-1)^k e^{-\pi ks}].$$ Using partial

fractions we have

$$H(s) = \frac{1}{s(s^2+.1s+1)} = \frac{1}{s} - \frac{s+.1}{s^2+.1s+1}$$

$$= \frac{1}{s} - \frac{s+.05}{(s+.05)^2+b^2} - \frac{.05}{(s+.05)^2+b^2},$$ where

$b^2 = [1-(.05)^2] = .9975.$ Now let

$$h(t) = \mathcal{L}^{-1}\{H(s)\} = 1 - e^{-.05t}\cos bt - \frac{.05}{b}e^{-.05t}\sin bt.$$ Hence,

$$y(t) = h(t) + 2\sum_{n=1}^{\infty}(-1)^k u_{\pi k}(t) h(t-\pi k),$$ and thus, for t > n the

solution will be approximated by

$\pm 1 - Ae^{-.05(t-n\pi)} \cos[b(t-n\pi) + \delta],$ and therefore converges as

$t\to\infty.$

20a. y(t) for n = 30   y(t) for n = 31

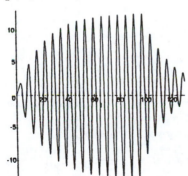
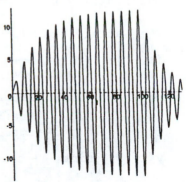

20b. From the graph of part a, A $\cong$ 12.5 and the frequency is 2π.

20c. From the graph
     (or analytically)
     A = 10 and the
     frequency is 2π.

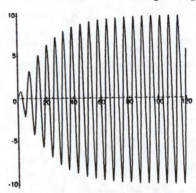

Section 6.5, Page 328

1. Proceeding as in Example 1, we take the Laplace transform
    of the D.E. and apply the I.C.:
    $(s^2 + 2s + 2)Y(s) = s + 2 + e^{-\pi s}$. Thus,
    $Y(s) = (s+2)/[(s+1)^2 + 1] + e^{-\pi s}/[(s+1)^2 + 1]$. We write
    the first term as $(s+1)/[(s+1)^2 + 1] + 1/[(s+1)^2 + 1]$.
    Applying Theorem 6.3.1 and using Table 6.2.1, we obtain
    the solution, $y = e^{-t}\cos t + e^{-t}\sin t - u_\pi(t)e^{-(t-\pi)}\sin t$.
    Note that $\sin(t-\pi) = -\sin t$.

3. Taking the Laplace transform and using the I.C. we have
    $$(s^2 + 3s+2)Y(s) = \frac{1}{2} + e^{-5s} + \frac{e^{-10s}}{s}. \quad \text{Thus}$$
    $$Y(s) = \frac{1/2}{s^2+3s+2} + \frac{e^{-5s}}{s^2+3s+2} + e^{-10s}(\frac{1/2}{s}+\frac{1/2}{s+2}-\frac{1}{s+1}) \text{ and}$$
    hence
    $$y(t) = \frac{1}{2}h(t) + u_5(t)h(t-5) + u_{10}(t)[\frac{1}{2}+\frac{1}{2}e^{-2(t-10)}-e^{-(t-10)}]$$
    where $h(t) = e^{-t} - e^{-2t}$.

5. The Laplace transform of the D.E. is
    $$(s^2+2s+3)Y(s) = \frac{1}{s^2+1} + e^{-3\pi s}, \text{ so}$$
    $$Y(s) = \frac{1}{(s^2+1)(s^2+2s+3)} + e^{-3\pi s}[\frac{1}{s^2+2s+3}]. \quad \text{Using partial}$$
    fractions or a computer algebra system we obtain
    $$y(t) = \frac{1}{4}\sin t - \frac{1}{4}\cos t + \frac{1}{4}e^{-t}\cos\sqrt{2}\,t + \frac{1}{\sqrt{2}}u_{3\pi}(t)h(t-3\pi),$$
    where $h(t) = e^{-t}\sin\sqrt{2}\,t$.

7. Taking the Laplace transform of the D.E. yields
    $$(s^2+1)Y(s) - y'(0) = \int_0^\infty e^{-st}\delta(t-2\pi)\cos t\,dt. \quad \text{Since}$$
    $\delta(t-2\pi) = 0$ for $t \neq 2\pi$ the integral on the right is equal
    to $\int_{-\infty}^\infty e^{-st}\delta(t-2\pi)\cos t\,dt$ which equals $e^{-2\pi s}\cos 2\pi$ from
    Eq.(16). Substituting for $y'(0)$ and solving for $Y(s)$
    gives $Y(s) = \dfrac{1}{s^2+1} + \dfrac{e^{-2\pi s}}{s^2+1}$ and hence
    $$y(t) = \sin t + u_{2\pi}(t)\sin(t-2\pi) = \begin{cases} \sin t & 0 \leq t < 2\pi \\ 2\sin t & 2\pi \leq t \end{cases}$$

10. See the solution for Problem 7.

13a. From Eq. (22) $y(t)$ will complete one cycle when $\sqrt{15}\,(t-5)/4 = 2\pi$ or $T = t - 5 = 8\pi/\sqrt{15}$, which is consistent with the plot in Fig. 6.5.3. Since an impulse causes a discontinuity in the first derivative, we need to find the value of y' at $t = 5$ and $t = 5 + T$. From Eq. (22) we have, for $t \geq 5$,

$$y' = e^{-(t-5)/4}[\frac{-1}{2\sqrt{15}}\sin\frac{\sqrt{15}}{4}(t-5) + \frac{1}{2}\cos\frac{\sqrt{15}}{4}(t-5)].$$ Thus

$y'(5) = \frac{1}{2}$ and $y'(5+T) = \frac{1}{2}e^{-T/4}$. Since the original impulse, $\delta(t-5)$, caused a discontinuity in y' of $1/2$, we must choose the impulse at $t = 5 + T$ to be $-e^{-T/4}$, which is equal and opposite to y' at $5 + T$.

13b. Now consider $2y'' + y' + 2y = \delta(t-5) + k\delta(t-5-T)$ with $y(0) = 0$, $y'(0) = 0$. Using the results of Example 1 we have

$$y(t) = \frac{2}{\sqrt{15}}u_5(t)e^{-(t-5)/4}\sin\frac{\sqrt{15}}{4}(t-5)$$

$$+ \frac{2k}{\sqrt{15}}u_{5+T}(t)e^{-(t-5-T)/4}\sin\frac{\sqrt{15}}{4}(t-5-T)$$

$$= \frac{2}{\sqrt{15}}e^{-(t-5)/4}[u_5(t)\sin\frac{\sqrt{15}}{4}(t-5)+ku_{5+T}(t)e^{T/4}\sin\frac{\sqrt{15}}{4}(t-5-T)]$$

$$= \frac{2}{\sqrt{15}}e^{-(t-5)/4}[u_5(t)+ke^{T/4}u_{5+T}(t)]\sin\frac{\sqrt{15}}{4}(t-5).$$ If

$y(t) \equiv 0$ for $t \geq 5 + T$, then $1 + ke^{T/4} = 0$, or $k = -e^{-T/4}$, as found in part (a).

17b. We have $(s^2+1)Y(s) = \sum\limits_{k=1}^{20} e^{-k\pi s}$ so that $Y(s) = \sum\limits_{k=1}^{20} \frac{e^{-ks}}{s^2+1}$

and hence $y(t) = \sum\limits_{k=1}^{20} u_{k\pi}(t)\sin(t-k\pi)$

$= u_\pi(t)\sin(t-\pi) + u_{2\pi}(t)\sin(t-2\pi) + \ldots + u_{10\pi}\sin(t-10\pi)$.
For $0 \leq t < \pi$, $y(t) \equiv 0$. For $\pi \leq t < 2\pi$, $y(t) = \sin(t-\pi)$ $= -\sin t$. For $2\pi \leq t < 3\pi$,
$y(t) = \sin(t-\pi) + \sin(t-2\pi) = -\sin t + \sin t \equiv 0$. Due to the periodicity of $\sin t$, the solution will exhibit this behavior in alternate intervals for $0 \leq t < 20\pi$. After $t = 20\pi$ the solution remains at zero.

21b. Taking the transform and using the I.C. we have

$$(s^2+1)Y(s) = \sum_{k=1}^{15} e^{-(2k-1)\pi} \text{ so that } Y(s) = \sum_{k=1}^{15} \frac{e^{-(2k-1)\pi}}{s^2+1}.$$

Thus $y(t) = \sum_{k=1}^{15} u_{(2k-1)\pi}(t)\sin[t-(2k-1)\pi]$

$$= \sin(t-\pi) + \sin(t-3\pi) \ldots + \sin(t-29\pi)$$
$$= -\sin t - \sin t \ldots -\sin t$$
$$= -15\sin t.$$

25b. Substituting for $f(t)$ we have

$y = \int_0^t e^{-(t-\tau)}\delta(\tau-\pi)\sin(t-\tau)d\tau.$ We know that the integration variable is always less than t (the upper limit) and thus for $t < \pi$ we have $\tau < \pi$ and thus $\delta(\tau-\pi) = 0$. Hence $y = 0$ for $t < \pi$. For $t > \pi$ utilize Eq.(16).

Section 6.6, Page 335

1c. Using the format of Eqs.(2) and (3) we have

$$f*(g*h) = \int_0^t f(t-\tau)(g*h)(\tau)d\tau$$

$$= \int_0^t f(t-\tau)[\int_0^\tau g(\tau-\eta)h(\eta)d\eta]d\tau$$

$$= \int_0^t [\int_\eta^t f(t-\tau)g(\tau-\eta)d\tau]h(\eta)(d\eta).$$

The last double integral is obtained from the previous line by interchanging the order of the η and τ integrations. Making the change of variable $\omega = \tau - \eta$ on the inside integral yields

$$f*(g*h) = \int_0^t [\int_0^{t-\eta} f(t-\eta-\omega)g(\omega)d\omega]h(\eta)d\eta$$

$$= \int_0^t (f*g)(t-\eta)h(\eta)d\eta = (f*g)*h.$$

4.   It is possible to determine $f(t)$ explicitly by using integration by parts and then find its transform $F(s)$. However, it is much more convenient to apply Theorem 6.6.1. Let us define $g(t) = t^2$ and $h(t) = \cos 2t$. Then, $f(t) = \int_0^t g(t-\tau)h(\tau)d\tau.$ Using Table 6.2.1, we have $G(s) \equiv \mathcal{L}\{g(t)\} = 2/s^3$ and $H(s) \equiv \mathcal{L}\{h(t)\} = s/(s^2+4)$. Hence, by Theorem 6.6.1, $\mathcal{L}\{f(t)\} = F(s) = G(s)H(s) = 2/s^2(s^2+4)$.

8.   As was done in Example 1 think of F(s) as the product of s^{-4} and $(s^2+1)^{-1}$ which, according to Table 6.2.1, are the transforms of $t^3/6$ and sint, respectively. Hence, by Theorem 6.6.1, the inverse transform of F(s) is

$$f(t) = (1/6)\int_0^t (t-\tau)^3 \sin\tau d\tau.$$

13. We take the Laplace transform of the D.E. and apply the I.C.: $(s^2 + 2s + 2)Y(s) = \alpha/(s^2 + \alpha^2)$. Solving for Y(s), we have $Y(s) = [\alpha/(s^2+\alpha^2)][(s+1)^2 + 1]^{-1}$, where the second factor has been written in a convenient way by completing the square. Thus Y(s) is seen to be the product of the transforms of $\sin\alpha t$ and $e^{-t}\sin t$ respectively. Hence, according to Theorem 6.6.1, $y = \int_0^t e^{-(t-\tau)}\sin(t-\tau)\sin\alpha\tau d\tau.$

15. Proceeding as in Problem 13 we obtain

$$Y(s) = \frac{s}{s^2+s+5/4} + \frac{1-e^{-s}}{s(s^2+s+5/4)}$$

$$= \frac{(s+1/2) - 1/2}{(s+1/2)^2+1} + \frac{1-e^{-s}}{s} \cdot \frac{1}{(s+1/2)^2+1},$$

where the first term is obtained by completing the square in the denominator and the second term is written as the product of two terms whose inverse transforms are known, so that Theorem 6.6.1 can be used. Note that $\mathcal{L}^{-1}\{(1-e^{-s})/s\} = 1 - u_\pi(t)$. Also note that a different form of the same solution would be obtained by writing the second term as $(1-e^{-\pi s})(\frac{a}{s} + \frac{bs + c}{(s+1/2)^2+1})$ and solving for a, b and c. In this case $\mathcal{L}^{-1}\{1-e^{-s}\} = \delta(t) - \delta(t-\pi)$ from Section 6.5.

17. Taking the Laplace transform, using the I.C. and solving, we have $Y(s) = (s+3)/(s+1)(s+2) + s/(s^2+\alpha^2)(s+1)(s+2)$. As in Problem 15, there are several correct ways the second term can be treated in order to use the convolution integral. In order to obtain the desired answer, write the second term as

$$\frac{s}{s^2+\alpha^2}(\frac{a}{s+1} + \frac{b}{s+2})$$ and solve for a and b.

20. To find $\Phi(s)$ you must recognize the integral that
     appears in the equation as a convolution integral.
     Taking the transform of both sides then yields

$$\Phi(s) + K(s)\Phi(s) = F(s), \text{ or } \Phi(s) = \frac{F(s)}{1+K(s)}.$$

CHAPTER 7

2. As in Example 1, let $x_1 = u$ and $x_2 = u'$, then $x_1' = x_2$ and $x_2' = u'' = 3\sin t - .5x_2 - 2x_1$.

4. In this case let $x_1 = u$, $x_2 = u'$, $x_3 = u''$, and $x_4 = u'''$.

5. Let $x_1 = u$ and $x_2 = u'$; then $x_1' = x_2$ is the first of the desired pair of equations. The second equation is obtained by substituting $u'' = x_2'$, $u' = x_2$, and $u = x_1$ in the given D.E. The I.C. become $x_1(0) = u_0$, $x_2(0) = u_0'$.

8. Follow the steps outlined in Problem 7. Solve the first D.E. for x_2 to obtain $x_2 = \dfrac{3}{2}x_1 - \dfrac{1}{2}x_1'$. Substitute this into the second D.E. to obtain $x_1'' - x_1' - 2x_1 = 0$, which has the solution $x_1 = c_1 e^{2t} + c_2 e^{-t}$. Differentiating this and substituting into the above equation for x_2 yields $x_2' = \dfrac{1}{2}c_1 e^{2t} + 2c_2 e^{-t}$. The I.C. then give

    $c_1 + c_2 = 3$ and $\dfrac{1}{2}c_1 + 2c_2 = \dfrac{1}{2}$, which yield

    $c_1 = \dfrac{11}{3}$, $c_2 = -\dfrac{2}{3}$. Thus $x_1 = \dfrac{11}{3}e^{2t} - \dfrac{2}{3}e^{-t}$ and

    $x_2 = \dfrac{11}{6}e^{2t} - \dfrac{4}{3}e^{-t}$. Note that for large t, the second term in each solution vanishes and we have $x_1 \cong \dfrac{11}{3}e^{2t}$ and

    $x_2 \cong \dfrac{11}{6}e^{2t}$, so that $x_1 \cong 2x_2$. This says that the graph will be asymptotic to the line $x_1 = 2x_2$ for large t.

9. Solving the first D.E. for x_2 gives $x_2 = \dfrac{4}{3}x_1' - \dfrac{5}{3}x_1$, which substituted into the second D.E. yields $x_1'' - 2.5x_1' + x_1 = 0$. Thus $x_1 = c_1 e^{t/2} + c_2 e^{2t}$ and $x_2 = -c_1 e^{t/2} + c_2 e^{2t}$. Using the I.C. yields $c_1 = -3/2$ and $c_2 = -1/3$. For large t, $x_1 \cong (-1/2)e^{2t}$ and $x_2 \cong (-1/2)e^{2t}$ and thus the graph is asymptotic to $x_1 = x_2$ in the third quadrant. The graph is shown on the right.

12.   9.

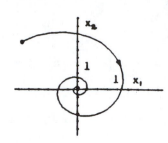   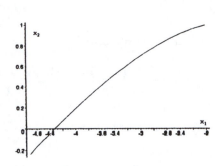

12. Solving the first D.E. for x_2 gives $x_2 = \frac{1}{2}x_1' + \frac{1}{4}x_1$ and substitution into the second D.E. gives $x_1'' + x_1' + \frac{17}{4}x_1 = 0$. Thus $x_1 = e^{-t/2}(c_1\cos 2t + c_2\sin 2t)$ and $x_2 = e^{-t/2}(c_2\cos 2t - c_1\sin 2t)$. The I.C. yields $c_1 = -2$ and $c_2 = 2$.

14. If $a_{12} \neq 0$, then solve the first equation for x_2, obtaining $x_2 = [x_1' - a_{11}x_1 - g_1(t)]/a_{12}$. Upon substituting this expression into the second equation, we have a second order linear O.D.E. for x_1. One I.C. is $x_1(0) = x_1^0$. The second I.C. is $x_2(0) = [x_1'(0) - a_{11}x_1(0) - g_1(0)]/a_{12} = x_2^0$. Solving for $x_1'(0)$ gives $x_1'(0) = a_{12}x_2^0 + a_{11}x_1^0 + g_1(0)$. These results hold when $a_{11}, \ldots, a_{22}$ are functions of t as long as the derivatives exist and $a_{12}(t)$ and $a_{21}(t)$ are not both zero on the interval. The initial conditions will involve $a_{11}(0)$ and $a_{12}(0)$.

19. Let us number the nodes 1,2, and 3 clockwise beginning with the top right node in Figure 7.1.4. Also let I_1, I_2, I_3, and I_4 denote the currents through the resistor $R = 1$, the inductor $L = 1$, the capacitor $C = \frac{1}{2}$, and the resistor $R = 2$, respectively. Let V_1, V_2, V_3, and V_4 be the corresponding voltage drops. Kirchhoff's first law applied to nodes 1 and 2, respectively, gives (i) $I_1 - I_2 = 0$ and (ii) $I_2 - I_3 - I_4 = 0$. Kirchhoff's second law applied to each loop gives (iii) $V_1 + V_2 + V_3 = 0$ and (iv) $V_3 - V_4 = 0$. The current-

voltage relation through each circuit element yields four more equations: (v) $V_1 = I_1$,   (vi) $I_2' = v_2$,

(vii) $(1/2)V_3' = I_3$ and (viii) $V_4 = 2I_4$. We thus have a system of eight equations in eight unknowns, and we wish to eliminate all of the variables except I_2 and V_3 from this system of equations. For example, we can use Eqs.(i) and (iv) to eliminate I_1 and V_4 in Eqs.(v) and (viii). Then use the new Eqs.(v) and (viii) to eliminate V_1 and I_4 in Eqs.(ii) and (iii). Finally, use the new Eqs. (ii) and (iii) in Eqs.(vi) and (vii) to obtain $I_2' = - I_2 - V_3$, $V_3' = 2I_2 - V_3$. These equations are identical (when subscripts on the remaining variables are dropped) to the equations given in the text.

21a. Note that the amount of water in each tank remains constant. Thus $Q_1(t)/30$ and $Q_2(t)/20$ represent oz./gal of salt in each tank. As in Example 1 of Section 2.3, we assume the mixture in each tank is well stirred. Then, for the first tank we have
$$\frac{dQ_1}{dt} = 1.5 - 3\frac{Q_1(t)}{30} + 1.5\frac{Q_2(t)}{20},$$ where the first term on the right represents the amount of salt per minute entering the mixture from an external source, the second term represents the loss of salt per minute going to Tank 2 and the third term represents the gain of salt per minute entering from Tank 2. Similarly, we have
$$\frac{dQ_2}{dt} = 3 + 3\frac{Q_1(t)}{30} - 4\frac{Q_2(t)}{20}$$ for Tank 2.

21b. Solve the second equation for $Q_1(t)$ to obtain $Q_1(t) = 10Q_2' + 2Q_2 - 30$. Substitution into the first equation then yields $10Q_2'' + 3Q_2' + \frac{1}{8}Q_2 = \frac{9}{2}$. The steady state solution for this is $Q_2^E = 8(9/2) = 36$. Substituting this value into the equation for Q_1 yields $Q_1^E = 72 - 30 = 42$.

21c. Substitute $Q_1 = x_1 + 42$ and $Q_2 = x_2 + 36$ into the equations found in part a.

Section 7.2, Page 355

1a. $2\mathbf{A} = \begin{pmatrix} 2 & -4 & 0 \\ 6 & 4 & -2 \\ -4 & 2 & 6 \end{pmatrix}$ so that

$2\mathbf{A} + \mathbf{B} = \begin{pmatrix} 2+4 & -4-2 & 0+3 \\ 6-1 & 4+5 & -2+0 \\ -4+6 & 2+1 & 6+2 \end{pmatrix} = \begin{pmatrix} 6 & -6 & 3 \\ 5 & 9 & -2 \\ 2 & 3 & 8 \end{pmatrix}$

1c. Using Eq.(9) and following Example 1 we have

$\mathbf{AB} = \begin{pmatrix} 4 + 2 + 0 & -2 - 10 + 0 & 3 + 0 + 0 \\ 12 - 2 - 6 & -6 + 10 - 1 & 9 + 0 - 2 \\ -8 - 1 + 18 & 4 + 5 + 3 & -6 + 0 + 6 \end{pmatrix}$,

which yields the correct answer.

6. $\mathbf{AB} = \begin{pmatrix} 6 & -5 & -7 \\ 1 & 9 & 1 \\ -1 & -2 & 8 \end{pmatrix}$ and $\mathbf{BC} = \begin{pmatrix} 5 & 3 & 3 \\ -1 & 7 & 3 \\ 2 & 3 & -2 \end{pmatrix}$ so that

$(\mathbf{AB})\mathbf{C} = \mathbf{A}(\mathbf{BC}) = \begin{pmatrix} 7 & -11 & -3 \\ 11 & 20 & 17 \\ -4 & 3 & -12 \end{pmatrix}$.

In problems 10 through 19 the method of row reduction, as illustrated in Example 2, can be used to find the inverse matrix or else to show that none exists. We start with the original matrix augmented by the indentity matrix, describe a suitable sequence of elementary row operations, and show the result of applying these operations.

10. Start with the given matrix augmented by the identity matrix. $\begin{pmatrix} 1 & 4 & . & 1 & 0 \\ & & . & & \\ -2 & 3 & . & 0 & 1 \end{pmatrix}$

Add 2 times the first row to the second row.
$\begin{pmatrix} 1 & 4 & . & 1 & 0 \\ & & . & & \\ 0 & 11 & . & 2 & 1 \end{pmatrix}$
Multiply the second row by (1/11).

$$\begin{pmatrix} 1 & 4 & . & 1 & 0 \\ & & . & & \\ 0 & 1 & . & 2/11 & 1/11 \end{pmatrix}$$

Add (-4) times the second row to the first row.

$$\begin{pmatrix} 1 & 0 & . & 3/11 & -4/11 \\ & & . & & \\ 0 & 1 & . & 2/11 & 1/11 \end{pmatrix}$$

Since we have performed the same operation on the given
matrix and the identity matrix, the 2 x 2 matric
appearing on the right side of this augmented matrix is
the desired inverse matrix. The answer can be checked by
multiplying it by the given matrix; the result should be
the indentity matrix.

12.   The augmented matrix in this case is:

$$\begin{pmatrix} 1 & 2 & 3 & . & 1 & 0 & 0 \\ & & & . & & & \\ 2 & 4 & 5 & . & 0 & 1 & 0 \\ & & & . & & & \\ 3 & 5 & 6 & . & 0 & 0 & 1 \end{pmatrix}$$

Add (-2) times the first row to the second row and (-3)
times the first row to the third row.

$$\begin{pmatrix} 1 & 2 & 3 & . & 1 & 0 & 0 \\ & & & . & & & \\ 0 & 0 & -1 & . & -2 & 1 & 0 \\ & & & . & & & \\ 0 & -1 & -3 & . & -3 & 0 & 1 \end{pmatrix}$$

Multiply the second and third rows by (-1) and
interchange them.

$$\begin{pmatrix} 1 & 2 & 3 & . & 1 & 0 & 0 \\ & & & . & & & \\ 0 & 1 & 3 & . & 3 & 0 & -1 \\ & & & . & & & \\ 0 & 0 & 1 & . & 2 & -1 & 0 \end{pmatrix}$$

Add (-3) times the third row to the first and second

$$\begin{pmatrix} 1 & 2 & 0 & . & -5 & 3 & 0 \\ & & & . & & & \\ 0 & 1 & 0 & . & -3 & 3 & -1 \\ & & & . & & & \\ 0 & 0 & 1 & . & 2 & -1 & 0 \end{pmatrix}$$

rows.

Add (-2) times the second row to the first row.

$$\begin{pmatrix} 1 & 0 & 0 & . & 1 & -3 & 2 \\ & & & . & & & \\ 0 & 1 & 0 & . & -3 & 3 & -1 \\ & & & . & & & \\ 0 & 0 & 1 & . & 2 & -1 & 0 \end{pmatrix}$$

The desired answer appears on the right side of this augmented matrix.

14. Again, start with the given matrix augmented by the identity matrix.

$$\begin{pmatrix} 1 & 2 & 1 & . & 1 & 0 & 0 \\ & & & . & & & \\ -2 & 1 & 8 & . & 0 & 1 & 0 \\ & & & . & & & \\ 1 & -2 & -7 & . & 0 & 0 & 1 \end{pmatrix}$$

Add (2) times the first row to the second row and add (-1) times the first row to the third row.

$$\begin{pmatrix} 1 & 2 & 1 & . & 1 & 0 & 0 \\ & & & . & & & \\ 0 & 5 & 10 & . & 2 & 1 & 0 \\ & & & . & & & \\ 0 & -4 & -8 & . & -1 & 0 & 1 \end{pmatrix}$$

Add (4/5) times the second row to the third row.

$$\begin{pmatrix} 1 & 2 & 1 & . & 1 & 0 & 0 \\ & & & . & & & \\ 0 & 5 & 10 & . & 2 & 1 & 0 \\ & & & . & & & \\ 0 & 0 & 0 & . & 3/5 & 4/5 & 0 \end{pmatrix}$$

Since the third row of the left matrix is all zeros, no further reduction can be performed, and the given matrix is singular.

22. $\mathbf{x}' = \begin{pmatrix} 4 \\ 2 \end{pmatrix} 2e^{2t} = \begin{pmatrix} 8 \\ 4 \end{pmatrix} e^{2t}$; and

$\begin{pmatrix} 3 & -2 \\ 2 & -2 \end{pmatrix} \mathbf{x} = \begin{pmatrix} 3 & -2 \\ 2 & -2 \end{pmatrix} \begin{pmatrix} 4 \\ 2 \end{pmatrix} e^{2t} = \begin{pmatrix} 12-4 \\ 8-4 \end{pmatrix} e^{2t} = \begin{pmatrix} 8 \\ 4 \end{pmatrix} e^{2t}.$

25. $\Psi' = \begin{pmatrix} -3e^{-3t} & 2e^{2t} \\ 12e^{-3t} & 2e^{2t} \end{pmatrix} = \begin{pmatrix} 1 & 1 \\ 4 & -2 \end{pmatrix} \begin{pmatrix} e^{-3t} & e^{2t} \\ -4e^{-3t} & e^{2t} \end{pmatrix}.$

Section 7.3, Page 366

1. Form the augmented matrix, as in Example 1, and use row
    reduction.

$$\begin{pmatrix} 1 & 0 & -1 & . & 0 \\ & & & . & \\ 3 & 1 & 1 & . & 1 \\ & & & . & \\ -1 & 1 & 2 & . & 2 \end{pmatrix}$$

Add (-3) times the first row to the second and add the
first row to the third.

$$\begin{pmatrix} 1 & 0 & -1 & . & 0 \\ & & & . & \\ 0 & 1 & 4 & . & 1 \\ & & & . & \\ 0 & 1 & 1 & . & 2 \end{pmatrix}$$

Add (-1) times the second row to the third.

$$\begin{pmatrix} 1 & 0 & -1 & . & 0 \\ & & & . & \\ 0 & 1 & 4 & . & 1 \\ & & & . & \\ 0 & 0 & -3 & . & 1 \end{pmatrix}$$

The third row is equivalent to $- 3x_3 = 1$ or $x_3 = - 1/3$.
Likewise the second row is equivalent to $x_2 + 4x_3 = 1$, so
$x_2 = 7/3$. Finally, from the first row, $x_1 - x_3 = 0$, so
$x_1 = - 1/3$. The answer can be checked by substituting
into the original equations.

2.   The augmented matrix is $\begin{pmatrix} 1 & 2 & -1 & . & 1 \\ & & & . & \\ 2 & 1 & 1 & . & 1 \\ & & & . & \\ 1 & -1 & 2 & . & 1 \end{pmatrix}$.   Row reduction then

yields $\begin{pmatrix} 1 & 2 & -1 & . & 1 \\ & & & . & \\ 0 & -3 & 3 & . & -1 \\ & & & . & \\ 0 & 0 & 0 & . & 1 \end{pmatrix}$.

The last row corresponds to the equation
$0x_1 + 0x_2 + 0x_3 = 1$, and there is no choice of x_1, x_2, and
x_3 that satisfies this equation.   Hence the given system
of equations has no solution.

3.   Form the augmented matrix and use row reduction.
$$\begin{pmatrix} 1 & 2 & -1 & . & 2 \\ & & & . & \\ 2 & 1 & 1 & . & 1 \\ & & & . & \\ 1 & -1 & 2 & . & -1 \end{pmatrix}$$
Add (-2) times the first row to the second and add (-1)
times the first row to the third.
$$\begin{pmatrix} 1 & 2 & -1 & . & 2 \\ & & & . & \\ 0 & -3 & 3 & . & -3 \\ & & & . & \\ 0 & -3 & 3 & . & -3 \end{pmatrix}$$
Add (-1) times the second row to the third row and then
multiply the second row by (-1/3).
$$\begin{pmatrix} 1 & 2 & -1 & . & 2 \\ & & & . & \\ 0 & 1 & -1 & . & 1 \\ & & & . & \\ 0 & 0 & 0 & . & 0 \end{pmatrix}$$

Since the last row has only zero entries, it may be dropped. The second row corresponds to the equation $x_2 - x_3 = 1$. We can assign an arbitrary value to either x_2 or x_3 and use this equation to solve for the other. For example, let $x_3 = c$, where c is arbitrary. Then $x_2 = 1 + c$. The first row corresponds to the equation $x_1 + 2x_2 - x_3 = 2$, so

$x_1 = 2 - 2x_2 + x_3 = 2 - 2(1+c) + c = -c$.

6. To determine whether the given set of vectors is linearly independent we must solve the system
$c_1 \mathbf{x}^{(1)} + c_2 \mathbf{x}^{(2)} + c_3 \mathbf{x}^{(3)} = \mathbf{0}$ for c_1, c_2, and c_3. Writing this in scalar form, we have
$$c_1 \qquad\quad + c_3 = 0$$
$$c_1 + c_2 \qquad\quad = 0, \text{ so the}$$
$$c_2 + c_3 = 0$$

augmented matrix is
$$\begin{pmatrix} 1 & 0 & 1 & . & 0 \\ & & & . & \\ 1 & 1 & 0 & . & 0 \\ & & & . & \\ 0 & 1 & 1 & . & 0 \end{pmatrix}$$

Row reduction yields
$$\begin{pmatrix} 1 & 0 & 1 & . & 0 \\ & & & . & \\ 0 & 1 & -1 & . & 0 \\ & & & . & \\ 0 & 0 & 2 & . & 0 \end{pmatrix}.$$

From the third row we have $c_3 = 0$. Then from the second row, $c_2 - c_3 = 0$, so $c_2 = 0$. Finally from the first row $c_1 + c_3 = 0$, so $c_1 = 0$. Since $c_1 = c_2 = c_3 = 0$, we conclude that the given vectors are linearly independent.

8. As in Problem 6 we wish to solve the system
$c_1 \mathbf{x}^{(1)} + c_2 \mathbf{x}^{(2)} + c_3 \mathbf{x}^{(3)} + c_4 \mathbf{x}^{(4)} = \mathbf{0}$ for c_1, c_2, c_3, and c_4. Form the augmented matrix and use row reduction.

$$\begin{pmatrix} 1 & -1 & -2 & -3 & . & 0 \\ & & & & . \\ 2 & 0 & -1 & 0 & . & 0 \\ & & & & . \\ 2 & 3 & 1 & -1 & . & 0 \\ & & & & . \\ 3 & 1 & 0 & 3 & . & 0 \end{pmatrix}$$

Add (-2) times the first row to the second, add (-2) times the first row to the third, and add (-3) times the first row to the fourth.

$$\begin{pmatrix} 1 & -1 & -2 & -3 & . & 0 \\ & & & & . \\ 0 & 2 & 3 & 6 & . & 0 \\ & & & & . \\ 0 & 5 & 5 & 5 & . & 0 \\ & & & & . \\ 0 & 4 & 6 & 12 & . & 0 \end{pmatrix}$$

Multiply the second row by $(1/2)$ and then add (-5) times the second row to the third and add (-4) times the second row to the fourth.

$$\begin{pmatrix} 1 & -1 & -2 & -3 & . & 0 \\ & & & & . \\ 0 & 1 & 3/2 & 3 & . & 0 \\ & & & & . \\ 0 & 0 & -5/2 & -10. & 0 \\ & & & & . \\ 0 & 0 & 0 & 0 & . & 0 \end{pmatrix}$$

The third row is equivalent to the equation $c_3 + 4c_4 = 0$. One way to satisfy this equation is by choosing $c_4 = -1$; then $c_3 = 4$.   From the second row we then have $c_2 = -(3/2)c_3 - 3c_4 = -6 + 3 = -3$.   Then, from the first row, $c_1 = c_2 + 2c_3 + 3c_4 = -3 + 8 - 3 = 2$.   Hence the given vectors are linearly dependent, and satisfy $2\mathbf{x}^{(1)} - 3\mathbf{x}^{(2)} + 4\mathbf{x}^{(3)} - \mathbf{x}^{(4)} = \mathbf{0}$.

14. Let $t = t_0$ be a fixed value of t in the interval

     $0 \le t \le 1$.   To determine whether $\mathbf{x}^{(1)}(t_0)$ and $\mathbf{x}^{(2)}(t_0)$ are

linearly dependent we must solve $c_1 \mathbf{x}^{(1)}(t_0) + c_2 \mathbf{x}^{(2)}(t_0) = \mathbf{0}$.
We have the augmented matrix

$$\left(\begin{matrix} e^{t_0} & 1 & . & 0 \\ \\ t_0 e^{t_0} & t_0 & . & 0 \end{matrix} \right).$$

Multiply the first row by $(-t_0)$ and add to the second row

to obtain $\left(\begin{matrix} e^{t_0} & 1 & . & 0 \\ \\ 0 & 0 & . & 0 \end{matrix} \right).$

Thus, for example, we can choose $c_1 = 1$ and $c_2 = -e^{t_0}$,
and hence the given vectors are linearly dependent at t_0.
Since t_0 is arbitrary the vectors are linearly dependent
at each point in the interval. However, there is no
linear relation between $\mathbf{x}^{(1)}$ and $\mathbf{x}^{(2)}$ that is valid
throughout the interval $0 \le t \le 1$. For example, if
$t_1 \ne t_0$, and if c_1 and c_2 are chosen as above, then

$$c_1 \mathbf{x}^{(1)}(t_1) + c_2 \mathbf{x}^{(2)}(t_1)$$

$$= \left(\begin{matrix} e^{t_1} \\ t_1 e^{t_1} \end{matrix} \right) + -e^{t_0} \left(\begin{matrix} 1 \\ t_1 \end{matrix} \right) = \left(\begin{matrix} e^{t_1} - e^{t_0} \\ t_1 e^{t_1} - t_1 e^{t_0} \end{matrix} \right) \ne \left(\begin{matrix} 0 \\ 0 \end{matrix} \right).$$

Hence the given vectors must be linearly independent on
$0 \le t \le 1$. In fact, the same argument applies to any
interval.

15. To find the eigenvalues and eigenvectors of the given

matrix we must solve $\left(\begin{matrix} 5-\lambda & -1 \\ 3 & 1-\lambda \end{matrix} \right) \left(\begin{matrix} x_1 \\ x_2 \end{matrix} \right) = \left(\begin{matrix} 0 \\ 0 \end{matrix} \right).$ The

determinant of coefficients is $(5-\lambda)(1-\lambda) - (-1)(3) = 0$,
or $\lambda^2 - 6\lambda + 8 = 0$. Hence $\lambda_1 = 2$ and $\lambda_2 = 4$ are the
eigenvalues. The eigenvector corresponding to λ_1 must

satisfy $\left(\begin{matrix} 3 & -1 \\ 3 & -1 \end{matrix} \right) \left(\begin{matrix} x_1 \\ x_2 \end{matrix} \right) = \left(\begin{matrix} 0 \\ 0 \end{matrix} \right)$, or $3x_1 - x_2 = 0$. If we let

$x_1 = 1$, then $x_2 = 3$ and the eigenvector is $\mathbf{x}^{(1)} = \left(\begin{matrix} 1 \\ 3 \end{matrix} \right)$, or

any constant multiple of this vector. Similarly, the
eigenvector corresponding to λ_2 must satisfy

$$\begin{pmatrix} 1 & -1 \\ 3 & -3 \end{pmatrix} \begin{pmatrix} x_1 \\ x_2 \end{pmatrix} = \begin{pmatrix} 0 \\ 0 \end{pmatrix}, \text{ or } x_1 - x_2 = 0. \text{ Hence } \mathbf{x}^{(2)} = \begin{pmatrix} 1 \\ 1 \end{pmatrix}, \text{ or}$$

a multiple thereof.

18. Since $\bar{a}_{12} = a_{21}$, the given matrix is Hermitian and we know in advance that its eigenvalues are real. To find the eigenvalues and eigenvectors we must solve

$$\begin{pmatrix} 1-\lambda & i \\ -i & 1-\lambda \end{pmatrix} \begin{pmatrix} x_1 \\ x_2 \end{pmatrix} = \begin{pmatrix} 0 \\ 0 \end{pmatrix}. \text{ The determinant of coefficients}$$

is $(1-\lambda)^2 - i(-i) = \lambda^2 - 2\lambda$, so the eigenvalues are $\lambda_1 = 0$ and $\lambda_2 = 2$; observe that they are indeed real even though the given matrix has imaginary entries. The eigenvector corresponding to λ_1 must satisfy

$$\begin{pmatrix} 1 & i \\ -i & 1 \end{pmatrix} \begin{pmatrix} x_1 \\ x_2 \end{pmatrix} = \begin{pmatrix} 0 \\ 0 \end{pmatrix}, \text{ or } x_1 + ix_2 = 0. \text{ Note that the second}$$

equation $-ix_1 + x_2 = 0$ is a multiple of the first. If $x_1 = 1$, then $x_2 = i$, and the eigenvector is

$$\mathbf{x}^{(1)} = \begin{pmatrix} 1 \\ i \end{pmatrix}. \text{ In a similar way we find that the}$$

eigenvector associated with λ_2 is $\mathbf{x}^{(2)} = \begin{pmatrix} 1 \\ -i \end{pmatrix}$.

21. The eigenvalues and eigenvectors satisfy

$$\begin{pmatrix} 1-\lambda & 0 & 0 \\ 2 & 1-\lambda & -2 \\ 3 & 2 & 1-\lambda \end{pmatrix} \begin{pmatrix} x_1 \\ x_2 \\ x_3 \end{pmatrix} = \begin{pmatrix} 0 \\ 0 \\ 0 \end{pmatrix}. \text{ The determinant of coefficients is}$$

$(1-\lambda)[(1-\lambda)^2 + 4] = 0$, which has roots $\lambda = 1$, $1 \pm 2i$. For $\lambda = 1$, we then have $2x_1 - 2x_3 = 0$ and $3x_1 + 2x_2 = 0$. Choosing

$x_1 = 2$ then yields $\begin{pmatrix} 2 \\ -3 \\ 2 \end{pmatrix}$ as the eigenvector corresponding to

$\lambda = 1$. For $\lambda = 1 + 2i$ we have
$-2ix_1 = 0$, $2x_1 - 2ix_2 - 2x_3 = 0$ and $3x_1 + 2x_2 - 2ix_3 = 0$,

yielding $x_1 = 0$ and $x_3 = -ix_2$. Thus $\begin{pmatrix} 0 \\ 1 \\ -i \end{pmatrix}$ is the eigenvector

corresponding to $\lambda = 1 + 2i$. A similar calculation shows that

$$\begin{pmatrix} 0 \\ 1 \\ i \end{pmatrix}$$ is the eigenvector corresponding to $\lambda = 1 - 2i$.

24. Since the given matrix is real and symmetric, we know that the eigenvalues are real. Further, even if there are repeated eigenvalues, there will be a full set of three linearly independent eigenvectors. To find the eigenvalues and eigenvectors we must solve

$$\begin{pmatrix} 3-\lambda & 2 & 4 \\ 2 & -\lambda & 2 \\ 4 & 2 & 3-\lambda \end{pmatrix} \begin{pmatrix} x_1 \\ x_2 \\ x_3 \end{pmatrix} = \begin{pmatrix} 0 \\ 0 \\ 0 \end{pmatrix}.$$ The determinant of

coefficients is $(3-\lambda)[-\lambda(3-\lambda)-4] - 2[2(3-\lambda) -8] + 4[4+4\lambda]$ $= -\lambda^3 + 6\lambda^2 + 15\lambda + 8$. Setting this equal to zero and solving we find $\lambda_1 = \lambda_2 = -1$, $\lambda_3 = 8$. The eigenvectors corresponding to λ_1 and λ_2 must satisfy

$$\begin{pmatrix} 4 & 2 & 4 \\ 2 & 1 & 2 \\ 4 & 2 & 4 \end{pmatrix} \begin{pmatrix} x_1 \\ x_2 \\ x_3 \end{pmatrix} = \begin{pmatrix} 0 \\ 0 \\ 0 \end{pmatrix};$$ hence there is only the single

relation $2x_1 + x_2 + 2x_3 = 0$ to be satisfied. Consequently, two of the variables can be selected arbitrarily and the third is then determined by this equation. For example, if $x_1 = 1$ and $x_3 = 1$, then $x_2 = -$ 4, and we obtain the eigenvector $x^{(1)} = \begin{pmatrix} 1 \\ -4 \\ 1 \end{pmatrix}$. Similarly,

if $x_1 = 1$ and $x_2 = 0$, then $x_3 = -1$, and we have the eigenvector $x^{(2)} = \begin{pmatrix} 1 \\ 0 \\ -1 \end{pmatrix}$, which is linearly independent of

$x^{(1)}$. There are many other choices that could have been made; however, by Eq.(38) there can be no more than two linearly independent eigenvectors corresponding to the eigenvalue -1. To find the eigenvector corresponding to λ_3 we must solve

$$\begin{pmatrix} -5 & 2 & 4 \\ 2 & -8 & 2 \\ 4 & 2 & -5 \end{pmatrix} \begin{pmatrix} x_1 \\ x_2 \\ x_3 \end{pmatrix} = \begin{pmatrix} 0 \\ 0 \\ 0 \end{pmatrix}.$$ Interchange the first and

second rows and use row reduction to obtain the equivalent system $x_1 - 4x_2 + x_3 = 0$, $2x_2 - x_3 = 0$. Since there are two equations to satisfy only one variable can be assigned an arbitrary value. If we let $x_2 = 1$, then

$x_3 = 2$ and $x_1 = 2$, so we find that $\mathbf{x}^{(3)} = \begin{pmatrix} 2 \\ 1 \\ 2 \end{pmatrix}$.

27. We are given that $\mathbf{Ax} = \mathbf{b}$ has solutions and thus we have $(\mathbf{Ax}, \mathbf{y}) = (\mathbf{b}, \mathbf{y})$. From Problem 26, though, $(\mathbf{Ax}, \mathbf{y}) = (\mathbf{x}, \mathbf{A^*y}) = 0$. Thus $(\mathbf{b}, \mathbf{y}) = 0$. For Example 2,

$\mathbf{A^*} = \overline{\mathbf{A}}^{-T} = \begin{pmatrix} 1 & -1 & 2 \\ -2 & 1 & -1 \\ 3 & -2 & 3 \end{pmatrix}$ and, using row reduction, the augmented

matrix for $\mathbf{A^*y} = \mathbf{0}$ becomes $\begin{pmatrix} 1 & -1 & 2 & 0 \\ 0 & 1 & -3 & 0 \\ 0 & 0 & 0 & 0 \end{pmatrix}$. Thus $\mathbf{y} = c\begin{pmatrix} 1 \\ 3 \\ 1 \end{pmatrix}$ and

hence $(\mathbf{b}, \mathbf{y}) = b_1 + 3b_2 + b_3 = 0$.

Section 7.4, Page 371

1. Use Mathematical Induction. It has already been proven that if $\mathbf{x}^{(1)}$ and $\mathbf{x}^{(2)}$ are solutions, then so is $c_1\mathbf{x}^{(1)} + c_2\mathbf{x}^{(2)}$. Assume that if $\mathbf{x}^{(1)}$, $\mathbf{x}^{(2)}$, ..., $\mathbf{x}^{(k)}$ are solutions, then $\mathbf{x} = c_1\mathbf{x}^{(1)} + \cdots + c_k\mathbf{x}^{(k)}$ is a solution. Then use Theorem 7.4.1 to conclude that $\mathbf{x} + c_{k+1}\mathbf{x}^{(k+1)}$ is also a solution and thus $c_1\mathbf{x}^{(1)} + \cdots + c_{k+1}\mathbf{x}^{(k+1)}$ is a solution if $\mathbf{x}^{(1)}$, ..., $\mathbf{x}^{(k+1)}$ are solutions.

2a. From Eq.(10) we have
$$W = \begin{vmatrix} x_1^{(1)} & x_1^{(2)} \\ x_2^{(1)} & x_2^{(2)} \end{vmatrix} = x_1^{(1)} x_2^{(2)} - x_2^{(1)} x_1^{(2)}.$$ Taking the

derivative of these two products yields four terms which may be written as

$$\frac{dW}{dt} = [\frac{dx_1^{(1)}}{dt}x_2^{(2)} - x_2^{(1)}\frac{dx_1^{(2)}}{dt}] + [x_1^{(1)}\frac{dx_2^{(2)}}{dt} - \frac{dx_2^{(1)}}{dt}x_1^{(2)}].$$

The terms in the square brackets can now be recognized as the respective determinants appearing in the desired solution. A similar result was mentioned in Problem 20 of Section 4.1.

2b. If $x^{(1)}$ is substituted into Eq.(3) we have

$$\frac{dx_1^{(1)}}{dt} = p_{11} x_1^{(1)} + p_{12} x_2^{(1)}$$

$$\frac{dx_2^{(1)}}{dt} = p_{21} x_1^{(1)} + p_{22} x_2^{(1)}.$$

Substituting the first equation above and its counterpart for $x^{(2)}$ into the first determinant appearing in dW/dt and evaluating the result yields $p_{11} \begin{vmatrix} x_1^{(1)} & x_1^{(2)} \\ x_2^{(1)} & x_2^{(2)} \end{vmatrix} = p_{11}W.$

Similarly, the second determinant in dW/dt is evaluated as $p_{22}W$, yielding the desired result.

2c. From prt b we have $\frac{dW}{W} = [p_{11}(t) + p_{22}(t)]dt$ which gives

$$W(t) = c \exp\int [p_{11}(t) + p_{22}(t)]dt.$$

6a. $W = \begin{vmatrix} t & t^2 \\ 1 & 2t \end{vmatrix} = 2t^2 - t^2 = t^2.$

6b. Pick $t = t_0$, then $c_1x^{(1)}(t_0) + c_2x^{(2)}(t_0) = 0$ implies

$$c_1\begin{pmatrix} t_0 \\ 1 \end{pmatrix} + c_2\begin{pmatrix} t_0^2 \\ 2t_0 \end{pmatrix} = \begin{pmatrix} 0 \\ 0 \end{pmatrix},$$ which has a non-zero solution

for c_1 and c_2 if and only if $\begin{vmatrix} t_0 & t_0^2 \\ 1 & 2t_0 \end{vmatrix} = 2t_0^2 - t_0^2 = t_0^2 = 0.$

Thus $x^{(1)}(t)$ and $x^{(2)}(t)$ are linearly independent at each point except $t = 0$. Thus they are linearly independent on every interval.

6c. From part a we see that the Wronskian vanishes at $t = 0$, but not at any other point. By Theorem 7.4.3, if $p(t)$, from Eq.(3), is continuous, then the Wronskian is either identically zero or else never vanishes. Hence, we conclude that the D.E. satisfied by $x^{(1)}(t)$ and $x^{(2)}(t)$ must have at least one discontinuous coefficient at $t = 0$.

6d. To obtain the system satisfied by $x^{(1)}$ and $x^{(2)}$ we consider

$$\mathbf{x} = c_1\mathbf{x}^{(1)} + c_2\mathbf{x}^{(2)}, \quad \text{or} \quad \begin{pmatrix} x_1 \\ x_2 \end{pmatrix} = c_1\begin{pmatrix} t \\ 1 \end{pmatrix} + c_2\begin{pmatrix} t^2 \\ 2t \end{pmatrix}.$$

Taking the derivative we obtain $\begin{pmatrix} x_1' \\ x_2' \end{pmatrix} = c_1\begin{pmatrix} 1 \\ 0 \end{pmatrix} + c_2\begin{pmatrix} 2t \\ 2 \end{pmatrix}.$

Solving this last system for c_1 and c_2 we find
$c_1 = x_1' - tx_2'$ and $c_2 = x_2'/2$. Thus

$$\begin{pmatrix} x_1 \\ x_2 \end{pmatrix} = (x_1' - tx_2')\begin{pmatrix} t \\ 1 \end{pmatrix} + \frac{x_2'}{2}\begin{pmatrix} t^2 \\ 2t \end{pmatrix}, \quad \text{which yields}$$

$x_1 = tx_1' - \dfrac{t^2}{2} x_2'$ and $x_2 = x_1'$. Writing this system in

matrix form we have $\mathbf{x} = \begin{pmatrix} t - t^2/2 \\ 1 \qquad 0 \end{pmatrix}\mathbf{x}'$. Finding the

inverse of the matrix multiplying $\mathbf{x}'$ yields the desired
solution.

Section 7.5, Page 381

1. Assuming that there are solutions of the form $\mathbf{x} = \xi e^{rt}$,
    we substitute into the D.E. to find
    $$r\xi e^{rt} = \begin{pmatrix} 3 & -2 \\ 2 & -2 \end{pmatrix}\xi e^{rt}. \quad \text{Since } \xi = I\xi = \begin{pmatrix} 1 & 0 \\ 0 & 1 \end{pmatrix}\xi, \text{ we can}$$
    write this equation as $\begin{pmatrix} 3 & -2 \\ 2 & -2 \end{pmatrix}\xi - r\begin{pmatrix} 1 & 0 \\ 0 & 1 \end{pmatrix}\xi = \mathbf{0}$ and

    thus we must solve $\begin{pmatrix} 3-r & -2 \\ 2 & -2-r \end{pmatrix}\begin{pmatrix} \xi_1 \\ \xi_2 \end{pmatrix} = \begin{pmatrix} 0 \\ 0 \end{pmatrix}$ for r, ξ_1, ξ_2.

    The determinant of the coefficients is
    $(3-r)(-2-r) + 4 = r^2 - r - 2$, so the eigenvalues are
    $r = -1, 2$. The eigenvector corresponding to $r = -1$

    satisfies $\begin{pmatrix} 4 & -2 \\ 2 & -1 \end{pmatrix}\begin{pmatrix} \xi_1 \\ \xi_2 \end{pmatrix} = \begin{pmatrix} 0 \\ 0 \end{pmatrix}$, which yields $2\xi_1 - \xi_2 = 0$.

    Thus $\mathbf{x}^{(1)}(t) = \xi^{(1)}e^{-t} = \begin{pmatrix} 1 \\ 2 \end{pmatrix}e^{-t}$, where we have set $\xi_1 = 1$.

    (Any other non zero choice would also work). In a

    similar fashion, for $r = 2$, we have $\begin{pmatrix} 1 & -2 \\ 2 & -4 \end{pmatrix}\begin{pmatrix} \xi_1 \\ \xi_2 \end{pmatrix} = \begin{pmatrix} 0 \\ 0 \end{pmatrix}$,

    or $\xi_1 - 2\xi_2 = 0$. Hence $\mathbf{x}^{(2)}(t) = \xi^{(2)}e^{2t} = \begin{pmatrix} 2 \\ 1 \end{pmatrix}e^{2t}$ by setting $\xi_2 = 1$. The general solution is then

$\mathbf{x} = c_1\mathbf{x}^{(1)}(t) + c_2\mathbf{x}^{(2)}(t)$. To sketch the trajectories we follow the steps illustrated in Examples 1 and 2.

Setting $c_2 = 0$ we have $\mathbf{x} = \begin{pmatrix} x_1 \\ x_2 \end{pmatrix} = c_1\begin{pmatrix} 1 \\ 2 \end{pmatrix} e^{-t}$ or $x_1 = c_1e^{-t}$

and $x_2 = 2c_1e^{-t}$ and thus one asymptote is given by

$x_2 = 2x_1$. In a similar
fashion $c_1 = 0$ gives
$x_2 = (1/2)x_1$ as a second
asymptote. Since the
roots differ in sign,
the trajectories for
this problem are similar
in nature to those in
Example 1. For $c_2 \neq 0$,
all solutions will be

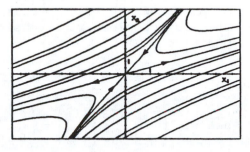

asymptotic to $x_2 = (1/2)x_1$ as $t \rightarrow \infty$. For $c_2 = 0$, the solution approaches the origin along the line $x_2 = 2x_1$.

5. Proceeding as in Problem 1 we assume a solution of the form $\mathbf{x} = \xi e^{rt}$, where r, ξ_1, ξ_2 must now satisfy

$\begin{pmatrix} -2-r & 1 \\ 1 & -2-r \end{pmatrix}\begin{pmatrix} \xi_1 \\ \xi_2 \end{pmatrix} = \begin{pmatrix} 0 \\ 0 \end{pmatrix}$. Evaluating the determinant of the

coefficients set equal to zero yields $r = -1$, -3 as the eigenvalues. For $r = -1$ we find $\xi_1 = \xi_2$ and thus

$\xi^{(1)} = \begin{pmatrix} 1 \\ 1 \end{pmatrix}$ and for $r = -3$ we find $\xi_2 = -\xi_1$ and hence

$\xi^{(2)} = \begin{pmatrix} 1 \\ -1 \end{pmatrix}$. The general solution is then

$\mathbf{x} = c_1\begin{pmatrix} 1 \\ 1 \end{pmatrix} e^{-t} + c_2\begin{pmatrix} 1 \\ -1 \end{pmatrix} e^{-3t}$. Since there are two negative

eigenvalues, we would expect the trajectories to be
similar to those of Example 2.
Setting $c_2 = 0$ and eliminating
t (as in Problem 1) we find

that $\begin{pmatrix} 1 \\ 1 \end{pmatrix} e^{-t}$ approaches the

origin along the line $x_2 = x_1$.

Similarly $\begin{pmatrix} 1 \\ -1 \end{pmatrix} e^{-3t}$ approaches

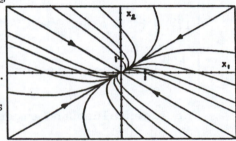

the origin along the line

$x_2 = -x_1$. As long as $c_1 \neq 0$ (since e^{-t} is the dominant term as $t \to 0$), all trajectories approach the origin asymptotic to $x_2 = x_1$. For $c_1 = 0$, the trajectory approaches the origin along $x_2 = -x_1$, as shown in the graph.

6. The characteristic equation is $(5/4 - r)^2 - 9/16 = 0$, so $r = 2, 1/2$. Since the roots are of the same size, the behavior of the solutions is similar to Problem 5, except the trajectories are reversed since the roots are positive.

7. Again assuming $\mathbf{x} = \xi e^{rt}$ we find that r, ξ_1, ξ_2 must

satisfy $\begin{pmatrix} 4-r & -3 \\ 8 & -6-r \end{pmatrix} \begin{pmatrix} \xi_1 \\ \xi_2 \end{pmatrix} = \begin{pmatrix} 0 \\ 0 \end{pmatrix}$. The determinant of the

coefficients set equal to zero yields $r = 0, -2$. For $r = 0$ we find $4\xi_1 = 3\xi_2$. Choosing $\xi_2 = 4$ we find $\xi_1 = 3$ and thus $\xi^{(1)} = \begin{pmatrix} 3 \\ 4 \end{pmatrix}$. Similarly for $r = -2$ we have

$\xi^{(2)} = \begin{pmatrix} 1 \\ 2 \end{pmatrix}$ and thus $\mathbf{x} = c_1 \begin{pmatrix} 3 \\ 4 \end{pmatrix} + c_2 \begin{pmatrix} 1 \\ 2 \end{pmatrix} e^{-2t}$. To sketch the

trajectories, note that the general solution is equivalent to the simultaneous equations $x_1 = 3c_1 + c_2 e^{-2t}$ and $x_2 = 4c_1 + 2c_2 e^{-2t}$. Solving the first equation for $c_2 e^{-2t}$ and substituting into the second yields $x_2 = 2x_1 - 2c_1$ and thus the trajectories are parallel straight lines.

9. The eigvalues are given by $\begin{vmatrix} 1-r & i \\ -i & 1-r \end{vmatrix} = (1-r)^2 + i^2 = $

$r(r-2) = 0$. For $r = 0$ we have $\begin{pmatrix} 1 & i \\ -i & 1 \end{pmatrix} \begin{pmatrix} \xi_1 \\ \xi_2 \end{pmatrix} = 0$ or

$-i\xi_1 + \xi_2 = 0$ and thus $\begin{pmatrix} 1 \\ i \end{pmatrix}$ is one eigenvector. Similarly

$\begin{pmatrix} 1 \\ -i \end{pmatrix}$ is the eigenvector for $r = 2$.

14. The eigenvalues and eigenvectors of the coefficient

matrix satisfy $\begin{pmatrix} 1-r & -1 & 4 \\ 3 & 2-r & -1 \\ 2 & 1 & -1-r \end{pmatrix} \begin{pmatrix} \xi_1 \\ \xi_2 \\ \xi_3 \end{pmatrix} = \begin{pmatrix} 0 \\ 0 \\ 0 \end{pmatrix}$. The determinant

of coefficients set equal to zero reduces to
$r^3 - 2r^2 - 5r + 6 = 0$, so the eigenvalues are
$r_1 = 1$, $r_2 = -2$, and $r_3 = 3$. The eigenvector

corresponding to r_1 must satisfy $\begin{pmatrix} 0 & -1 & 4 \\ 3 & 1 & -1 \\ 2 & 1 & -2 \end{pmatrix} \begin{pmatrix} \xi_1 \\ \xi_2 \\ \xi_3 \end{pmatrix} = \begin{pmatrix} 0 \\ 0 \\ 0 \end{pmatrix}$.

Using row reduction we obtain the equivalent system
$\xi_1 + \xi_3 = 0$, $\xi_2 - 4\xi_3 = 0$. Letting $\xi_1 = 1$, it follows
that

$\xi_3 = -1$ and $\xi_2 = -4$, so $\xi^{(1)} = \begin{pmatrix} 1 \\ -4 \\ -1 \end{pmatrix}$. In a similar way the

eigenvectors corresponding to r_2 and r_3 are found to be

$\xi^{(2)} = \begin{pmatrix} 1 \\ -1 \\ -1 \end{pmatrix}$ and $\xi^{(3)} = \begin{pmatrix} 1 \\ 2 \\ 1 \end{pmatrix}$, respectively. Thus the

general solution of the given D.E. is

$\mathbf{x} = c_1 \begin{pmatrix} 1 \\ -4 \\ -1 \end{pmatrix} e^t + c_2 \begin{pmatrix} 1 \\ -1 \\ -1 \end{pmatrix} e^{-2t} + c_3 \begin{pmatrix} 1 \\ 2 \\ 1 \end{pmatrix} e^{3t}$. Notice that the

"trajectories" of this solution would lie in the $x_1 x_2 x_3$
three dimensional space.

16. The eigenvalues and eigenvectors of the coefficient

matrix are found to be $r_1 = -1$, $\xi^{(1)} = \begin{pmatrix} 1 \\ 1 \end{pmatrix}$ and $r_2 = 3$,

$\xi^{(2)} = \begin{pmatrix} 1 \\ 5 \end{pmatrix}$. Thus the general solution of the given D.E.

is $\mathbf{x} = c_1 \begin{pmatrix} 1 \\ 1 \end{pmatrix} e^{-t} + c_2 \begin{pmatrix} 1 \\ 5 \end{pmatrix} e^{3t}$. The I.C. yields the

system of equations $c_1 \begin{pmatrix} 1 \\ 1 \end{pmatrix} + c_2 \begin{pmatrix} 1 \\ 5 \end{pmatrix} = \begin{pmatrix} 1 \\ 3 \end{pmatrix}$. The augmented matrix of this system is $\begin{pmatrix} 1 & 1 & . & 1 \\ & & . & \\ 1 & 5 & . & 3 \end{pmatrix}$ and by row reduction

we obtain $\begin{pmatrix} 1 & 1 & . & 1 \\ & & . & \\ 0 & 1 & .1/2 \end{pmatrix}$. Thus $c_2 = 1/2$ and $c_1 = 1/2$.

Substituting these values in the general solution gives the solution of the I.V.P. As $t \to \infty$, the solution

becomes asymptotic to $\mathbf{x} = \dfrac{1}{2}\begin{pmatrix} 1 \\ 5 \end{pmatrix} e^{3t}$, or $x_2 = 5x_1$.

20. Substituting $\mathbf{x} = \xi t^r$ into the D.E. we obtain

$r\xi t^r = \begin{pmatrix} 2 & -1 \\ 3 & -2 \end{pmatrix} \xi t^r$. For $t \neq 0$ this equation can be

written as $\begin{pmatrix} 2-r & -1 \\ 3 & -2-r \end{pmatrix} \begin{pmatrix} \xi_1 \\ \xi_2 \end{pmatrix} = \begin{pmatrix} 0 \\ 0 \end{pmatrix}$. The eigenvalues and

eigenvectors are $r_1 = 1$, $\xi^{(1)} = \begin{pmatrix} 1 \\ 1 \end{pmatrix}$ and $r_2 = -1$,

$\xi^{(2)} = \begin{pmatrix} 1 \\ 3 \end{pmatrix}$. Substituting these in the assumed form we

obtain the general solution $\mathbf{x} = c_1 \begin{pmatrix} 1 \\ 1 \end{pmatrix} t + c_2 \begin{pmatrix} 1 \\ 3 \end{pmatrix} t^{-1}$.

25.

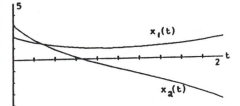

27.

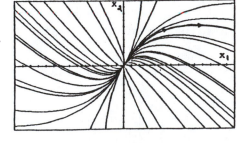
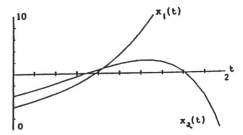

31c. The eigevalues are given by

$$\begin{vmatrix} -1-r & -1 \\ -\alpha & -1-r \end{vmatrix} = r^2 + 2r + 1 - \alpha = 0. \quad \text{Thus } r_{1,2} = -1\pm\sqrt{\alpha}.$$

Note that in Part (a) the eigenvalues are both negative while in Part (b) they differ in sign. Thus, in this part, if we choose $\alpha = 1$, then one eigenvalue is zero, which is the transition of the one root from negative to positive. This is the desired bifurcation point.

Section 7.6, Page 390

1. We assume a solution of the form $\mathbf{x} = \xi e^{rt}$ thus r and ξ are solutions of $\begin{pmatrix} 3-r & -2 \\ 4 & -1-r \end{pmatrix}\begin{pmatrix} \xi_1 \\ \xi_2 \end{pmatrix} = \begin{pmatrix} 0 \\ 0 \end{pmatrix}$. The determinant of coefficients is $(r^2-2r-3) + 8 = r^2 - 2r + 5$, so the eigenvalues are $r = 1 \pm 2i$. The eigenvector corresponding to $1 + 2i$ satisfies $\begin{pmatrix} 2-2i & -2 \\ 4 & -2-2i \end{pmatrix}\begin{pmatrix} \xi_1 \\ \xi_2 \end{pmatrix} = \begin{pmatrix} 0 \\ 0 \end{pmatrix}$, or $(2-2i)\xi_1 - 2\xi_2 = 0$. If $\xi_1 = 1$, then $\xi_2 = 1-i$ and

$\xi^{(1)} = \begin{pmatrix} 1 \\ 1-i \end{pmatrix}$ and thus one complex-valued solution of the D.E. is

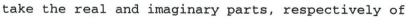

$\mathbf{x}^{(1)}(t) = \begin{pmatrix} 1 \\ 1-i \end{pmatrix} e^{(1+2i)t}$.

To find real-valued solutions (see Eqs.8 and 9) we take the real and imaginary parts, respectively of

$\mathbf{x}^{(1)}(t)$. Thus $\mathbf{x}^{(1)}(t) = \begin{pmatrix} 1 \\ 1-i \end{pmatrix}e^t(\cos 2t + i\sin 2t)$

$$= e^t\begin{pmatrix} \cos 2t + i\sin 2t \\ \cos 2t + \sin 2t + i(\sin 2t - \cos 2t) \end{pmatrix}$$

$$= e^t\begin{pmatrix} \cos 2t \\ \cos 2t + \sin 2t \end{pmatrix} + ie^t\begin{pmatrix} \sin 2t \\ \sin 2t - \cos 2t \end{pmatrix}$$

Hence the general solution of the D.E. is

$$\mathbf{x} = c_1 e^t\begin{pmatrix} \cos 2t \\ \cos 2t + \sin 2t \end{pmatrix} + c_2 e^t\begin{pmatrix} \sin 2t \\ \sin 2t - \cos 2t \end{pmatrix}. \quad \text{The}$$

solutions spiral to ∞ as $t \to \infty$ due to the e^t terms.

7.   The eigenvalues and eigenvectors of the coefficient

matrix satisfy $\begin{pmatrix} 1-r & 0 & 0 \\ 2 & 1-r & -2 \\ 3 & 2 & 1-r \end{pmatrix} \begin{pmatrix} \xi_1 \\ \xi_2 \\ \xi_3 \end{pmatrix} = \begin{pmatrix} 0 \\ 0 \\ 0 \end{pmatrix}$.   The

determinant of coefficients reduces to $(1-r)(r^2 - 2r + 5)$
so the eigenvalues are $r_1 = 1$, $r_2 = 1 + 2i$, and
$r_3 = 1 - 2i$.   The eigenvector corresponding to r_1
satisfies

$\begin{pmatrix} 0 & 0 & 0 \\ 2 & 0 & -2 \\ 3 & 2 & 0 \end{pmatrix} \begin{pmatrix} \xi_1 \\ \xi_2 \\ \xi_3 \end{pmatrix} = \begin{pmatrix} 0 \\ 0 \\ 0 \end{pmatrix}$; hence $\xi_1 - \xi_3 = 0$ and

$3\xi_1 + 2\xi_2 = 0$.   If we let $\xi_2 = -3$ then $\xi_1 = 2$ and $\xi_3 = 2$,

so one solution of the D.E. is $\begin{pmatrix} 2 \\ -3 \\ 2 \end{pmatrix} e^t$.   The eigenvector

corresponding to r_2 satisfies $\begin{pmatrix} -2i & 0 & 0 \\ 2 & -2i & -2 \\ 3 & 2 & -2i \end{pmatrix} \begin{pmatrix} \xi_1 \\ \xi_2 \\ \xi_3 \end{pmatrix} = \begin{pmatrix} 0 \\ 0 \\ 0 \end{pmatrix}$.

Hence $\xi_1 = 0$ and $i\xi_2 + \xi_3 = 0$.   If we let $\xi_2 = 1$, then
$\xi_3 = -i$. Thus a complex-valued solution is

$\begin{pmatrix} 0 \\ 1 \\ -i \end{pmatrix} e^t(\cos 2t + i \sin 2t)$.   Taking the real and imaginary

parts, see prob. 1, we obtain $\begin{pmatrix} 0 \\ \cos 2t \\ \sin 2t \end{pmatrix} e^t$ and $\begin{pmatrix} 0 \\ \sin 2t \\ -\cos 2t \end{pmatrix} e^t$,

respectively.   Thus the general solution is

$\mathbf{x} = c_1 \begin{pmatrix} 2 \\ -3 \\ 2 \end{pmatrix} e^t + c_2 e^t \begin{pmatrix} 0 \\ \cos 2t \\ \sin 2t \end{pmatrix} + c_3 e^t \begin{pmatrix} 0 \\ \sin 2t \\ -\cos 2t \end{pmatrix}$, which spirals

to ∞ about the x_1 axis in the $x_1x_2x_3$ space as $t \to \infty$.

9.   The eigenvalues and eigenvectors of the coefficient

matrix satisfy $\begin{pmatrix} 1-r & -5 \\ 1 & -3-r \end{pmatrix} \begin{pmatrix} \xi_1 \\ \xi_2 \end{pmatrix} = \begin{pmatrix} 0 \\ 0 \end{pmatrix}$.   The determinant of

coefficients is $r^2 + 2r + 2$ so that the eigenvalues are
$r = -1 \pm i$.   The eigenvector corresponding to $r = -1 + i$

is given by $\begin{pmatrix} 2-i & -5 \\ 1 & -2-i \end{pmatrix} \begin{pmatrix} \xi_1 \\ \xi_2 \end{pmatrix} = 0$ so that $\xi_1 = (2+i)\xi_2$ and

thus one complex-valued solution is

$x^{(1)}(t) = \begin{pmatrix} 2+i \\ 1 \end{pmatrix} e^{(-1+i)t}$. Finding the real and complex

parts of $x^{(1)}$ leads to the general solution

$x = c_1 e^{-t} \begin{pmatrix} 2\cos t - \sin t \\ \cos t \end{pmatrix} + c_2 e^{-t} \begin{pmatrix} 2\sin t + \cos t \\ \sin t \end{pmatrix}$.   Setting

$t = 0$ we find $x(0) = \begin{pmatrix} 1 \\ 1 \end{pmatrix} = c_1 \begin{pmatrix} 2 \\ 1 \end{pmatrix} + c_2 \begin{pmatrix} 1 \\ 0 \end{pmatrix}$, which is

equivalent to the system $\begin{matrix} 2c_1 + c_2 = 1 \\ c_1 + 0 = 1 \end{matrix}$.   Thus $c_1 = 1$ and

$c_2 = -1$ and

$$x(t) = e^{-t} \begin{pmatrix} 2\cos t - \sin t \\ \cos t \end{pmatrix} - e^{-t} \begin{pmatrix} 2\sin t + \cos t \\ \sin t \end{pmatrix}$$

$$= e^{-t} \begin{pmatrix} \cos t - 3\sin t \\ \cos t - \sin t \end{pmatrix}, \text{ which spirals to zero as}$$

$t \to \infty$ due to the e^{-t} term.

11a. The eigenvalues are given by

$$\begin{vmatrix} 3/4-r & -2 \\ 1 & -5/4-r \end{vmatrix} = r^2 + r/2 + 17/16 = 0.$$

11d. The trajectory starts at
$(5,5)$ in the x_1x_2 plane
and spirals around and
converges to the t axis
as $t \to \infty$.

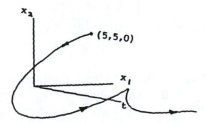

15a. The eigenvalues satisfy $\begin{vmatrix} 2-r & -5 \\ \alpha & -2-r \end{vmatrix} = r^2 - 4 + 5\alpha = 0$, so

$r_1, r_2 = \pm\sqrt{4-5\alpha}$.

15b. The critical value of α yields $r_1 = r_2 = 0$, or $\alpha = 4/5$.

15c.

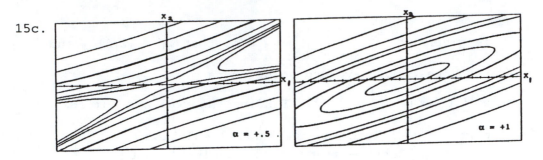

16a. $\begin{vmatrix} 5/4-r & 3/4 \\ \alpha & 5/4-r \end{vmatrix} = r^2 - 5r/2 + (25/16 - 3\alpha/4) = 0$, so

$r_{1,2} = 5/4 \pm \sqrt{3\alpha}/2$.

16b. There are two critical values of α. For $\alpha < 0$ the eigenvalues are complex, while for $\alpha > 0$ they are real. There will be a second critical value of α when $r_2 = 0$, or $\alpha = 25/12$. In this case the second real eigenvalue goes from positive to negative.

16c.

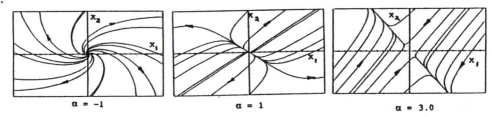

18a. We have $\begin{vmatrix} 3-r & \alpha \\ -6 & -4-r \end{vmatrix} = r^2 + r - 12 + 6\alpha = 0$, so

$r_1, r_2 = -1/2 \pm \sqrt{49-24\alpha}/2$.

18b. The critical values occur when $49 - 24\alpha = 1$ (in which case $r_2 = 0$) and when $49 - 24\alpha = 0$, in which case $r_1 = r_2 = -1/2$. Thus $\alpha = 2$ and $\alpha = 49/24 \cong 2.04$.

18c.

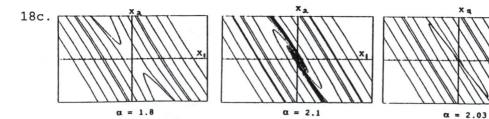

21. If we seek solutions of the form $\mathbf{x} = \xi t^r$, then r must be an eigenvalue and ξ a corresponding eigenvector of the coefficient matrix. Thus r and ξ satisfy

$$\begin{pmatrix} -1-r & -1 \\ 2 & -1-r \end{pmatrix}\begin{pmatrix} \xi_1 \\ \xi_2 \end{pmatrix} = \begin{pmatrix} 0 \\ 0 \end{pmatrix}.$$ The determinant of coefficients

is $(-1-r)^2 + 2 = r^2 + 2r + 3$, so the eigenvalues are $r = -1 \pm \sqrt{2}\,i$. The eigenvector corresponding to

$-1 + \sqrt{2}\,i$ satisfies $\begin{pmatrix} -\sqrt{2}\,i & -1 \\ 2 & -\sqrt{2}\,i \end{pmatrix}\begin{pmatrix} \xi_1 \\ \xi_2 \end{pmatrix} = \begin{pmatrix} 0 \\ 0 \end{pmatrix}$ or

$\sqrt{2}\,i\xi_1 + \xi_2 = 0$. If we let $\xi_1 = 1$, then $\xi_2 = -\sqrt{2}\,i$, and

$\xi^{(1)} = \begin{pmatrix} 1 \\ -\sqrt{2}\,i \end{pmatrix}$. Thus a complex-valued solution of the

given D.E. is $\begin{pmatrix} 1 \\ -\sqrt{2}\,i \end{pmatrix} t^{-1+\sqrt{2}\,i}$. From Eq. (15) of

Section 5.5 we have (since $t^{\sqrt{2}\,i} = e^{\ln t^{\sqrt{2}\,i}} = e^{\sqrt{2}\,i\,\ln t}$)
$t^{-1+\sqrt{2}\,i} = t^{-1}[\cos(\sqrt{2}\,\ln t) + i\sin(\sqrt{2}\,\ln t)]$ for $t > 0$.
Separating the complex valued solution into real and imaginary parts, we obtain the two real-valued solutions

$$\mathbf{u} = t^{-1}\begin{pmatrix} \cos(\sqrt{2}\,\ln t) \\ \sqrt{2}\ \sin(\sqrt{2}\,\ln t) \end{pmatrix} \text{ and } \mathbf{v} = t^{-1}\begin{pmatrix} \sin(\sqrt{2}\,\ln t) \\ -\sqrt{2}\cos(\sqrt{2}\,\ln t) \end{pmatrix}.$$

23a. The eigenvalues are given by $(r+1/4)[(r+1/4)^2 + 1] = 0$.

23b.

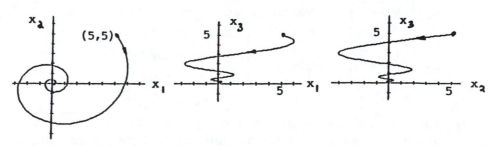

23c. Graph starts in the first octant and spirals around the x_3 axis, converging to zero.

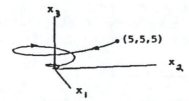

29a. We have $y_1' = x_1' = y_2$, $y_3' = x_2' = y_4$, $y_2' = -2y_1 + y_3$, and $y_4' = y_1 - 2y_3$. Thus

$$\mathbf{y}' = \begin{pmatrix} 0 & 1 & 0 & 0 \\ -2 & 0 & 1 & 0 \\ 0 & 0 & 0 & 1 \\ 1 & 0 & -2 & 0 \end{pmatrix} \mathbf{y}.$$

29b. The eigenvalues are given by $r^4 + 4r^2 + 3 = 0$, which yields $r^2 = -1, -\sqrt{3}$, so $r = \pm i, \pm\sqrt{3}\,i$.

29c. For $r = \pm i$ the eigenvectors are given by

$$\begin{pmatrix} -i & 1 & 0 & 0 \\ -2 & -i & 1 & 0 \\ 0 & 0 & -i & 1 \\ 1 & 0 & -2 & -i \end{pmatrix} \begin{pmatrix} \xi_1 \\ \xi_2 \\ \xi_3 \\ \xi_4 \end{pmatrix} = 0.$$

Choosing $\xi_1 = 1$ yields $\xi_2 = i$ and choosing $\xi_3 = 1$ yields $\xi_4 = i$, so $(1, i, 1, i)^T(\cos t + i\sin t)$ is a solution. Finding the real and imaginary parts yields

$\mathbf{w}_1 = (\cos t, -\sin t, \cos t, -\sin t)^T$ and

$\mathbf{w}_2 = (\sin t, \cos t, \sin t, \cos t)^T$ as two real solutions.

In a similar fashion, for $r = \pm\sqrt{3}\,i$, we obtain

$\xi = (1, \sqrt{3}\,i, -1, -\sqrt{3}\,i)$ and

$\mathbf{w}_3 = (\cos\sqrt{3}\,t, -\sqrt{3}\sin\sqrt{3}\,t, -\cos\sqrt{3}\,t, \sqrt{3}\sin\sqrt{3}\,t)^T$ and

$\mathbf{w}_4 = (\sin\sqrt{3}\,t, \sqrt{3}\cos\sqrt{3}\,t, -\sin\sqrt{3}\,t, -\sqrt{3}\cos\sqrt{3}\,t)^T$.

Thus $\mathbf{y} = c_1\mathbf{w}_1 + c_2\mathbf{w}_2 + c_3\mathbf{w}_3 + c_4\mathbf{w}_4$, so $\mathbf{y}^T(0) = (2, 1, 2, 1)$ yields $c_1 + c_3 = 2$, $c_2 + \sqrt{3}\,c_4 = 1$, $c_1 - c_3 = 2$, and $c_2 - \sqrt{3}\,c_4 = 1$, which yields $c_1 = 2$, $c_2 = 1$, and

$c_3 = c_4 = 0$. Hence $\mathbf{y} = \begin{pmatrix} 2\cos t + \sin t \\ -2\sin t + \cos t \\ 2\cos t + \sin t \\ -2\sin t + \cos t \end{pmatrix}$.

29e. The natural frequencies are $\omega_1 = 1$ and $\omega_2 = \sqrt{3}$, which are the absolute value of the eigenvalues. For any other choice of I.C., both frequencies will be present, and thus another mode of oscillation with a different frequency (depending on the I.C.) will be present.

Section 7.7, Page 400

Each of the Problems 1 through 10, except 2 and 8, has been solved in one of the previous sections. Thus a fundamental matrix for the given systems can be readily written down. The fundamental matrix $\Phi(t)$ satisfying $\Phi(0) = \mathbf{I}$ can then be found, as shown in the following problems.

2. The characteristic equation is given by $\begin{vmatrix} -3/4-r & 1/2 \\ 1/8 & -3/4-r \end{vmatrix} =$

$r^2 + 3r/2 + 1/2 = 0$, so $r = 1, 1/2$. For $r = 1$ we have

$\begin{pmatrix} 1/4 & 1/2 \\ 1/8 & 1/4 \end{pmatrix} \begin{pmatrix} \xi_1 \\ \xi_2 \end{pmatrix} = \begin{pmatrix} 0 \\ 0 \end{pmatrix}$, and thus $\xi^{(1)} = \begin{pmatrix} -2 \\ 1 \end{pmatrix}$. Likewise

$\xi^{(2)} = \begin{pmatrix} 2 \\ 1 \end{pmatrix}$ and thus $\mathbf{x}^{(1)}(t) = \begin{pmatrix} -2 \\ 1 \end{pmatrix} e^{-t}$ and $\mathbf{x}^{(2)}(t) = \begin{pmatrix} 2 \\ 1 \end{pmatrix} e^{-t/2}$. To

find the first column of Φ we choose c_1 and c_2 so that

$c_1 \mathbf{x}^{(1)}(0) + c_2 \mathbf{x}^{(2)}(0) = \begin{pmatrix} 1 \\ 0 \end{pmatrix}$, which yields $-2c_1 + 2c_2 = 1$ and

$c_1 + c_2 = 0$. Thus $c_1 = -1/4$ and $c_2 = 1/4$ and the first column

of Φ is $\begin{pmatrix} 1/2e^{-t/2} + 1/2e^t \\ 1/4e^{-t/2} - 1/4e^{-t/2} \end{pmatrix}$. The second colunm of Φ is

determined by $d_1 \mathbf{x}^{(1)}(0) + d_2 \mathbf{x}^{(2)}(0) = \begin{pmatrix} 0 \\ 1 \end{pmatrix}$ which yields $d_1 = d_2 =$

$1/2$ and thus the second column of Φ is $\begin{pmatrix} e^{-t/2} - e^{-t} \\ 1/2e^{-t/2} + 1/2e^{-t} \end{pmatrix}$.

4. From Problem 4 of Section 7.5 we have the two linearly

independent solutions $\mathbf{x}^{(1)}(t) = \begin{pmatrix} 1 \\ -4 \end{pmatrix} e^{-3t}$ and

$\mathbf{x}^{(2)}(t) = \begin{pmatrix} 1 \\ 1 \end{pmatrix} e^{2t}$. Hence a fundamental matrix Ψ is given

by $\Psi(t) = \begin{pmatrix} e^{-3t} & e^{2t} \\ -4e^{-3t} & e^{2t} \end{pmatrix}$. To find the fundamental matrix

$\Phi(t)$ satisfying the I.C. $\Phi(0) = \mathbf{I}$ we can proceed in
either of two ways. One way is to find $\Psi(0)$, invert it
to obtain $\Psi^{-1}(0)$, and then to form the product
$\Psi(t)\Psi^{-1}(0)$, which is $\Phi(t)$. Alternatively, we can find
the first column of Φ by determining the linear
combination

$c_1 \mathbf{x}^{(1)}(t) + c_2 \mathbf{x}^{(2)}(t)$ that satisfies the I.C. $\begin{pmatrix} 1 \\ 0 \end{pmatrix}$. This

requires that $c_1 + c_2 = 1$, $-4c_1 + c_2 = 0$, so we obtain
$c_1 = 1/5$ and $c_2 = 4/5$. Thus the first column of $\Phi(t)$ is

$$\begin{pmatrix} (1/5)e^{-3t} + (4/5)e^{2t} \\ -(4/5)e^{-3t} + (4/5)e^{2t} \end{pmatrix}.$$ Similarly, the second column of

Φ is that linear combination of $\mathbf{x}^{(1)}(t)$ and $\mathbf{x}^{(2)}(t)$ that

satisfies the I.C. $\begin{pmatrix} 0 \\ 1 \end{pmatrix}$. Thus we must have

$c_1 + c_2 = 0$, $-4c_1 + c_2 = 1$; therefore $c_1 = -1/5$ and
$c_2 = 1/5$. Hence the second column of $\Phi(t)$ is

$$\begin{pmatrix} -(1/5)e^{-3t} + (1/5)e^{2t} \\ (4/5)e^{-3t} + (1/5)e^{2t} \end{pmatrix}.$$

6. Two linearly independent real-valued solutions of the
    given D.E. were found in Problem 2 of Section 7.6. Using
    the result of that problem, we have

$$\Psi(t) = \begin{pmatrix} -2e^{-t}\sin 2t & 2e^{-t}\cos 2t \\ e^{-t}\cos 2t & e^{-t}\sin 2t \end{pmatrix}.$$ To find $\Phi(t)$

we determine the linear combinations of the columns of

$\Psi(t)$ that satisfy the I.C. $\begin{pmatrix} 1 \\ 0 \end{pmatrix}$ and $\begin{pmatrix} 0 \\ 1 \end{pmatrix}$, respectively.

In the first case c_1 and c_2 satisfy $0c_1 + 2c_2 = 1$ and
$c_1 + 0c_2 = 0$. Thus $c_1 = 0$ and $c_2 = 1/2$. In the second
case we have $0c_1 + 2c_2 = 0$ and $c_1 + 0c_2 = 1$, so $c_1 = 1$
and $c_2 = 0$. Using these values of c_1 and c_2 to form the
first and second columns of $\Phi(t)$ respectively, we obtain

$$\Phi(t) = \begin{pmatrix} e^{t}\cos 2t & -2e^{-t}\sin 2t \\ (1/2)e^{-t}\sin 2t & e^{-t}\cos 2t \end{pmatrix}.$$

10. From Problem 14 Section 7.5 we have $\mathbf{x}^{(1)} = \begin{pmatrix} 1 \\ -4 \\ -1 \end{pmatrix}e^{t}$,

$\mathbf{x}^{(2)} = \begin{pmatrix} 1 \\ -1 \\ 1 \end{pmatrix}e^{-2t}$ and $\mathbf{x}^{(3)} = \begin{pmatrix} 1 \\ 2 \\ 1 \end{pmatrix}e^{3t}$. For the first column

of Φ we want to choose c_1, c_2, c_3 such that $c_1\mathbf{x}^{(1)}(0) +$

$c_2\mathbf{x}^{(2)}(0) + c_3\mathbf{x}^{(3)}(0) = \begin{pmatrix} 1 \\ 0 \\ 0 \end{pmatrix}$. Thus $c_1 + c_2 + c_3 = 1$,

$-4c_1 - c_2 + 2c_3 = 0$ and $-c_1 - c_2 + c_3 = 0$, which yield
$c_1 = 1/6$, $c_2 = 1/3$ and $c_3 = 1/2$. The first column of Φ
is then $(1/6e^{t} + 1/3e^{-2t} + 1/2e^{3t}, -2/3e^{t} - 1/3e^{-2t} + e^{3t}$,
$-1/6e^{t} - 1/3e^{-2t} + 1/2e^{3t})^{T}$. Likewise, for the second

column we have $d_1\mathbf{x}^{(1)}(0) + d_2\mathbf{x}^{(2)}(0) + d_3\mathbf{x}^{(3)}(0) = \begin{pmatrix} 0 \\ 1 \\ 0 \end{pmatrix}$,

which yields $d_1 = -1/3$, $d_2 = 1/3$ and $d_3 = 0$ and thus $(-1/3e^t + 1/3e^{-2t}, 4/3e^t - 1/3e^{-2t}, 1/3e^t - 1/3e^{-2t})^T$ is the second column of $\Phi(t)$. Finally, for the third column

we have $e_1\mathbf{x}^{(1)}(0) + e_2\mathbf{x}^{(2)}(0) + e_3\mathbf{x}^{(3)}(0) = \begin{pmatrix} 0 \\ 0 \\ 1 \end{pmatrix}$, which

gives $e_1 = 1/2$, $e_2 = -1$ and $e_3 = 1/2$ and hence $(1/2e^t - e^{-2t} + 1/2e^{3t}, -2e^t + e^{-2t} + e^{3t}, -1/2e^t + e^{-2t} + 1/2e^{3t})^T$ is the third column of $\Phi(t)$.

11. From Eq. (14) the solution is given by $\Phi(t)\mathbf{x}^0$. Thus

$$\mathbf{x} = \begin{pmatrix} 3/2e^t - 1/2e^{-t} & -1/2e^t + 1/2e^{-t} \\ 3/2e^t - 3/2e^{-t} & -1/2e^t + 3/2e^{-t} \end{pmatrix} \begin{pmatrix} 2 \\ 1 \end{pmatrix}$$

$$= \begin{pmatrix} 7/2e^t - 3/2e^t \\ 7/2e^t - 9/2e^{-t} \end{pmatrix} = \frac{7}{2}\begin{pmatrix} 1 \\ 1 \end{pmatrix}e^t - \frac{3}{2}\begin{pmatrix} 1 \\ 3 \end{pmatrix}e^{-t}.$$

Section 7.8, Page 407

1. The eigenvalues and eigenvectors of the given coefficient matrix satisfy $\begin{pmatrix} 3-r & -4 \\ 1 & -1-r \end{pmatrix} \begin{pmatrix} \xi_1 \\ \xi_2 \end{pmatrix} = \begin{pmatrix} 0 \\ 0 \end{pmatrix}$. The determinant of coefficients is $(3-r)(-1-r) + 4 = r^2 - 2r + 1 = (r-1)^2$ so $r_1 = 1$ and $r_2 = 1$. The eigenvectors corresponding to this double eigenvalue satisfy $\begin{pmatrix} 2 & -4 \\ 1 & -2 \end{pmatrix} \begin{pmatrix} \xi_1 \\ \xi_2 \end{pmatrix} = \begin{pmatrix} 0 \\ 0 \end{pmatrix}$, or $\xi_1 - 2\xi_2 = 0$. Thus the only eigenvectors are multiples of $\xi^{(1)} = \begin{pmatrix} 2 \\ 1 \end{pmatrix}$. One solution of the given D.E. is

$\mathbf{x}^{(1)}(t) = \begin{pmatrix} 2 \\ 1 \end{pmatrix} e^t$, but there is no second solution of this form. To find a second solution we assume, as in Eq. (13), that $\mathbf{x} = \xi t e^t + \eta e^t$ and substitute this expression into the D.E. As in Example 2 we find that ξ is an eigenvector, so we choose $\xi = \begin{pmatrix} 2 \\ 1 \end{pmatrix}$. Then η must satisfy $\begin{pmatrix} 2 & -4 \\ 1 & -2 \end{pmatrix}\begin{pmatrix} \eta_1 \\ \eta_2 \end{pmatrix} = \begin{pmatrix} 2 \\ 1 \end{pmatrix}$, which verifies Eq.(16).

Solving these equations yields $\eta_1 - 2\eta_2 = 1$. If $\eta_2 = k$, where k is an arbitrary constant, then $\eta_1 = 1 + 2k$. Hence the second solution that we obtain is

$$\mathbf{x}^{(2)}(t) = \begin{pmatrix} 2 \\ 1 \end{pmatrix}te^t + \begin{pmatrix} 1 + 2k \\ k \end{pmatrix}e^t = \begin{pmatrix} 2 \\ 1 \end{pmatrix}te^t + \begin{pmatrix} 1 \\ 0 \end{pmatrix}e^t + k\begin{pmatrix} 2 \\ 1 \end{pmatrix}e^t.$$

The last term is a multiple of the first solution $\mathbf{x}^{(1)}(t)$ and may be neglected, that is, we may set $k = 0$. Thus

$$\mathbf{x}^{(2)}(t) = \begin{pmatrix} 2 \\ 1 \end{pmatrix}te^t + \begin{pmatrix} 1 \\ 0 \end{pmatrix}e^t \text{ and the general solution is}$$

$\mathbf{x} = c_1\mathbf{x}^{(1)}(t) + c_2\mathbf{x}^{(2)}(t)$. All solutions diverge to infinity as $t \to \infty$. The graph is shown on the right.

3. The origin is attracting   1.

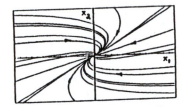

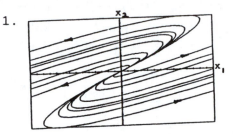

5.   Substituting $\mathbf{x} = \xi e^{rt}$ into the given system, we find that the eigenvalues and eigenvectors satisfy

$$\begin{pmatrix} 1-r & 1 & 1 \\ 2 & 1-r & -1 \\ 0 & -1 & 1-r \end{pmatrix}\begin{pmatrix} \xi_1 \\ \xi_2 \\ \xi_3 \end{pmatrix} = \begin{pmatrix} 0 \\ 0 \\ 0 \end{pmatrix}.$$ The determinant of coefficients

is $-r^3 + 3r^2 - 4$ and thus $r_1 = -1$, $r_2 = 2$ and $r_3 = 2$. The eigenvector corresponding to r_1 satisfies

$$\begin{pmatrix} 2 & 1 & 1 \\ 2 & 2 & -1 \\ 0 & -1 & 2 \end{pmatrix}\begin{pmatrix} \xi_1 \\ \xi_2 \\ \xi_3 \end{pmatrix} = \begin{pmatrix} 0 \\ 0 \\ 0 \end{pmatrix} \text{ which yields } \xi^{(1)} = \begin{pmatrix} -3 \\ 4 \\ 2 \end{pmatrix} \text{ and}$$

$$\mathbf{x}^{(1)} = \begin{pmatrix} -3 \\ 4 \\ 2 \end{pmatrix}e^{-t}.$$ The eigenvectors corresponding to the

double eigenvalue must satsify $\begin{pmatrix} -1 & 1 & 1 \\ 2 & -1 & -1 \\ 0 & -1 & -1 \end{pmatrix}\begin{pmatrix} \xi_1 \\ \xi_2 \\ \xi_3 \end{pmatrix} = \begin{pmatrix} 0 \\ 0 \\ 0 \end{pmatrix}$, which yields the single eigenvector $\xi^{(2)} = \begin{pmatrix} 0 \\ 1 \\ -1 \end{pmatrix}$ and hence

$\mathbf{x}^{(2)}(t) = \begin{pmatrix} 0 \\ 1 \\ -1 \end{pmatrix} e^{2t}$. The second solution corresponding to the double eigenvalue will have the form specified by Eq.(13), which yields $\mathbf{x}^{(3)} = \begin{pmatrix} 0 \\ 1 \\ -1 \end{pmatrix} te^{2t} + \eta e^{2t}$.

Substituting this into the given system, or using Eq.(16), we find that η satisfies $\begin{pmatrix} -1 & 1 & 1 \\ 2 & -1 & -1 \\ 0 & -1 & -1 \end{pmatrix} \begin{pmatrix} \eta_1 \\ \eta_2 \\ \eta_3 \end{pmatrix} = \begin{pmatrix} 0 \\ 1 \\ -1 \end{pmatrix}$.

Using row reduction we find that $\eta_1 = 1$ and $\eta_2 + \eta_3 = 1$, where either η_2 or η_3 is arbitrary. If we choose $\eta_2 = 0$, then $\eta = \begin{pmatrix} 1 \\ 0 \\ 1 \end{pmatrix}$ and thus $\mathbf{x}^{(3)} = \begin{pmatrix} 0 \\ 1 \\ -1 \end{pmatrix} te^{2t} + \begin{pmatrix} 1 \\ 0 \\ 1 \end{pmatrix} e^{2t}$. The general solution is then $\mathbf{x} = c_1\mathbf{x}^{(1)} + c_2\mathbf{x}^{(2)} + c_3\mathbf{x}^{(3)}$.

9. We have $\begin{vmatrix} 2-r & 3/2 \\ -3/2 & -1-r \end{vmatrix} = (r-1/2)^2 = 0$. For $r = 1/2$, the eigenvector is given by $\begin{pmatrix} 3/2 & 3/2 \\ -3/2 & -3/2 \end{pmatrix} \begin{pmatrix} \xi_1 \\ \xi_2 \end{pmatrix} = 0$, so $\xi = \begin{pmatrix} 1 \\ -1 \end{pmatrix}$ and $\begin{pmatrix} 1 \\ -1 \end{pmatrix} e^{t/2}$ is one solution. For the second solution we have $\mathbf{x} = \xi te^{t/2} + \eta e^{t/2}$, where $(\mathbf{A} - \frac{1}{2}\mathbf{I})\eta = \xi$, $\mathbf{A}$ being the coefficient matrix for this problem. This last equation reduces to $3\eta_1/2 + 3\eta_2/2 = 1$ and $-3\eta_1/2 - 3\eta_2/2 = -1$. Choosing $\eta_2 = 0$ yields $\eta_1 = 2/3$ and hence

$\mathbf{x} = c_1 \begin{pmatrix} 1 \\ -1 \end{pmatrix} e^{t/2} + c_2 \begin{pmatrix} 2/3 \\ 0 \end{pmatrix} e^{t/2} + c_2 \begin{pmatrix} 1 \\ -1 \end{pmatrix} te^{t/2}$. $\mathbf{x}(0) = \begin{pmatrix} 3 \\ -2 \end{pmatrix}$

gives $c_1 + 2c_2/3 = 3$ and $-c_1 = -2$, and hence $c_1 = 2$, $c_2 = 3/2$. Substituting these into the above $\mathbf{x}$ yields the solution.

11. The eigenvalues are $r = 1, 1, 2$. For $r = 2$, we have

$$\begin{pmatrix} -1 & 0 & 0 \\ -4 & -1 & 0 \\ 3 & 6 & 0 \end{pmatrix} \begin{pmatrix} \xi_1 \\ \xi_2 \\ \xi_3 \end{pmatrix} = \begin{pmatrix} 0 \\ 0 \\ 0 \end{pmatrix},$$ which yields $\xi = \begin{pmatrix} 0 \\ 0 \\ 1 \end{pmatrix}$, so one

solution is $\mathbf{x}^{(1)} = \begin{pmatrix} 0 \\ 0 \\ 1 \end{pmatrix} e^{2t}$. For $r = 1$, we have

$$\begin{pmatrix} 0 & 0 & 0 \\ -4 & 0 & 0 \\ 3 & 6 & 1 \end{pmatrix} \begin{pmatrix} \xi_1 \\ \xi_2 \\ \xi_3 \end{pmatrix} = \begin{pmatrix} 0 \\ 0 \\ 0 \end{pmatrix},$$ which yields the second solution

$\mathbf{x}^{(2)} = \begin{pmatrix} 0 \\ 1 \\ -6 \end{pmatrix} e^t$. The third solution is of the form

$\mathbf{x}^{(3)} = \begin{pmatrix} 0 \\ 1 \\ -6 \end{pmatrix} te^t + \eta e^t$, where $\begin{pmatrix} 0 & 0 & 0 \\ -4 & 0 & 0 \\ 3 & 6 & 1 \end{pmatrix} \eta = \begin{pmatrix} 0 \\ 1 \\ -6 \end{pmatrix}$ and thus

$\eta_1 = -1/4$ and $6\eta_2 + \eta_3 = -21/4$. Choosing $\eta_2 = 0$ gives
$\eta_3 = -21/4$ and hence

$$\mathbf{x}(t) = c_1 \begin{pmatrix} 0 \\ 1 \\ -6 \end{pmatrix} e^t + c_2 \left[\begin{pmatrix} -1/4 \\ 0 \\ -21/4 \end{pmatrix} e^t + \begin{pmatrix} 0 \\ 1 \\ -6 \end{pmatrix} te^t \right] + c_3 \begin{pmatrix} 0 \\ 0 \\ 1 \end{pmatrix} e^{2t}.$$ The

I.C. then yield $c_1 = 2$, $c_2 = 4$ and $c_3 = 3$ and hence

$$\mathbf{x} = \begin{pmatrix} -1 \\ 2 \\ -33 \end{pmatrix} e^t + 4 \begin{pmatrix} 0 \\ 1 \\ -6 \end{pmatrix} te^t + 3 \begin{pmatrix} 0 \\ 0 \\ 1 \end{pmatrix} e^{2t},$$ which become unbounded

as $t \to \infty$.

12.

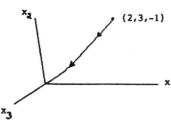
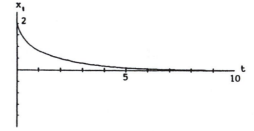

14. Assuming $\mathbf{x} = \xi t^r$ and substituting into the given system,
we find r and ξ must satisfy $\begin{pmatrix} 1-r & -4 \\ 4 & -7-r \end{pmatrix} \begin{pmatrix} \xi_1 \\ \xi_2 \end{pmatrix} = \begin{pmatrix} 0 \\ 0 \end{pmatrix}$, which
has the double eigenvalue $r = -3$ and single eigenvector

$\begin{pmatrix} 1 \\ 1 \end{pmatrix}$. Hence one solution of the given D.E. is

$\mathbf{x}^{(1)}(t) = \begin{pmatrix} 1 \\ 1 \end{pmatrix} t^{-3}$. By analogy with the scalar case

considered in Section 5.5 and Example 2 of this section, we seek a second solution of the form $\mathbf{x} = \eta t^{-3}\ln t + \zeta t^{-3}$. Substituting this expression into the D.E. we find that η and ζ satisfy the equations $(\mathbf{A} + 3\mathbf{I})\eta = \mathbf{0}$ and

$(\mathbf{A} + 3\mathbf{I})\zeta = \eta$, where $\mathbf{A} = \begin{pmatrix} 1 & -4 \\ 4 & -7 \end{pmatrix}$ and $\mathbf{I}$ is the identity

matrix. Thus $\eta = \begin{pmatrix} 1 \\ 1 \end{pmatrix}$, from above, and ζ is found to be

$\begin{pmatrix} 0 \\ -1/4 \end{pmatrix}$. Thus a second solution is

$\mathbf{x}^{(2)}(t) = \begin{pmatrix} 1 \\ 1 \end{pmatrix} t^{-3}\ln t + \begin{pmatrix} 1 \\ -1/4 \end{pmatrix} t^{-3}$.

15. All solutions of the given system approach zero as $t \to \infty$ if and only if the eigenvalues of the coefficient matrix either are real and negative or else are complex with negative real part. Write down the determinantal equation satisfied by the eigenvalues and determine when the eigenvalues are as stated.

17a. The eigenvalues and eigenvectors of the coefficient

matrix satisfy $\begin{pmatrix} 1-r & 1 & 1 \\ 2 & 1-r & -1 \\ -3 & 2 & 4-r \end{pmatrix} \begin{pmatrix} \xi_1 \\ \xi_2 \\ \xi_3 \end{pmatrix} = \begin{pmatrix} 0 \\ 0 \\ 0 \end{pmatrix}$. The determinant

of coefficients is $8 - 12r + 6r^2 - r^3 = (2-r)^3$, so the eigenvalues are $r_1 = r_2 = r_3 = 2$. The eigenvectors corresponding to this triple eigenvalue satisfy

$\begin{pmatrix} -1 & 1 & 1 \\ 2 & -1 & -1 \\ -3 & 2 & 2 \end{pmatrix} \begin{pmatrix} \xi_1 \\ \xi_2 \\ \xi_3 \end{pmatrix} = \begin{pmatrix} 0 \\ 0 \\ 0 \end{pmatrix}$. Using row reduction we can reduce

this to the equivalent system $\xi_1 - \xi_2 - \xi_3 = 0$, and $\xi_2 + \xi_3 = 0$. If we let $\xi_2 = 1$, then $\xi_3 = -1$ and $\xi_1 = 0$,

so the only eigenvectors are multiples of $\xi = \begin{pmatrix} 0 \\ 1 \\ -1 \end{pmatrix}$.

17b. From part a, one solution of the given D.E. is

$$\mathbf{x}^{(1)}(t) = \begin{pmatrix} 0 \\ 1 \\ -1 \end{pmatrix} e^{2t}, \text{ but there are no other linearly}$$

independent solutions of this form.

17c. We now seek a second solution of the form
$\mathbf{x} = \xi t e^{2t} + \eta e^{2t}$. Thus $\mathbf{A}\mathbf{x} = \mathbf{A}\xi t e^{2t} + \mathbf{A}\eta e^{2t}$ and
$\mathbf{x}' = 2\xi t e^{2t} + \xi e^{2t} + 2\eta e^{2t}$. Equating like terms, we then
have $(\mathbf{A}-2\mathbf{I})\xi = 0$ and $(\mathbf{A}-2\mathbf{I})\eta = \xi$. Thus ξ is as in part
a and the second equation yields
$$\begin{pmatrix} -1 & 1 & 1 \\ 2 & -1 & -1 \\ -3 & 2 & 2 \end{pmatrix} \begin{pmatrix} \eta_1 \\ \eta_2 \\ \eta_3 \end{pmatrix} = \begin{pmatrix} 0 \\ 1 \\ -1 \end{pmatrix}. \text{ By row reduction this is}$$

equivalent to the system $\begin{pmatrix} 1 & -1 & -1 \\ 0 & 1 & 1 \\ 0 & 0 & 0 \end{pmatrix} \begin{pmatrix} \eta_1 \\ \eta_2 \\ \eta_3 \end{pmatrix} = \begin{pmatrix} 0 \\ 1 \\ 0 \end{pmatrix}.$ If we

choose $\eta_3 = 0$, then $\eta_2 = 1$ and $\eta_1 = 1$, so $\eta = \begin{pmatrix} 1 \\ 1 \\ 0 \end{pmatrix}.$ Hence

a second solution of the D.E. is

$$\mathbf{x}^{(2)}(t) = \begin{pmatrix} 0 \\ 1 \\ -1 \end{pmatrix} t e^{2t} + \begin{pmatrix} 1 \\ 1 \\ 0 \end{pmatrix} e^{2t}.$$

17d. Assuming $\mathbf{x} = \xi(t^2/2)e^{2t} + \eta t e^{2t} + \zeta e^{2t}$, we have
$\mathbf{A}\mathbf{x} = \mathbf{A}\xi(t^2/2)e^{2t} + \mathbf{A}\eta t e^{2t} + \mathbf{A}\xi e^{2t}$ and
$\mathbf{x}' = \xi t e^{2t} + 2\xi(t^2/2)e^{2t} + \eta e^{2t} + 2\eta t e^{2t} + 2\xi e^{2t}$ and thus
$(\mathbf{A}-2\mathbf{I})\xi = 0$, $(\mathbf{A}-2\mathbf{I})\eta = \xi$ and $(\mathbf{A}-2\mathbf{I})\zeta = \eta$. Again, ξ and η
are as found previously and the last equation is
equivalent to
$$\begin{pmatrix} -1 & 1 & 1 \\ 2 & -1 & -1 \\ -3 & 2 & 2 \end{pmatrix} \begin{pmatrix} \zeta_1 \\ \zeta_2 \\ \zeta_3 \end{pmatrix} = \begin{pmatrix} 1 \\ 1 \\ 0 \end{pmatrix}. \text{ By row reduction we find the}$$

equivalent system $\begin{pmatrix} 1 & -1 & -1 \\ 0 & 1 & 1 \\ 0 & 0 & 0 \end{pmatrix} \begin{pmatrix} \zeta_1 \\ \zeta_2 \\ \zeta_3 \end{pmatrix} = \begin{pmatrix} -1 \\ 3 \\ 0 \end{pmatrix}.$ If we let

$\zeta_2 = 0$, then $\zeta_3 = 3$ and $\zeta_1 = 2$, so $\zeta = \begin{pmatrix} 2 \\ 0 \\ 3 \end{pmatrix}$ and

$$\mathbf{x}^{(3)}(t) = \begin{pmatrix} 0 \\ 1 \\ -1 \end{pmatrix} (t^2/2)e^{2t} + \begin{pmatrix} 1 \\ 1 \\ 0 \end{pmatrix} te^{2t} + \begin{pmatrix} 2 \\ 0 \\ 3 \end{pmatrix} e^{2t}.$$

17e. Ψ is the matrix with $\mathbf{x}^{(1)}$ as the first column, $\mathbf{x}^{(2)}$ as the second column and $\mathbf{x}^{(3)}$ as the third column.

17f. $\mathbf{T} = \begin{pmatrix} 0 & 1 & 2 \\ 1 & 1 & 0 \\ -1 & 0 & 3 \end{pmatrix}$ and using row operations on $\mathbf{T}$ and $\mathbf{I}$, or a

computer algebra system, $\mathbf{T}^{-1} = \begin{pmatrix} -3 & 3 & 2 \\ 3 & -2 & -2 \\ -1 & 1 & 1 \end{pmatrix}$ and thus

$$\mathbf{T}^{-1}\mathbf{A}\mathbf{T} = \begin{pmatrix} 2 & 1 & 0 \\ 0 & 2 & 1 \\ 0 & 0 & 2 \end{pmatrix} = \mathbf{J}.$$

19a. $\mathbf{J}^2 = \mathbf{J}\mathbf{J} = \begin{pmatrix} \lambda & 1 \\ 0 & \lambda \end{pmatrix}\begin{pmatrix} \lambda & 1 \\ 0 & \lambda \end{pmatrix} = \begin{pmatrix} \lambda^2 & 2\lambda \\ 0 & \lambda^2 \end{pmatrix}$

$\mathbf{J}^3 = \mathbf{J}\mathbf{J}^2 = \begin{pmatrix} \lambda & 1 \\ 0 & \lambda \end{pmatrix}\begin{pmatrix} \lambda^2 & 2\lambda \\ 0 & \lambda^2 \end{pmatrix} = \begin{pmatrix} \lambda^3 & 3\lambda^2 \\ 0 & \lambda^3 \end{pmatrix}$

19b. Based upon the results of part a, assume

$\mathbf{J}^n = \begin{pmatrix} \lambda^n & n\lambda^{n-1} \\ 0 & \lambda^n \end{pmatrix}$, then

$\mathbf{J}^{n+1} = \mathbf{J}\mathbf{J}^n = \begin{pmatrix} \lambda & 1 \\ 0 & \lambda \end{pmatrix}\begin{pmatrix} \lambda^n & n\lambda^{n-1} \\ 0 & \lambda^n \end{pmatrix}$

$= \begin{pmatrix} \lambda^{n+1} & (n+1)\lambda^n \\ 0 & \lambda^{n+1} \end{pmatrix}$, which is the same as $\mathbf{J}^n$ with n

replaced by n+1. Thus, by mathematical induction, $\mathbf{J}^n$ has the desired form.

19c. From Eq.(23), Section 7.7, we have

$$\exp(\mathbf{J}t) = \mathbf{I} + \sum_{n=1}^{\infty} \frac{\mathbf{J}^n t^n}{n!}$$

$$= \mathbf{I} + \sum_{n=1}^{\infty} \begin{pmatrix} \dfrac{\lambda^n t^n}{n!} & \dfrac{n\lambda^{n-1} t^n}{n!} \\ 0 & \dfrac{\lambda^n t^n}{n!} \end{pmatrix}$$

$$= \begin{pmatrix} 1 + \sum_{n=1}^{\infty} \dfrac{\lambda^n t^n}{n!} & \sum_{n=1}^{\infty} \dfrac{\lambda^{n-1} t^n}{(n-1)!} \\ 0 & 1 + \sum_{n=1}^{\infty} \dfrac{\lambda^n t^n}{n!} \end{pmatrix}$$

$$= \begin{pmatrix} e^{\lambda t} & t e^{\lambda t} \\ 0 & e^{\lambda t} \end{pmatrix}, \text{ since}$$

$$\sum_{n=1}^{\infty} \frac{\lambda^{n-1} t^n}{(n-1)!} = t\left(1 + \sum_{n=1}^{\infty} \frac{\lambda^n t^n}{n!}\right) = t e^{\lambda t}.$$

19d. From Eq. (28), Section 7.7, we have

$$\mathbf{x} = \exp(\mathbf{J}t)\mathbf{x}^0 = \begin{pmatrix} e^{\lambda t} & t e^{\lambda t} \\ 0 & e^{\lambda t} \end{pmatrix}\begin{pmatrix} x_1^0 \\ x_2^0 \end{pmatrix} = \begin{pmatrix} x_1^0 e^{\lambda t} + x_2^0 t e^{\lambda t} \\ x_2^0 e^{\lambda t} \end{pmatrix}$$

$$= \begin{pmatrix} x_1^0 \\ x_2^0 \end{pmatrix} e^{\lambda t} + \begin{pmatrix} x_2^0 \\ 0 \end{pmatrix} t e^{\lambda t}.$$

Section 7.9, Page 417

1. From Section 7.5 Problem 3 we have

$$\mathbf{x}^{(c)} = c_1 \begin{pmatrix} 1 \\ 1 \end{pmatrix} e^t + c_2 \begin{pmatrix} 1 \\ 3 \end{pmatrix} e^{-t}.$$ Note that

$$\mathbf{g}(t) = \begin{pmatrix} 1 \\ 0 \end{pmatrix} e^t + \begin{pmatrix} 0 \\ 1 \end{pmatrix} t$$ and that $r = 1$ is an eigenvalue of
the coefficient matrix. Thus if the method of
undetermined coefficients is used, the assumed form is
given by Eq. (18).

2. Using methods of previous sections, we find that the
eigenvalues are $r_1 = 2$ and $r_2 = -2$, with corresponding
eigenvectors $\begin{pmatrix} \sqrt{3} \\ 1 \end{pmatrix}$ and $\begin{pmatrix} 1 \\ -\sqrt{3} \end{pmatrix}$. Thus

$$\mathbf{x}^{(c)} = c_1 \begin{pmatrix} \sqrt{3} \\ 1 \end{pmatrix} e^{2t} + c_2 \begin{pmatrix} 1 \\ -\sqrt{3} \end{pmatrix} e^{-2t}.$$ Writing the

nonhomogeneous term as $\begin{pmatrix} 1 \\ 0 \end{pmatrix} e^t + \begin{pmatrix} 0 \\ \sqrt{3} \end{pmatrix} e^{-t}$ we see that we

can assume $\mathbf{x}^{(p)} = \mathbf{a}e^t + \mathbf{b}e^{-t}$. Substituting this in the D.E., we obtain

$$\mathbf{a}e^t - \mathbf{b}e^{-t} = \mathbf{A}\mathbf{a}e^t + \mathbf{A}\mathbf{b}e^{-t} + \begin{pmatrix} 1 \\ 0 \end{pmatrix} e^t + \begin{pmatrix} 0 \\ \sqrt{3} \end{pmatrix} e^{-t},$$ where $\mathbf{A}$

is the given coefficient matrix. All the terms involving

e^t must add to zero and thus we have $\mathbf{A}\mathbf{a} - \mathbf{a} + \begin{pmatrix} 1 \\ 0 \end{pmatrix} = \begin{pmatrix} 0 \\ 0 \end{pmatrix}.$

This is equivalent to the system
$\sqrt{3} a_2 = -1$ and $\sqrt{3} a_1 - 2a_2 = 0$, or $a_1 = -2/3$ and
$a_2 = -1/\sqrt{3}$. Likewise the terms involving e^{-t} must add

to zero, which yields $\mathbf{A}\mathbf{b} + \mathbf{b} + \begin{pmatrix} 0 \\ \sqrt{3} \end{pmatrix} = \begin{pmatrix} 0 \\ 0 \end{pmatrix}.$ The solution

of this system is $b_1 = -1$ and $b_2 = 2/\sqrt{3}$. Substituting these values for $\mathbf{a}$ and $\mathbf{b}$ into $\mathbf{x}^{(p)}$ and adding $\mathbf{x}^{(p)}$ to $\mathbf{x}^{(c)}$ yields the desired solution.

3.   The method of undetermined coefficients is not straight
     forward since the assumed form of $\mathbf{x}^{(p)} = \mathbf{a}\cos t + \mathbf{b}\sin t$
     leads to singular equations for $\mathbf{a}$ and $\mathbf{b}$. From Problem 3
     of Section 7.6 we find that a fundamental matrix is

$$\Psi(t) = \begin{pmatrix} 5\cos t & 5\sin t \\ 2\cos t + \sin t & -\cos t + 2\sin t \end{pmatrix}.$$ The inverse
matrix is

$$\Psi^{-1}(t) = \begin{pmatrix} \dfrac{\cos t - 2\sin t}{5} & \sin t \\ \dfrac{2\cos t + \sin t}{5} & -\cos t \end{pmatrix},$$ which may be found as

in Section 7.2 or by using a computer algebra system.
Thus we may use the method of variation of parameters
where $\mathbf{x} = \Psi(t)\mathbf{u}(t)$ and $\mathbf{u}(t)$ is given by
$\mathbf{u}'(t) = \Psi^{-1}(t)\mathbf{g}(t)$ from Eq.(27). For this problem
$$\mathbf{g}(t) = \begin{pmatrix} -\cos t \\ \sin t \end{pmatrix}$$ and thus

$$\mathbf{u}'(t) = \begin{pmatrix} \dfrac{\cos t - 2\sin t}{5} & \cdot \\[2mm] & \sin t \\[2mm] \dfrac{2\cos t + \sin t}{5} & -\cos t \end{pmatrix} \begin{pmatrix} -\cos t \\ \sin t \end{pmatrix}$$

$$= \frac{1}{5} \begin{pmatrix} 2 - 3\cos t2t + \sin 2t \\ -1 - \cos 2t - 3\sin 2t \end{pmatrix},$$

after multiplying and using appropriate trigonometric identities. Integration and multiplication by Ψ yields the desired solution.

4.   In this problem we use the method illustrated in Example 1. From Problem 4 of Section 7.5 we have the transformation matrix $\mathbf{T} = \begin{pmatrix} 1 & 1 \\ -4 & 1 \end{pmatrix}$. Inverting $\mathbf{T}$ we find that $\mathbf{T}^{-1} = \dfrac{1}{5}\begin{pmatrix} 1 & -1 \\ 4 & 1 \end{pmatrix}$. If we let $\mathbf{x} = \mathbf{Ty}$ and substitute into the D.E., we obtain

$$\mathbf{y}' = \frac{1}{5}\begin{pmatrix} 1 & -1 \\ 4 & 1 \end{pmatrix}\begin{pmatrix} 1 & 1 \\ 4 & -2 \end{pmatrix}\begin{pmatrix} 1 & 1 \\ -4 & 1 \end{pmatrix}\mathbf{y} + \frac{1}{5}\begin{pmatrix} 1 & -1 \\ 4 & 1 \end{pmatrix}\begin{pmatrix} e^{-2t} \\ -2e^{t} \end{pmatrix}$$

$$= \begin{pmatrix} -3 & 0 \\ 0 & 2 \end{pmatrix}\mathbf{y} + \frac{1}{5}\begin{pmatrix} e^{-2t} + 2e^{t} \\ 4e^{-2t} - 2e^{t} \end{pmatrix}.$$ This corresponds to the two scalar equations

$$y_1' + 3y_1 = (1/5)e^{-2t} + (2/5)e^{t},$$
$$y_2' - 2y_2 = (4/5)e^{-2t} - (2/5)e^{t},$$

which may be solved by the methods of Section 2.1. For the first equation the integrating factor is e^{3t} and we obtain $(e^{3t}y_1)' = (1/5)e^{t} + (2/5)e^{4t}$, so $e^{3t}y_1 = (1/5)e^{t} + (1/10)e^{4t} + c_1$. For the second equation the integrating factor is e^{-2t}, so $(e^{-2t}y_2)' = (4/5)e^{-4t} - (2/5)e^{-t}$. Hence $e^{-2t}y_2 = -(1/5)e^{-4t} + (2/5)e^{-t} + c_2$. Thus

$$\mathbf{y} = \begin{pmatrix} 1/5 \\ -1/5 \end{pmatrix}e^{-2t} + \begin{pmatrix} 1/10 \\ 2/5 \end{pmatrix}e^{t} + \begin{pmatrix} c_1 e^{-3t} \\ c_2 e^{2t} \end{pmatrix}.$$ Finally, multiplying by $\mathbf{T}$, we obtain

$$\mathbf{x} = \mathbf{Ty} = \begin{pmatrix} 0 \\ -1 \end{pmatrix}e^{-2t} + \begin{pmatrix} 1/2 \\ 0 \end{pmatrix}e^{t} + c_1\begin{pmatrix} 1 \\ -4 \end{pmatrix}e^{-3t} + c_2\begin{pmatrix} 1 \\ 1 \end{pmatrix}e^{2t}.$$

The last two terms are the general solution of the corresponding homogeneous system, while the first two terms constitute a particular solution of the nonhomogeneous system.

12. Since the coefficient matrix is the same as that of Problem 3, use the same procedure as done in that problem, including the Ψ^{-1} found there. In the interval $\pi/2 < t < \pi$ $\sin t > 0$ and $\cos t < 0$; hence $|\sin t| = \sin t$, but $|\cos t| = -\cos t$.

14. To verify that the given vector is the general solution of the corresponding system, it is sufficient to substitute it into the D.E. Note also that the two terms in $\mathbf{x}^{(c)}$ are linearly independent. If we seek a solution of the form $\mathbf{x} = \Psi(t)\mathbf{u}(t)$ then we find that the equation corresponding to Eq.(26) is $t\Psi(t)\mathbf{u}'(t) = \mathbf{g}(t)$, where

$$\Psi(t) = \begin{pmatrix} t & 1/t \\ t & 3/t \end{pmatrix} \text{ and } \mathbf{g}(t) = \begin{pmatrix} 1-t^2 \\ 2t \end{pmatrix}. \text{ Thus}$$

$\mathbf{u}' = (1/t)\Psi^{-1}(t)\mathbf{g}(t)$. Using a computer algebra system or row operations on Ψ and $\mathbf{I}$, we find that

$$\Psi^{-1} = \begin{pmatrix} 3/2t & -1/2t \\ -t/2 & t/2 \end{pmatrix} \text{ and hence } u_1' = \frac{3}{2t^2} - \frac{3}{2} - \frac{1}{t} \text{ and}$$

$u_2' = \dfrac{-1}{2} + \dfrac{t^2}{2} + t$, which yields $u_1 = \dfrac{-3}{2t} - \dfrac{3t}{2} - \ln t + c_1$

and $u_2 = -\dfrac{1}{2}t + \dfrac{t^3}{6} + \dfrac{t^2}{2} + c_2$. Multiplication of $\mathbf{u}$ by

$\Psi(t)$ yields the desired solution.

CHAPTER 8

Section 8.1, Page 427

In the following problems that ask for a large number of numerical calculations the first few steps are shown. It is then necessary to use these samples as a model to format a computer program or calculator to find the remaining values.

1a. The Euler formulas is $y_{n+1} = y_n + h(3 + t_n - y_n)$ for
$n = 0,1,2,3...$ and with $t_0 = 0$ and $y_0 = 1$. Thus
$y_1 = 1 + .05(3 + 0 - 1) = 1.1$
$y_2 = 1.1 + .05(3 + .05 - 1.1) = 1.1975 \cong y(.1)$
$y_3 = 1.1975 + .05(3 + .1 - 1.1975) = 1.29263$
$y_4 = 1.29263 + .05(3 + .15 - 1.29263) = 1.38549 \cong y(.2)$.

1c. The backward Euler formula is $y_{n+1} = y_n + h(3 + t_{n+1} - y_{n+1})$.
Solving this for y_{n+1} we find $y_{n+1} = [y_n + h(3 + t_{n+1})]/(1+n)$.
Thus $y_1 = \dfrac{1 + .05(3.05)}{1.05} = 1.097619$ and

$y_2 = \dfrac{1.097619 + .05(3.1)}{1.05} = 1.192971$.

5a. $y_1 = y_0 + h\dfrac{y_0^2 + 2t_0 y_0}{3 + t_0^2} = .5 + .05\dfrac{(.5)^2 + 0}{3 + 0} = .504167$

$y_2 = .504167 + .05\dfrac{(.504167)^2 + 2(.05)(.504167)}{3 + (.05)^2} = .509239$

5c. $y_1 = .5 + .05\dfrac{y_1^2 + 2(.05)y_1}{3 + (.05)^2}$, which is a quadratic
equation in y_1. Using the quadratic formula, or an
equation solver, we obtain $y_1 = .5050895$. Thus

$y_2 = .5050895 + .05\dfrac{y_2^2 + 2(.1)y_2}{3 + (.1)^2}$ which is again quadratic
in y_2, yielding $y_2 = .5111273$.

7a. For part a eighty steps must be taken, that is,
$n = 0,1,...79$ and for part b 160 steps must taken with
$n = 0,1,...159$. Thus use of a programmable calculator or
a computer is required.

7c. We have $y_{n+1} = y_n + h(.5 - t_{n+1} + 2y_{n+1})$, which is linear in
y_{n+1} and thus we have $y_{n+1} = \dfrac{y_n + .5h - ht_{n+1}}{1 - 2h}$. Again, 80

steps are needed here and 160 steps in part d. In This
case a spreadsheet is very useful. The first few, the
middle three and last two lines are shown for h = .025:

n	y_n	t_n	y_{n+1}
0	1	0	1.06513
1	1.06513	.025	1.13303
2	1.13303	.050	1.20381
⋮			
38	7.49768	.950	7.87980
39	7.87980	.975	8.28137
40	8.28137	1.000	8.70341
⋮			
78	55.62105	1.950	58.50966
79	58.50966	1.975	61.54964
80	61.54964	2.000	

At least eight decimal places were used in all
calculations.

9c. The backward Euler formula gives
$y_{n+1} = y_n + h\sqrt{t_{n+1} + y_{n+1}}$. Subtracting y_n from both
sides, squaring both sides, and solving for y_{n+1} yields
$$y_{n+1} = y_n + \frac{h^2}{2} + h\sqrt{y_n + t_{n+1} + h^2/4} .$$ Alternately, an
equation solver can be used to solve
$y_{n+1} = y_n + h\sqrt{t_{n+1} + y_{n+1}}$ for y_{n+1}. The first few values, for
h = 0.25, are y_1 = 3.043795, y_2 = 2.088082, y_3 = 3.132858 and
y_4 = 3.178122 ≅ y(.1).

15. If $y' = 1 - t + 4y$ then
$y'' = -1 + 4y' = -1 + 4(1-t+4y) = 3 - 4t + 16y$. In
Eq.(12) we let y_n, y_n' and y_n'' denote the approximate
values of $\phi(t_n)$, $\phi'(t_n)$, and $\phi''(t_n)$, respectively.
Keeping the first three terms in the Taylor series we
have
$y_{n+1} = y_n + y_n'h + y_n'' h^2/2$
$\quad = y_n + (1 - t_n + 4y_n)h + (3 - 4t_n + 16y_n)h^2/2$. For n = 0,
$t_0 = 0$ and $y_0 = 1$ we have
$$y_1 = 1 + (1 - 0 + 4)(.1) + (3 - 0 + 16)\frac{(.1)^2}{2} = 1.595.$$

16. If $y = \phi(t)$ is the exact solution of the I.V.P., then
$\phi'(t) = 2\phi(t) - 1$ and $\phi''(t) = 2\phi'(t) = 4\phi(t) - 2$. From
Eq.(21), $e_{n+1} = [2\phi(\bar{t}_n) - 1]h^2$, $t_n < \bar{t}_n < t_n + h$. Thus a

bound for e_{n+1} is $|e_{n+1}| \leq [1 + 2 \max_{0 \leq t \leq 1} |\phi(t)|]h^2$. Since the

exact solution is $y = \phi(t) = [1 + \exp(2t)]/2$,

$e_{n+1} = h^2 \exp(2\bar{t}_n)$. Therefore

$|e_1| \leq (0.1)^2 \exp(0.2) = 0.012$ and

$|e_4| \leq (0.1)^2 \exp(0.8) = 0.022$, since the maximum value of

$\exp(2\bar{t}_n)$ occurs at $t = .1$ and $t = .4$ respectively. From

Problem 2 of Section 2.7, the actual error in the first

step is .0107.

19. The local truncation error is $e_{n+1} = \phi''(\bar{t}_n)h^2/2$. For this

problem $\phi'(t) = 5t - 3\phi^{1/2}(t)$ and thus

$\phi''(t) = 5 - (3/2)\phi^{-1/2}\phi' = 19/2 - (15/2)t\phi^{-1/2}$.

Substituting this last expression into e_{n+1} yields the

desired answer.

22d. Since $y'' = -5\pi\sin 5\pi t$, Eq.(21) gives $e_{n+1} = -(5\pi/2)\sin(5\pi\bar{t}_n)h^2$.

Thus $|e_{n+1}| < \dfrac{5\pi}{2}h^2 < .05$, or $h < \dfrac{1}{\sqrt{50\pi}} \cong .08$.

23a. From Eq.(14) we have $E_n = \phi(t_n) - y_n$. Using this in

Eq.(20) we obtain

$E_{n+1} = E_n + h\{f[t_n,\phi(t_n)] - f(t_n,y_n)\} + \phi''(\bar{t}_n)h^2/2$. Using

the given inequality involving L we have

$|f[t_n,\phi(t_n)] - f(t_n,y_n)| \leq L |\phi(t_n) - y_n| = L|E_n|$ and thus

$|E_{n+1}| \leq |E_n| + hL|E_n| + \max_{t_0 \leq t \leq t_n} |\phi''(t)|h^2/2 = \alpha|E_n| + \beta h^2$.

23b. Since $\alpha = 1 + hL$, $\alpha - 1 = hL$. Hence $\beta h^2(\alpha^n-1)/(\alpha-1) = \beta h^2[(1+hL)^n - 1]/hL = \beta h[(1+hL)^n - 1]/L$.

23c. $(1+hL)^n \leq \exp(nhL)$ follows from the observation that

$\exp(nhL) = [\exp(nL)]^n = (1 + hL + h^2L^2/2! + ...)^n$.

Noting that $nh = t_n - t_0$, the rest follows from Eq.(ii).

24. The Taylor series for $\phi(t)$ about $t = t_{n+1}$ is

$\phi(t) = \phi(t_{n+1}) + \phi'(t_{n+1})(t-t_{n+1}) + \phi''(t_{n+1})\dfrac{(t-t_{n+1})^2}{2} + ...$

Letting $\phi'(t) = f(t,\phi(t))$, $t = t_n$ and $h = t_{n+1} - t_n$ we

have $\phi(t_n) = \phi(t_{n+1}) - f(t_{n+1},\phi(t_{n+1}))h + \phi''(\bar{t}_n)h^2/2$, where

$t_n < \bar{t}_n < t_{n+1}$. Thus

$\phi(t_{n+1}) = \phi(t_n) + f(t_{n+1},\phi(t_{n+1}))h - \phi''(\bar{t}_n)h^2/2$. Comparing

this to Eq. 13 we then have $e_{n+1} = -\phi''(\bar{t}_n)h^2/2$.

25b. From Problem 1 we have $y_{n+1} = y_n + h(3 + t_n - y_n)$, so
$$y_1 = 1 + .05(3 + 0 - 1) = 1.1$$
$$y_2 = 1.1 + .05(3 + .05 - 1.1) = 1.20 \cong y(.1)$$
$$y_3 = 1.20 + .05(3 + .1 - 1.20) = 1.30$$
$$y_4 = 1.30 + .05(3 + .15 - 1.30) = 1.39 \cong y(.2).$$

Section 8.2, Page 434

1a. The improved Euler formula is
$$y_{n+1} = y_n + [y_n' + f(t_n + h, y_n + hy_n')]h/2 \text{ where}$$
$y' = f(t,y) = 3 + t - y$. Hence $y_n' = 3 + t_n - y_n$ and
$f(t_n + h, y_n + hy_n') = 3 + t_{n+1} - (y_n + hy_n')$. Thus we obtain
$$y_{n+1} = y_n + (3 + t_n - y_n)h + \frac{h^2}{2}(1 - y_n')$$
$$= y_n + (3 + t_n - y_n)h + \frac{h^2}{2}(-2 - t_n + y_n). \text{ Thus}$$
$$y_1 = 1 + (3-1)(.05) + \frac{(.05)^2}{2}(-2+1) = 1.098750 \text{ and}$$
$$y_2 = y_1 + (3 +.05 - y_1)(.05) + \frac{(.05)^2}{2}(-2 -.05 + y_1) = 1.19512$$
are the first two steps. In this case, the equation
specifying y_{n+1} is somewhat more complicated when
$y_n' = 3 + t_n - y_n$ was substituted. When designing the steps to
calculate y_{n+1} on a computer, y_n' can be calculated first and
thus the simpler formula for y_{n+1} can be used. The exact
solution is $y(t) = 2 + t - e^{-t}$, so $y(.1) = 1.19516$,
$y(.2) = 1.38127$, $y(.3) = 1.55918$ and $y(.04) = 1.72968$, so the
approximations using $h = .0125$ are quite accurate to five
decimal places.

4. In this case $y_n' = 2t_n + e^{-t_n y_n}$ and thus the improved Euler
formula is
$$y_{n+1} = y_n + \frac{[(2t_n + e^{-t_n y_n}) + 2t_{n+1} + e^{-t_{n+1}(y_n + hy_n')}]h}{2}. \text{ For}$$
$n = 0, 1, 2$ we get $y_1 = 1.05122$, $y_2 = 1.10483$ and $y_3 = 1.16072$
for $h = .05$.

10. See Problem 4.

11. The improved Euler formula is
$$y_{n+1} = y_n + \frac{f(t_n,y_n) + f(t_{n+1}, y_n + hf(t_n,y_n))}{2} h. \text{ As suggested}$$
in the text, it's best to perform the following steps when

implementing this formula: let $k_1 = (4 - t_n y_n)/(1 + y_n^2)$,
$k_2 = y_n + hk_1$ and $k_3 = (4 - t_{n+1}k_2)/(1 + k_2^2)$.   Then
$y_{n+1} = y_n + (k_1 + k_3)h/2$.

14a. Since $\phi(t_n + h) = \phi(t_{n+1})$ we have, using the first part of
Eq.(5) and the given equation,
$$e_{n+1} = \phi(t_{n+1}) - y_{n+1} = [\phi(t_n) - y_n] + [\phi'(t_n) -$$
$$\frac{y_n' + f(t_n + h, y_n + hy_n')}{2}]h + \phi''(t_n)h^2/2! + \phi'''(\bar{t}_n)h^3/3!.$$
Since $y_n = \phi(t_n)$ and $y_n' = \phi'(t_n) = f(t_n, y_n)$ this reduces
to
$$e_{n+1} = \phi''(t_n)h^2/2! - \{f[t_n + h, y_n + hf(t_n, y_n)]$$
$$- f(t_n, y_n)\}h/2! + \phi'''(\bar{t}_n)h^3/3!,$$
which can be written in the form of Eq.(i).

14b. First observe that $y' = f(t, y)$ and $y'' = f_t(t, y) +$
$f_y(t, y)y'$.   Hence $\phi''(t_n) = f_t(t_n, y_n) + f_y(t_n, y_n)f(t_n, y_n)$.
Using the given Taylor series, with $a = t_n$, $h = h$, $b = y_n$
and $k = hf(t_n, y_n)$ we have

$$f[t_n + h, y_n + hf(t_n, y_n)] = f(t_n, y_n) + f_t(t_n, y_n)h + f_y(t_n, y_n)hf(t_n, y_n)$$
$$+ [f_{tt}(\xi, \eta)h^2 + 2f_{ty}(\xi, \eta)h^2f(t_n, y_n) + f_{yy}(\xi, \eta)h^2f^2(t_n, y_n)]/2!$$

where $t_n < \xi < t_n + h$ and $|\eta - y_n| < h|f(t_n, y_n)|$.
Substituting this in Eq.(i) and using the earlier
expression for $\phi''(t_n)$ we find that the first term on the
right side of Eq.(i) reduces to
$-[f_{tt}(\xi, \eta) + 2f_{ty}(\xi, \eta)f(t_n, y_n) + f_{yy}(\xi, \eta)f^2(t_n, y_n)]h^3/4$,
which is proportional to h^3 plus, possibly, higher order
terms.   The reason that there may be higher order terms
is because ξ and η will, in general, depend upon h.

14c. If $f(t, y)$ is linear in t and y, then $f_{tt} = f_{ty} = f_{yy} = 0$
and the terms appearing in the last formula of part (b)
are all zero.

15. Since $\phi(t) = [4t - 3 + 19\exp(4t)]/16$ we have
$\phi'''(t) = 76\exp(4t)$ and thus from Problem 14c, since f is
linear in t and y, we find
$$e_{n+1} = 38[\exp(4\bar{t}_n)]h^3/3.   \text{ Thus}$$
$|e_{n+1}| \leq (38h^3/3)\exp(8) = 37,758.8h^3$ on $0 \leq t \leq 2$. For $n = 1$,
we have $|e_1| = |\phi(t_1) - y_1| \leq (38/3)\exp(0.2)(.05)^3 = .001934$,
which is approximately 1/15 of the error indicated in Eq.(27)
of the previous section.

19. The Euler method gives
$$y_1 = y_0 + h(5t_0 - 3\sqrt{y_0}) = 2 + .1(-3\sqrt{2}) = 1.57574$$ and the
improved Euler method gives

$$y_1 = y_0 + \frac{f(t_0,y_0) + f(t_1,y_1)}{2} h$$

$$= 2 + [-3\sqrt{2} + (.5 - 3\sqrt{1.57574})].05 = 1.62458.$$

Thus, the estimated error in using the Euler method is
$1.62458 - 1.57574 = .04884$. Since we want our error tolerance
to be no greater than .0025 we need to adjust the step size
downward by a factor of $\sqrt{.0025/.04884} \cong .226$. Thus a step
size of $h = (.1)(.23) = .023$ would be needed for the required
local truncation error bound of .0025.

24. The modified Euler formula is
$y_{n+1} = y_n + hf[t_n + h/2, y_n + (h/2)f(t_n,y_n)]$ where
$f(t,y) = 5t - 3\sqrt{y}$. Thus
$$y_1 = 2 + .05[5(t_0 + .025) - 3\text{sqrt}(2 + .025(5t_0 - 3\sqrt{2}))]$$
$$= 1.79982 \text{ for } t_0 = 0.$$ The values obtained here are
between the values for $h = .05$ and for $h = .025$ using the
Euler method in Problem 2.

Section 8.3, Page 438

4.   The Runge-Kutta formula is
$y_{n+1} = y_n + h(k_{n1} + 2k_{n2} + 2k_{n3} + k_{n4})/6$ where k_{n1}, k_{n2}
etc. are given by Eqs.(3). Thus for
$f(t,y) = 2t + e^{-ty}$, $(t_0,y_0) = (0,1)$ and $h = .1$ we have
$k_{01} = 0 + e^0 = 1$
$k_{02} = 2(0 + .05) + e^{-(0+.05)(1+.05k_{01})} = 1.048854$
$k_{03} = 2(.05) + e^{-(.05)(1+.05k_{02})} = 1.048738$
$k_{04} = 2(.1) + e^{-(.1)(1+.1k_{03})} = 1.095398$ and hence
$y(.1) \cong y_1 = 1 + .1(k_{01} + 2k_{02} + 2k_{03} + k_{04})/6 = 1.104843$.

11. We have $f(t_n,y_n) = (4 - t_n y_n)/(1 + y_n^2)$. Thus for $t_0 = 0$,
$y_0 = -2$ and $h = .1$ we have
$k_{01} = f(0,-2) = .8$
$k_{02} = f(.05, -2 + .05(.8)) = f(.05, -1.96) = .846414$,
$k_{03} = f(.05, -2 + .05k_{02}) = f(.05, -1.957679) = .847983$,
$k_{04} = f(.1, -2 + .1k_{03}) = f(.1, -1.915202) = .897927$, and
$y_1 = -2 + .1(k_{01} + 2k_{02} + 2k_{03} + k_{04})/6 = -1.915221$. For
comparison, see Problem 11 in Sections 8.1 and 8.2.

14a.

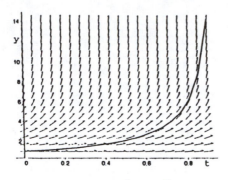

14b. We have $f(t_n, y_n) = t_n^2 + y_n^2$, $t_0 = 0$, $y_0 = 1$ and $h = .1$ so

$k_{01} = 0^2 + 1^2 = 1$

$k_{02} = (.05)^2 + (1 + .05)^2 = 1.105$

$k_{03} = (.05)^2 + [1 + .05(1.105)]^2 = 1.11605$

$k_{04} = (.1)^2 + [1 + .1(1.11605)]^2 = 1.245666$ and thus

$y_1 = 1 + .1(k_{01} + 2k_{02} + 2k_{03} + k_{04})/6 = 1.11146$. Using these steps in a computer program, we obtain the following values for y:

t	h = .1	h = .05	h = .025	h = 0.125
.8	5.842	5.8481	5.8483	5.8486
.9	14.0218	14.2712	14.3021	14.3046
.95		46.578	49.757	50.3935

14c. No accurate solution can be obtained for y(1), as the values at t = .975 for h = .025 and h = .0125 are 1218 and 23,279 respectively. These are caused by the slope field becoming vertical as t → 1.

Section 8.4, Page 444

4a. The predictor formula is

$Y_{n+1} = Y_n + h(55f_n - 59f_{n-1} + 37f_{n-2} - 9f_{n-3})/24$

and the corrector formula is

$Y_{n+1} = Y_n + h(9f_{n+1} + 19f_n - 5f_{n-1} + f_{n-2})/24$, where

$f_n = 2t_n + \exp(-t_n y_n)$. Using the Range-Kutta method, from Section 8.3, Problem 4a, we have for $t_0 = 0$ and $y_0 = 1$, $y_1 = 1.1048431$, $y_2 = 1.2188411$ and $y_3 = 1.3414680$. Thus the predictor formula gives $y_4 = 1.4725974$, so $f_4 = 1.3548603$ and the corrector formula then gives $y_4 = 1.4726173$, which is the desired value. These results, and the next step, are summarized in the following table:

n	Y_n	f_n	Y_{n+1}	f_{n+1}	Y_{n+1} Corrected
0	1	1			
1	1.1048431	1.0954004			
2	1.2188411	1.1836692			
3	1.3414680	1.2686862	1.4725974	1.3548603	1.4726173
4	1.4726173	1.3548559	1.6126246	1.4465016	1.6126215
5	1.6126215				

formula, and the corrected y_{n+1} is given by the corrector formula. Note that the value for f_4 on the line for $n = 4$ uses the corrected value for y_4, and differs slightly from the f_4 on the line for $n = 3$, which uses the predicted value for y_4.

4b. The fourth order Adams-Moulton method is given by Eq. (10): $y_{n+1} = y_n + (h/24)(9f_{n+1} + 19f_n + 5f_{n-1} + f_{n-2})$. Substituting $h = .1$ we obtain
$y_{n+1} = y_n + (.0375)(19f_n - 5f_{n-1} + f_{n-2}) + .0375f_{n+1}$.
For $n = 2$ we then have

$y_3 = y_2 + (.0375)(19f_2 - 5f_1 + f_0) + .0375f_3$
$\quad = 1.293894103 + .0375(.6 + e^{-.3y_3})$, using values for
y_2, f_0, f_1, f_2 from part a. An equation solver then yields
$y_3 = 1.341469821$. Likewise

$y_4 = y_3 + (.0375)(19f_3 - 5f_2 + f_1) + .0375f_4$
$\quad = 1.421811841 + .0375(.8 + e^{-.4y_4})$, where f_3 is calculated
using the y_3 found above. This last equation yields
$y_4 = 1.472618922$. Finally

$y_5 = y_4 + (.0375)(19f_4 - 5f_3 + f_2) + .0375f_5$
$\quad = 1.558379316 + .0375(1.0 + e^{-.5y_5})$, which gives
$y_5 = 1.612623138$.

4c. We use Eq. (16):
$y_{n+1} = (1/25)(48y_n - 36y_{n-1} + 16y_{n-2} - 3y_{n-3} + 12hf_{n+1})$.
Thus $y_4 = .04(48y_3 - 36y_2 + 16y_1 - 3y_0) + .048f_4$
$\quad\quad = 1.40758686 + .048(.8 + e^{-.4y_4})$, using values
for y_0, y_1. y_2, y_3 from part a. An equation solver then
yields $y_4 = 1.472619913$. Likewise
$y_5 = .04(48y_4 - 36y_3 + 16y_2 - 3y_1) + .048f_5$
$\quad = 1.54319349 + .048(1 + e^{-.5y_5})$, which gives
$y_5 = 1.612625556$.

7a. Using the predictor and corrector formulas (Eqs. 6 and 10) with $f_n = .5 - t_n + 2y_n$ and using the Runge-Kutta method to calculate y_1, y_2 and y_3, we obtain the following table for $h = .05$, $t_0 = 0$, $y_0 = 1$:

n	y_n	f_n	y_{n+1}	f_{n+1}	y_{n+1} corrected
0	1	2.5			
1	1.130171	2.710342			
2	1.271403	2.9420805			
3	1.424858	3.199717	1.591820	3.483640	1.591825
4	1.591825	3.483649	1.773716	3.797433	1.773721
5	1.773721	3.797443	1.972114	4.144227	1.972119
6	1.972119	4.144238	2.188747	4.527495	2.188753
7	2.188753	4.527507	2.425535	4.951070	2.425542
8	2.425542	4.951084	2.684597	5.419194	2.684604
9	2.684604	5.419209	2.968276	5.936551	2.968284
10	2.968284				

7b. From Eq.(10) we have

$$y_{n+1} = y_n + \frac{h}{24}(9f_{n+1} + 19f_n - 5f_{n-1} + f_{n-2})$$

$$= y_n + \frac{h}{24}[9(.5 - t_{n+1} + 2y_{n+1}) + 19f_n - 5f_{n-1} + f_{n-2}].$$

Solving for y_{n+1} we obtain

$$y_{n+1} = [y_n + \frac{h}{24}(19f_n - 5f_{n-1} + f_{n-2} + 4.5 - 9t_{n+1})]/(1-.75h).$$

For h = .05, $t_0 = 0$, $y_0 = 1$ and using y_1 and y_2 as calculated using the Runge-Kutta formula, we obtain the following table:

n	y_n	f_n	y_{n+1}
0	1	2.5	
1	1.130171	2.710342	
2	1.271403	2.942805	1.424859
3	1.424859	3.199718	1.591825
4	1.591825	3.483650	1.773722
5	1.773722	3.797444	1.972120
6	1.972120	4.144241	2.188755
7	2.188755	4.527510	2.425544
8	2.425544	4.951088	2.684607
9	2.684607	5.419214	2.968287
10	2.968287		

7c. From Eq (16) we have

$$y_{n+1} = (48y_n - 36y_{n-1} + 16y_{n-2} - 3y_{n-3} + 12hf_{n+1})/25$$

$$= [48y_n - 36y_{n-1} + 16y_{n-2} - 3y_{n-3} + 12h(.5-t_{n+1})]/25+(24/25)hy_{n+1}.$$

Solving for y_{n+1} we have

$$y_{n+1} = [48y_n - 36y_{n-1} + 16y_{n-2} - 3y_{n-3} + 12h(.5-t_{n+1})]/(25-24h).$$

Again, using Runge-Kutta to find y_1 and y_2, we then obtain the following table:

n	y_n	y_{n+1}
0	1	
1	1.130170833	
2	1.271402571	
3	1.424858497	1.591825573
4	1.591825573	1.773724801
5	1.773724801	1.972125968
6	1.972125968	2.188764173
7	2.188764173	2.425557376
8	2.425557376	2.684625416
9	2.684625416	2.968311063
10	2.968311063	

The exact solution is $y(t) = e^t + t/2$ so $y(.5) = 2.9682818$ and $y(2) = 55.59815$, so we see that the predictor-corrector method in part a is accurate through three decimal places.

16. Let $P_2(t) = At^2 + Bt + C$. As in Eqs. (12) and (13) let $P_2(t_{n-1}) = y_{n-1}$, $P_2(t_n) = y_n$, $P_2(t_{n+1}) = y_{n+1}$ and $P_2'(t_{n+1}) = f(t_{n+1}, y_{n+1}) = f_{n+1}$. Recall that $t_{n-1} = t_n - h$ and $t_{n+1} = t_n + h$ and thus we have the four equations:

$$A(t_n-h)^2 + B(t_n-h) + C = y_{n-1} \quad \text{(i)}$$
$$At_n^2 + Bt_n + C = y_n \quad \text{(ii)}$$
$$A(t_n+h)^2 + B(t_n+h) + C = y_{n+1} \quad \text{(iii)}$$
$$2A(t_n+h) + B = f_{n+1} \quad \text{(iv)}$$

Subtracting Eq. (i) from Eq. (ii) to get Eq. (v) (not shown) and subtracting Eq. (ii) from Eq. (iii) to get Eq. (vi) (not shown), then subtracting Eq. (v) from Eq. (vi) yields $y_{n+1} - 2y_n + y_{n-1} = 2Ah^2$, which can be solved for A. Thus $B = f_{n+1} - 2A(t_n+h)$ [from Eq. (iv)] and $C = y_n - t_nf_{n+1} + At_n^2 + 2At_nh$ [from Eq. (ii)]. Using these values for A, B and C in Eq. (iv) yields $y_{n+1} = (1/3)(4y_n-y_{n-1}+2hf_{n+1})$, which is Eq. (15).

Section 8.5, Page 454

2a. If $0 \le t \le 1$ then we know $0 \le t^2 \le 1$ and hence $e^y \le t^2 + e^y \le 1 + e^y$. Since each of these terms represents a slope, we may conclude that the solution of Eq.(i) is bounded above by the solution of Eq.(iii) and is bounded below by the solution of Eq.(iv).

2b. $\phi_1(t)$ and $\phi_2(t)$ can each be found by separation of variables. For $\phi_1(t)$ we have $\dfrac{1}{1+e^y}dy = dt$, or

$\dfrac{e^{-y}}{e^{-y}+1}dy = dt$. Integrating both sides yields

$-\ln(e^{-y}+1) = t + c$. Solving for y we find
$y = \ln[1/(c_1e^{-t}-1)]$. Setting t = 0 and y = 0, we obtain
$c_1 = 2$ and thus $\phi_1(t) = \ln[e^t/(2-e^t)]$. As t → ln 2, we
see that $\phi_1(t)$ → ∞. A similar analysis shows that
$\phi_2(t) = \ln[1/(c_2-t)]$, where $c_2 = 1$ when the I.C. are
used. Thus $\phi_2(t)$ → ∞ as t → 1 and thus we conclude
that $\phi(t)$ → ∞ for some t such that $\ln 2 \le t \le 1$.

2c. From Part b: $\phi_1(.9) = \ln[1/(c_1e^{-.9}-1)] = 3.4298$ yields
$c_1 = 2.5393$ and thus $\phi_1(t)$ → ∞ when t ≅ .9319.
Similarly for $\phi_2(t)$ we have $c_2 = .9324$ and thus
$\phi_2(t)$ → ∞ when t ≅ .932.

4a. The D.E. is $y' + 10y = 2.5t^2 + .5t$. So $y_h = ce^{-10t}$ is the
solution of the related homogeneous equation and the
particular solution, using undetermined coefficients, is
$y_p = At^2 + Bt + C$.
Substituting this
into the D.E. yields
A = 1/4, B = C = 0.
To satisfy the I.C.,
c = 4, so
$y(t) = 4e^{-10t} + (t^2/4)$,
which is shown in the
graph.

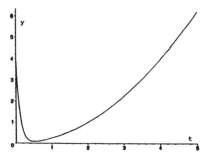

4b. From the discussion following Eq (15), we see that h must
be less than $\dfrac{2}{|r|}$ for the Euler method to be stable.
Thus, for r = 10, h < .2. For h = .2 we obtain the
following values:

t =	4	4.2	4.4	4.6	4.8	5.0
y =	8	.4	8.84	1.28	9.76	2.24

and for h = .18 we obtain:

t =	4.14	4.32	4.50	4.68	4.86	5.04
y =	4.26	4.68	5.04	5.48	5.89	6.35.

Clearly the second set of values is stable, although far
from accurate.

4c. For a step size of .25 we find

              t = 4   4.25   4.75   5.00
              y = 4.018 4.533   5.656   6.205,

     for a step size of .28 we find

              t = 4.2   4.48   4.76   5.00
              y = 10.14   10.89   11.68   12.51,

     and for a step size of .3 we find

              t = 4.2   4.5   4.8   5.1
              y = 353   484   664   912.

     Thus instability appears to occur about h = .28 and
     certainly by h = .3. Note that the exact solution for
     t = 5 is y = 6.2500, so for h = .25 we do obtain a good
     approximation.

4d. For h = .5 the error at t = 5 is .013, while for h = .385,
     the error at t = 5.005 is .01.

5a. The general solution of the D.E. is $y(t) = t + ce^{\lambda t}$,
     where $y(0) = 0 \rightarrow c = 0$ and thus $y(t) = t$, which is
     independent of λ.

5c. Your result in Part b will depend upon the particular
     computer system and software that you use. If there is
     sufficient accuracy, you will obtain the solution y = t
     for t on $0 \le t \le 1$ for each value of λ that is given,
     since there is no discretization error. If there is not
     sufficient accuracy, then round-off error will affect
     your calculations. For the larger values of λ, the
     numerical solution will quickly diverge from the exact
     solution, y = t, to the general solution $y = t + ce^{\lambda t}$,
     where the value of c depends upon the round-off error.
     If the latter case does not occur, you may simulate it by
     computing the numerical solution to the I.V.P.
     $y' - \lambda y = 1 - \lambda t$, y(.1) = .10000001. Here we have
     assumed that the numerical solution is exact up to the
     point t = .09 [i.e. y(.09) = .09] and that at t = .1
     round-off error has occurred as indicated by the slight
     error in the I.C. It has also been found that a larger
     step size (h = .05 or h = .1) may also lead to round-off
     error.

Section 8.6, Page 457

2a. The Euler formula is $\mathbf{x}_{n+1} = \mathbf{x}_n + h\mathbf{f}_n$, where

$\mathbf{f}_n = \begin{pmatrix} 2x_n + t_n y_n \\ x_n y_n \end{pmatrix}$, $x_0 = 1$ and $y_0 = 1$. Thus $f_0 = \begin{pmatrix} 2-0 \\ (1)(1) \end{pmatrix} =$

$\begin{pmatrix} 2 \\ 1 \end{pmatrix}$, $\mathbf{x}_1 = \begin{pmatrix} 1+.1(2) \\ 1+.1(1) \end{pmatrix} = \begin{pmatrix} 1.2 \\ 1.1 \end{pmatrix}$,

$f_1 = \begin{pmatrix} 2.4+.1(1.1) \\ (1.2)(1.1) \end{pmatrix} = \begin{pmatrix} 2.51 \\ 1.32 \end{pmatrix}$

and $\mathbf{x}_2 = \begin{pmatrix} 1.2+.1(2.51) \\ 1.1+.1(1.32) \end{pmatrix} = \begin{pmatrix} 1.451 \\ 1.232 \end{pmatrix} \cong \begin{pmatrix} \phi(.2) \\ \psi(.2) \end{pmatrix}$

2b. Eqs. (7) give:

$\mathbf{k}_{01} = \begin{pmatrix} f(0,1,1) \\ g(0,1,1) \end{pmatrix} = \begin{pmatrix} 2+0 \\ (1)(1) \end{pmatrix} = \begin{pmatrix} 2 \\ 1 \end{pmatrix}$

$\mathbf{k}_{02} = \begin{pmatrix} 2.4+.1(1.1) \\ (1.2)(1.1) \end{pmatrix} = \begin{pmatrix} 2.51 \\ 1.32 \end{pmatrix}$

$\mathbf{k}_{03} = \begin{pmatrix} 2.502+.1(1.132) \\ (1.251)(1.132) \end{pmatrix} = \begin{pmatrix} 2.6152 \\ 1.41613 \end{pmatrix}$

$\mathbf{k}_{04} = \begin{pmatrix} 3.04608+.2(1.28323) \\ (1.52304)(1.28323) \end{pmatrix} = \begin{pmatrix} 3.30273 \\ 1.95441 \end{pmatrix}$

Using Eq. (6) in scalar form, we then have
$x_1 = 1+(.2/6)[2+2(2.51)+2(2.6152)+3.30273] = 1.51844$
$y_1 = 1+(.2/6)[1+2(1.32)+2(1.41613)+1.95440] = 1.28089$.

7. Write a computer program to do this problem as there are
    twenty steps or more for $h \le .05$.

8. If we let $y = x'$, then $y' = x''$ and thus we obtain the
    system $x' = y$ and $y' = t-3x-t^2y$, with $x(0) = 1$ and
    $y(0) = x'(0) = 2$. Thus $f(t,x,y) = y$,
    $g(t,x,y) = t - 3x - t^2y$, $t_0 = 0$, $x_0 = 1$ and $y_0 = 2$. If a
    program has been written for an earlier problem, then its
    best to use that. Otherwise, the first two steps are as
    follows:

$\mathbf{k}_{01} = \begin{pmatrix} 2 \\ -3 \end{pmatrix}$

$$k_{02} = \begin{pmatrix} 2+(-.15) \\ .05-3(1.1)-(.05)^2(1.85) \end{pmatrix} = \begin{pmatrix} 1.85 \\ -3.25463 \end{pmatrix}$$

$$k_{03} = \begin{pmatrix} 2+(-.16273) \\ .05-3(1.0925)-(.05)^2(1.83727) \end{pmatrix} = \begin{pmatrix} 1.83727 \\ -3.23209 \end{pmatrix}$$

$$k_{04} = \begin{pmatrix} 2+(-.32321) \\ .1-3(1.18373)-(.1)^2(1.67679) \end{pmatrix} = \begin{pmatrix} 1.67679 \\ -3.46796 \end{pmatrix}$$

and thus

x_1 = 1+(.1/6)[2 + 2(1.85)+2(1.83727)+(1.67679)]=1.18419,
y_1 = 2+(.1/6)[-3-2(3.25463)-2(3.23209)-3.46796]=1.67598,

which are approximations to $x(.1)$ and $y(.1) = x'(.1)$.
In a similar fashion we find

$$k_{11} = \begin{pmatrix} 1.67598 \\ -3.46933 \end{pmatrix} \qquad k_{12} = \begin{pmatrix} 1.50251 \\ -3.68777 \end{pmatrix}$$

$$k_{13} = \begin{pmatrix} 1.49159 \\ -3.66151 \end{pmatrix} \qquad k_{14} = \begin{pmatrix} 1.30983 \\ -3.85244 \end{pmatrix}$$

and thus

x_2=x_1+(.1/6)[1.67598+2(1.50251)+2(1.49159)+1.30983]=1.33376
y_2=y_1-(.1/6)[3.46933+2(3.68777)+2(3.66151)+3.85244]=1.30897.

Three more steps must be taken in order to approximate
$x(.5)$ and $y(.5) = x'(.5)$. The intermediate steps yield
$x(.3) \cong 1.44489$, $y(.3) \cong .9093062$ and $x(.4) \cong 1.51499$,
$y(.4) \cong .4908795$.

CHAPTER 9

For Problems 1 through 16, once the eigenvalues have been found, Table 9.1.1 will, for the most part, quickly yield the type of critical point and the stability. In all cases it can be easily verified that **A** is nonsingular.

1a. The eigenvalues are found from the equation $\det(\mathbf{A}-r\mathbf{I})=0$.
Substituting the values for **A** we have $\begin{vmatrix} 3-r & -2 \\ 2 & -2-r \end{vmatrix} =$
$r^2 - r - 2 = 0$ and thus the eigenvalues are $r_1 = -1$ and
$r_2 = 2$. For $r_1 = -1$, we have $\begin{pmatrix} 4 & -2 \\ 2 & -1 \end{pmatrix}\begin{pmatrix} \xi_1 \\ \xi_2 \end{pmatrix} = \begin{pmatrix} 0 \\ 0 \end{pmatrix}$ and thus
$\xi^{(1)} = \begin{pmatrix} 1 \\ 2 \end{pmatrix}$ and for r_2 we have $\begin{pmatrix} 1 & -2 \\ 2 & -4 \end{pmatrix}\begin{pmatrix} \xi_1 \\ \xi_2 \end{pmatrix} = \begin{pmatrix} 0 \\ 0 \end{pmatrix}$ and thus
$\xi^{(2)} = \begin{pmatrix} 2 \\ 1 \end{pmatrix}$.

1b. Since the eigenvalues differ in sign, the critical point is a saddle point and is unstable.

1d.
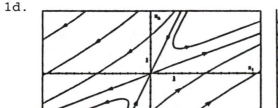
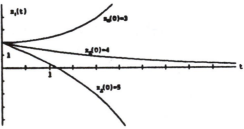

4a. Again the eigenvalues are given by $\begin{vmatrix} 1-r & -4 \\ 4 & -7-r \end{vmatrix} =$
$r^2 + 6r + 9 = 0$ and thus $r_1 = r_2 = -3$. The eigenvectors
are solutions of $\begin{pmatrix} 4 & -4 \\ 4 & -4 \end{pmatrix}\begin{pmatrix} \xi_1 \\ \xi_2 \end{pmatrix} = \begin{pmatrix} 0 \\ 0 \end{pmatrix}$ and hence there is
just one eigenvector $\xi = \begin{pmatrix} 1 \\ 1 \end{pmatrix}$.

4b. Since the eigenvalues are negative, $(0,0)$ is an improper node which is asymptotically stable. If we had found that there were two independent eigenvectors then $(0,0)$ would have been a proper node, as indicated in Case 3a.

4d.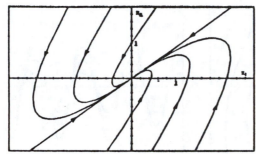

7a. In this case det$(\mathbf{A} - r\mathbf{I}) = r^2 - 2r + 5$ and thus the
     eigenvalues are $r_{1,2} = 1 \pm 2i$. For $r_1 = 1 + 2i$ we have
     $$\begin{pmatrix} 2-2i & -2 \\ 4 & -2-2i \end{pmatrix} \begin{pmatrix} \xi_1 \\ \xi_2 \end{pmatrix} = \begin{pmatrix} 2-2i & -2 \\ 8-8i & -8 \end{pmatrix} \begin{pmatrix} \xi_1 \\ \xi_2 \end{pmatrix} = \begin{pmatrix} 0 \\ 0 \end{pmatrix} \text{ and thus}$$
     $\xi^{(1)} = \begin{pmatrix} 1 \\ 1-i \end{pmatrix}$. Similarly for $r_2 = 1-2i$ we have
     $$\begin{pmatrix} 2+2i & -2 \\ 4 & -2+2i \end{pmatrix} \begin{pmatrix} \xi_1 \\ \xi_2 \end{pmatrix} = \begin{pmatrix} 0 \\ 0 \end{pmatrix} \text{ and hence } \xi^{(2)} = \begin{pmatrix} 1 \\ 1+i \end{pmatrix}.$$

7b. Since the eigenvalues are complex with positive real
     part, we conclude that the critical point is a spiral
     point and is unstable.

7d.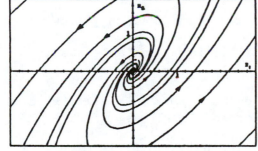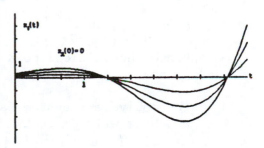

10a. Again, det$(\mathbf{A}-r\mathbf{I}) = r^2 + 9$ and thus we have $r_{1,2} = \pm 3i$.
     For $r_1 = 3i$ we have $\begin{pmatrix} 1-3i & 2 \\ -5 & -1-3i \end{pmatrix} \begin{pmatrix} \xi_1 \\ \xi_2 \end{pmatrix} = \begin{pmatrix} 0 \\ 0 \end{pmatrix}$ and thus
     $\xi^{(1)} = \begin{pmatrix} 2 \\ -1+3i \end{pmatrix}$. Likewise for $r_2 = -3i$,
     $\begin{pmatrix} 1+3i & 2 \\ -5 & -1+3i \end{pmatrix} \begin{pmatrix} \xi_1 \\ \xi_2 \end{pmatrix} = \begin{pmatrix} 0 \\ 0 \end{pmatrix}$ so that $\xi^{(2)} = \begin{pmatrix} 2 \\ -1-3i \end{pmatrix}$.

10b. Since the eigenvalues are pure imaginary the critical
     point is a center, which is stable.

10d.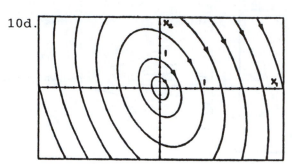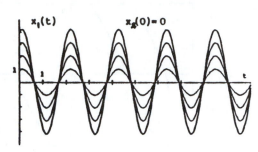

13. If we let $\mathbf{x} = \mathbf{x}^0 + \mathbf{u}$ then $\mathbf{x}' = \mathbf{u}'$ and thus the system becomes $\mathbf{u}' = \begin{pmatrix} 1 & 1 \\ 1 & -1 \end{pmatrix} \mathbf{x}^0 + \begin{pmatrix} 1 & 1 \\ 1 & -1 \end{pmatrix} \mathbf{u} - \begin{pmatrix} 2 \\ 0 \end{pmatrix}$ which will be in the form of Eq.(2) if $\begin{pmatrix} 1 & 1 \\ 1 & -1 \end{pmatrix} \mathbf{x}^0 = \begin{pmatrix} 2 \\ 0 \end{pmatrix}$. Using row operations, this last set of equations is equivalent to $\begin{pmatrix} 1 & 1 \\ 0 & -2 \end{pmatrix} \mathbf{x}^0 = \begin{pmatrix} 2 \\ -2 \end{pmatrix}$ and thus $x_1^0 = 1$ and $x_2^0 = 1$. Since $\mathbf{u}' = \begin{pmatrix} 1 & 1 \\ 1 & -1 \end{pmatrix} \mathbf{u}$ has $(0,0)$ as the critical point, we conclude that $(1,1)$ is the critical point of the original system. As in the earlier problems, the eigenvalues are given by $\begin{vmatrix} 1-r & 1 \\ 1 & -1-r \end{vmatrix} = r^2 - 2 = 0$ and thus $r_{1,2} = \pm\sqrt{2}$.
Hence the critical point $(1,1)$ is an unstable saddle point.

17. The equivalent system is $dx/dt = y, dy/dt = -(k/m)x-(c/m)y$ which is written in the form of Eq.(2) as $\dfrac{d}{dt}\begin{pmatrix} x \\ y \end{pmatrix} = \begin{pmatrix} 0 & 1 \\ -k/m & -c/m \end{pmatrix}\begin{pmatrix} x \\ y \end{pmatrix}$. The point $(0,0)$ is clearly a critical point, and since $\mathbf{A}$ is nonsingular, it is the only one. The characteristic equation is $r^2+(c/m)r+k/m=0$ so $r_1, r_2 = [-c \pm (c^2 - 4km)^{1/2}]/2m$. In the underdamped case $c^2 - 4km < 0$, the characteristic roots are complex with negative real parts and thus the critical point $(0,0)$ is an asymptotically stable spiral point. In the overdamped case $c^2 - 4km > 0$, the characteristic roots are real, unequal, and negative and hence the critical point $(0,0)$ is asymptotically stable node. In the critically damped case $c^2 - 4km = 0$, the characteristic roots are equal and negative. As indicated in the solution to Problem 4, to determine whether this is an improper or proper node we must determine whether there

are one or two linearly independent eigenvectors. The
eigenvectors satisfy the equation

$$\begin{pmatrix} c/2m & 1 \\ -k/m & -c/2m \end{pmatrix} \begin{pmatrix} \xi_1 \\ \xi_2 \end{pmatrix} = \begin{pmatrix} 0 \\ 0 \end{pmatrix},$$ which has just one solution if

$c^2 - 4km = 0$. Thus the critical point $(0,0)$ is an
asymptotically stable improper node.

18a. If **A** has one zero eigenvalue then for $r = 0$ we have
$\det(\mathbf{A}-r\mathbf{I}) = \det\mathbf{A} = 0$. Hence **A** is singular which means
Ax = **0** has infinitely many solutions and consequently
there are infinitely many critical points.

18b. From Chapter 7, the solution is $\mathbf{x}(t) = c_1\xi^{(1)} + c_2\xi^{(2)}e^{r_2 t}$,
which can be written in scalar form as
$x_1 = c_1\xi_1^{(1)} + c_2\xi_1^{(2)}e^{r_2 t}$ and $x_2 = c_1\xi_2^{(1)} + c_2\xi_2^{(2)}e^{r_2 t}$.
Assuming $\xi_1^{(2)} \neq 0$, the first equation can be solved for
$c_2 e^{r_2 t}$, which is then substituted into the second
equation to yield $x_2 = c_1\xi_2^{(1)} + [\xi_2^{(2)}/\xi_1^{(2)}][x_1 - c_1\xi_1^{(1)}]$.
These are straight lines parallel to the vector $\xi^{(2)}$.
Note that the family of lines is independent of c_2. If
$\xi_1^{(2)} = 0$, then the lines are vertical. If $r_2 > 0$, the
direction of motion will be in the same direction as
indicated for $\xi^{(2)}$. If $r_2 < 0$, then it will be in the
opposite direction.

19a. $\det(A-rI) = r^2 - (a_{11}+a_{22})r + a_{11}a_{22} - a_{21}a_{12} = 0$. If
$a_{11} + a_{22} = 0$, then $r^2 = -(a_{11}a_{22} - a_{21}a_{12}) < 0$ if
$a_{11}a_{22} - a_{21}a_{12} > 0$.

19b. Eq.(i) can be written in scalar form as $dx/dt = a_{11}x +$
$a_{12}y$ and $dy/dt = a_{21}x + a_{22}y$, which then yields Eq.(iii).
Ignoring the middle quotient in Eq.(iii), we can rewrite
that equation as $(a_{21}x + a_{22}y)dx - (a_{11}x + a_{12}y)dy = 0$,
which is exact since $a_{22} = -a_{11}$ from Eq.(ii)..

19c. Integrating $\phi_x = a_{21}x + a_{22}y$ we obtain $\phi = a_{21}x^2/2 + a_{22}xy$
$+ g(y)$ and thus $a_{22}x + g' = -a_{11}x - a_{12}y$ or $g' = -a_{12}y$
using Eq.(ii). Hence $a_{21}x^2/2 + a_{22}xy - a_{12}y^2/2 = k/2$ is
the solution to Eq.(iii). The quadratic equation $Ax^2 +$
$Bxy + Cy^2 = D$ is an ellipse provided $B^2 - 4AC < 0$. Hence
for our problem if

$a_{22}^2 + a_{21}a_{12} < 0$ then Eq.(iv) is an ellipse. Using

$a_{11} + a_{22} = 0$ we have $a_{22}^2 = -a_{11}a_{22}$ and hence

$-a_{11}a_{22} + a_{21}a_{12} < 0$ or $a_{11}a_{22} - a_{21}a_{12} > 0$, which is true by Eqs.(ii). Thus Eq.(iv) is an ellipse under the conditions of Eqs.(ii).

20. The given system can be written as $\dfrac{d}{dt}\begin{pmatrix} x \\ y \end{pmatrix} = \begin{pmatrix} a_{11} & a_{12} \\ a_{21} & a_{22} \end{pmatrix}\begin{pmatrix} x \\ y \end{pmatrix}$.

Thus the eigenvalues are given by
$r^2 - (a_{11} + a_{22})r + a_{11}a_{22} - a_{12}a_{21} = 0$ and using the given definitions we rewrite this as $r^2 - pr + q = 0$ and thus $r_{1,2} = (p \pm \sqrt{p^2 - 4q})/2 = (p \pm \sqrt{\Delta})/2$. The results are now obtained using Table 9.1.1.

Section 9.2, Page 477

1. Solutions of the D.E. for x
    are y are $x = Ae^{-t}$ and
    $y = Be^{-2t}$ respectively.
    $x(0) = 4$ and $y(0) = 2$ yield
    $A = 4$ and $B = 2$, so $x = 4e^{-t}$
    and $y = 2e^{-2t}$. Solving the
    first equation for e^{-t} and
    then substituting into the

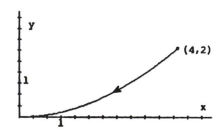

second yields $y = 2[x/4]^2 = x^2/8$, which is a parabola.
From the original D.E., or from the parametric solutions,
we find that $0 < x \le 4$ and $0 < y \le 2$ for $t \ge 0$ and thus
only the portion of the parabola shown is the trajectory,
with the direction of motion indicated.

3. Utilizing the approach indicated in Eq.(14), we have
    $dy/dx = -x/y$, which separates into $xdx + ydy = 0$.
    Integration then yields the circle $x^2 + y^2 = c^2$, where
    $c^2 = 16$ for both sets of I.C. The direction of motion
    can be found from the original D.E. and is
    counterclockwise for both I.C. To obtain the parametric
    equations, we write the system in the form
    $\dfrac{d}{dt}\begin{pmatrix} x \\ y \end{pmatrix} = \begin{pmatrix} 0 & -1 \\ 1 & 0 \end{pmatrix}\begin{pmatrix} x \\ y \end{pmatrix}$, which has the characteristic

    equation $\begin{vmatrix} -r & -1 \\ 1 & -r \end{vmatrix} = r^2 + 1 = 0$, or $r = \pm i$. Following the procedures of Section 7.6, we find that one solution of the above system is $\begin{pmatrix} 1 \\ -i \end{pmatrix} e^{it} = \begin{pmatrix} \cos t + i\sin t \\ \sin t - i\cos t \end{pmatrix}$ and thus two real solutions are $\mathbf{u}(t) = \begin{pmatrix} \cos t \\ \sin t \end{pmatrix}$ and $\mathbf{v}(t) = \begin{pmatrix} \sin t \\ -\cos t \end{pmatrix}$. The general solution of the system is then $\begin{pmatrix} x \\ y \end{pmatrix} = c_1\mathbf{u}(t) + c_2\mathbf{v}(t)$ and hence the first I.C. yields $c_1 = 4$, $c_2 = 0$, or $x = 4\cos t$, $y = 4\sin t$. The second I.C. yields $c_1 = 0$, $c_2 = -4$, or $x = -4\sin t$, $y = 4\cos t$. Note that both these parametric representations satsify the form of the trajectories found in the first part of this problem.

7a. The critical points are given by the solutions of $x(1-x-y) = 0$ and $y(1/2 - y/4 - 3x/4) = 0$. The solutions corresponding to either $x = 0$ or $y = 0$ are seen to be $x = 0$, $y = 0$; $x = 0$, $y = 2$; $x = 1$, $y = 0$. In addition, there is a solution corresponding to the intersection of the lines $1 - x - y = 0$ and $1/2 - y/4 - 3x/4 = 0$ which is the point $x = 1/2$, $y = 1/2$. Thus the critical points are $(0,0)$, $(0,2)$, $(1,0)$, and $(1/2,1/2)$.

7b.

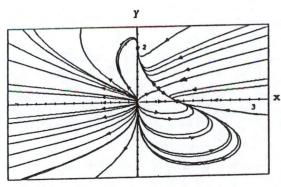

7c. For $(0,0)$ since all trajectories leave this point, this is an unstable node. For $(0,2)$ and $(1,0)$ since the trajectories tend to these points, respectively, they are asymptotically stable nodes. For $(1/2,1/2)$, one trajectory tends to $(1/2,1/2)$ while all others tend to infinity, so this is an unstable saddle point.

12a. The critical points are given by $y = 0$ and $x(1 - x^2/6 - y/5) = 0$, so $(0,0)$, $(\sqrt{6},0)$ and $(-\sqrt{6},0)$ are the only critical points.

12b.

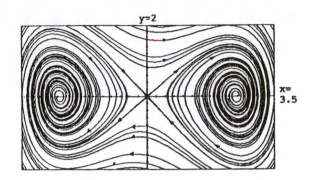

12c. Clearly $(\sqrt{6},0)$ and $(-\sqrt{6},0)$ are spiral points, and are asymptotically stable since the trajectories tend to each point, respectively. $(0,0)$ is a saddle point, which is unstable, since the trajectories behave like the ones for $(1/2,1/2)$ in Problem 7.

15a. $\dfrac{dy}{dx} = \dfrac{dy/dt}{dx/dt} = \dfrac{8x}{2y}$, so $4xdx - ydy = 0$ and thus $4x^2 - y^2 = c$, which are hyperbolas for $c \neq 0$ and straight lines $y = \pm 2x$ for $c = 0$.

15b.   19b.

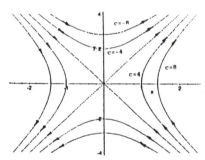   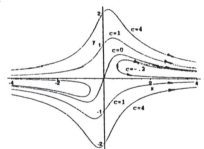

19a. $\dfrac{dy}{dx} = \dfrac{y-2xy}{-x+y+x^2}$, so $(y-2xy)dx + (x-y-x^2)dy = 0$, which is an exact D.E. Therefore $\phi(x,y) = xy - x^2y + g(y)$ and hence $\dfrac{\partial \phi}{\partial y} = x - x^2 + g'(y) = x - y - x^2$, so $g'(y) = -y$ and $g(y) = -y^2/2$. Thus $2x^2y - 2xy + y^2 = c$ (after multiplying by -2) is the desired solution.

21a. $\dfrac{dy}{dx} = \dfrac{-sinx}{y}$, so $ydy + sinxdx = 0$ and thus $y^2/2 - cosx = c$.

21b.

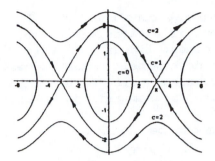

23. We know that $\phi'(t) = F[\phi(t), \psi(t)]$ and
     $\psi'(t) = G[\phi(t), \psi(t)]$ for $\alpha < t < \beta$. By direct
     substitution we have
     $\Phi'(t) = \phi'(t-s) = F[\phi(t-s), \psi(t-s)] = F[\Phi(t), \Psi(t)]$ and
     $\Psi'(t) = \psi'(t-s) = G[\phi(t-s), \psi(t-s)] = G[\Phi(t), \Psi(t)]$ for
     $\alpha < t-s < \beta$ or $\alpha+s < t < \beta+s$.

24. Suppose that $t_1 > t_0$. Let $s = t_1 - t_0$. Since the system
     is autonomous, the result of Problem 23, with s replaced
     by $-s$ shows that $x = \phi_1(t+s)$ and $y = \psi_1(t+s)$ generates
     the same trajectory (C_1) as $x = \phi_1(t)$ and $y = \psi_1(t)$. But
     at $t = t_0$ we have $x = \phi_1(t_0+s) = \phi_1(t_1) = x_0$ and
     $y = \psi_1(t_0+s) = \psi_1(t_1) = y_0$. Thus the solution
     $x = \phi_1(t+s)$, $y = \psi_1(t+s)$ satisfies <u>exactly</u> the same
     initial conditions as the solution $x = \phi_0(t)$, $y = \psi_0(t)$
     which generates the trajectory C_0. Hence C_0 and C_1 are
     the same.

25. From the existence and uniqueness theorem we know that if
     the two solutions $x = \phi(t)$, $y = \psi(t)$ and $x = x_0$, $y = y_0$
     satisfy $\phi(a) = x_0$, $\psi(a) = y_0$ and $x = x_0$, $y = y_0$ at $t = a$,
     then these solutions are identical. Hence $\phi(t) = x_0$ and
     $\psi(t) = y_0$ for all t contradicting the fact that the
     trajectory generated by $[\phi(t), \psi(t)]$ started at a
     noncritical point.

26. By direct substitution
     $\Phi'(t) = \phi'(t+T) = F[\phi(t+T), \psi(t+T)] = F[\Phi(t), \Psi(t)]$ and
     $\Psi'(t) = \psi'(t+T) = G[\phi(t+T), \psi(t+T)], G[\Phi(t), \Psi(t)]$.
     Furthermore $\Phi(t_0) = x_0$ and $\Psi(t_0) = y_0$. Thus by the
     existence and uniqueness theorem $\Phi(t) = \phi(t)$ and
     $\Psi(t) = \psi(t)$ for all t.

<u>Section 9.3, Page 487</u>

In Problems 1 through 4, write the system in the form of

Eq.(4). Then if $\mathbf{g(0)} = \mathbf{0}$ we may conclude that $(0,0)$ is a critical point. In addition, if $\mathbf{g}$ satisfies Eq.(5) or Eq.(6), then the system is almost linear. In this case the linear system, Eq.(1), will determine, in most cases, the type and stability of the critical point $(0,0)$ of the almost linear system. These results are summarized in Table 9.3.1.

3. In this case the system can be written as

$$\frac{d}{dt}\begin{pmatrix} x \\ y \end{pmatrix} = \begin{pmatrix} 0 & 0 \\ -1 & 0 \end{pmatrix}\begin{pmatrix} x \\ y \end{pmatrix} + \begin{pmatrix} (1+x)\sin y \\ 1 - \cos y \end{pmatrix}.$$ However, the

coefficient matrix is singular and $g_1(x,y) = (1+x)\sin y$

does not satisfy Eq.(6). However, if we consider the Taylor series for $\sin y$, we see that $(1+x)\sin y - y =$

$\sin y - y + x\sin y = -y^3/3! + y^5/5! + \cdots + x(y - y^3/3! + \cdots)$,

which does satisfy Eq.(6), using $x = r\cos\theta$, $y = r\sin\theta$. Thus the first equation now becomes

$$\frac{dx}{dt} = y + [(1+x)\sin y - y] \text{ and hence}$$

$$\frac{d}{dt}\begin{pmatrix} x \\ y \end{pmatrix} = \begin{pmatrix} 0 & 1 \\ -1 & 0 \end{pmatrix}\begin{pmatrix} x \\ y \end{pmatrix} + \begin{pmatrix} (1+x)\sin y - y \\ 1-\cos y \end{pmatrix}, \text{ where the}$$

coefficient matrix is now nonsingular and

$$\mathbf{g}(x,y) = \begin{pmatrix} (1+x)\sin y - y \\ 1-\cos y \end{pmatrix} \text{ satisfies Eq.(6).}$$

4. In this case the system can be written as

$$\frac{d}{dt}\begin{pmatrix} x \\ y \end{pmatrix} = \begin{pmatrix} 1 & 0 \\ 1 & 1 \end{pmatrix}\begin{pmatrix} x \\ y \end{pmatrix} + \begin{pmatrix} y^2 \\ 0 \end{pmatrix} \text{ and thus } \mathbf{A} = \begin{pmatrix} 1 & 0 \\ 1 & 1 \end{pmatrix} \text{ and}$$

$$\mathbf{g} = \begin{pmatrix} y^2 \\ 0 \end{pmatrix}. \text{ Since } \mathbf{g(0)} = \begin{pmatrix} 0 \\ 0 \end{pmatrix} \text{ we conclude that } (0,0) \text{ is a}$$

critical point. Following the procedure of Example 1, we let $x = r\cos\theta$ and $y = r\sin\theta$ and thus

$$g_1(x,y)/r = \frac{r^2\sin^2\theta}{r} \to 0 \text{ as } r \to 0 \text{ and thus the system is}$$

almost linear. Since $\det(\mathbf{A}-r\mathbf{I}) = (r-1)^2$, we find that the eigenvalues are $r_1 = r_2 = 1$. Since the roots are equal, we must determine whether there are one or two eigenvectors to classify the type of critical point. The

eigenvectors are determined by $\begin{pmatrix} 0 & 0 \\ 1 & 0 \end{pmatrix}\begin{pmatrix} \xi_1 \\ \xi_2 \end{pmatrix} = \begin{pmatrix} 0 \\ 0 \end{pmatrix}$ and hence there is only one eigenvector $\xi = \begin{pmatrix} 0 \\ 1 \end{pmatrix}$. Thus the critical point for the linear system is an unstable improper node. From Table 9.3.1 we then conclude that the given system, which is almost linear, has a critical point near (0,0) which is either a node or spiral point (depending on how the roots bifurcate) which is unstable.

6a. The critical points are the solutions of x(1-x-y) = 0 and y(3-x-2y) = 0. Solutions are x = 0, y = 0; x = 0, 3 - 2y = 0 which gives y = 3/2; y = 0 and 1 - x = 0 which give x = 1; and 1 - x - y = 0, 3 - x - 2y = 0 which give x = -1, y = 2. Thus the critical points are (0,0), (0,3/2), (1,0) and (-1,2).

6b, 6c. For the critical point (0,0) the D.E. is already in the form of an almost linear system; and the corresponding linear system is du/dt = u, dv/dt = 3v which has the eigenvalues r_1 = 1 and r_2 = 3. Thus the critical point (0,0) is an unstable node. Each of the other three critical points is dealt with in the same manner; we consider only the critical point (-1,2). In order to translate this critical point to the origin we set x(t) = -1 + u(t), y(t) = 2 + v(t) and substitute in the D.E. to obtain

$du/dt = -1 + u - (-1+u)^2 - (-1+u)(2+v) = u + v - u^2 - uv$ and

$dv/dt = 3(2+v) - (-1+u)(2+v) - 2(2+v)^2 = -2u - 4v - uv - 2v^2$. Writing this in the form of Eq.(4) we find that

$$\mathbf{A} = \begin{pmatrix} 1 & 1 \\ -2 & -4 \end{pmatrix} \text{ and } \mathbf{g} = -\begin{pmatrix} u^2 + uv \\ uv + 2v^2 \end{pmatrix}$$

which is an almost linear system. The eigenvalues of the corresponding linear system are $r = (-3 \pm \sqrt{9 + 8})/2$ and hence the critical point (-1,2), of the original system, is an unstable saddle point.

10a. The critical points are solutions of $x + x^2 + y^2 = 0$ and y(1-x) = 0, which yield (0,0) and (-1,0).

10b. For (0,0) the D.E. is already in the form of an almost linear system and thus du/dt = u and dv/dt = v. For (-1,0) we let u = x+1, v = y so that substituting x = u-1 and y = v into the D.E. we obtain $\dfrac{du}{dt} = -u + u^2 + v^2$ and

$\dfrac{dv}{dt}$ = 2v - uv. Thus the corresponding linear system is

u' = -u and v' = -2v.

10c. For (0,0) $\mathbf{A}$ = $\begin{pmatrix} 1 & 0 \\ 0 & 1 \end{pmatrix}$ which has r_1 = r_2 = 1, so that

(0,0), for the nonlinear system, will be either a node or
spiral point, depending on how the roots bifurcate. In
any case, since r_1 and r_2 are positive, the system will

be unstable. For (-1,0) $\mathbf{A}$ = $\begin{pmatrix} -1 & 0 \\ 0 & 2 \end{pmatrix}$ and thus

r_1 = -1 and r_2 = 2, and hence the nonlinear system, from
Table 9.3.1, has an unstable saddle point at (-1,0).

18a. The system is $\dfrac{d}{dt}\begin{pmatrix} x \\ y \end{pmatrix}$ = $\begin{pmatrix} 1 & 0 \\ 0 & -2 \end{pmatrix}\begin{pmatrix} x \\ y \end{pmatrix}$ + $\begin{pmatrix} 0 \\ x^3 \end{pmatrix}$ and thus is

almost linear using the procedures outlined in the
earlier problems. The corresponding linear system has
the eigenvalues r_1 = 1, r_2 = -2 and thus (0,0) is an
unstable saddle point for both the linear and almost
linear systems.

18b. The trajectories of the linear system are the solutions
of dx/dt = x and dy/dt = -2y and thus x(t) = $c_1 e^t$ and
y(t) = $c_2 e^{-2t}$. To sketch these, solve the first equation
for e^t and substitute into the second to obtain
y = $c_1^2 c_2 / x^2$, $c_1 \neq 0$.
Several trajectories
are shown in the figure.
Since x(t) = $c_1 e^t$, we
must pick c_1 = 0 for
x → 0 and t → ∞. Thus
x = 0, y = $c_2 e^{-2t}$ (the
vertical axis) is the
only trajectory for
which x → 0, y → 0 as t → ∞.

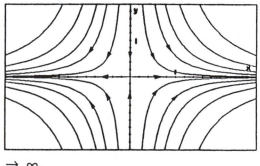

18c. For x ≠ 0 we have dy/dx = (dy/dt)/(dx/dt) = $(-2y+x^3)/x$.
This is a linear equation, and the general solution is
y = $x^3/5$ + k/x^2, where k is an arbitrary constant. In
addition the system of equations has the solution x = 0,
y = Be^{-2t}. Any solution with its initial point on the
y-axis (x=0) is given by the latter solution. The
trajectories corresponding to these solutions approach

the origin as t → ∞. The
trajectory that passes
through the origin and
divides the family of
curves is given by k = 0,
namely $y = x^3/5$. This
trajectory corresponds to
the trajectory y = 0 for
the linear problem.
Several trajectories
are sketched in the figure.

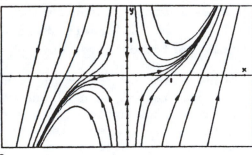

22a.

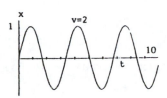 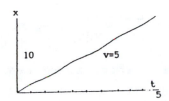 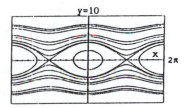

22b. From the graphs in part a, we see that v_c is between v = 2
and v = 5. Using several values for v, we estimate $v_c \cong 4.00$.

23a.

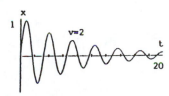 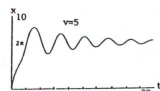 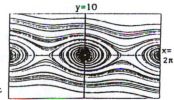

For v = 2, the motion is damped oscillatory about x = 0.
For v = 5, the pendulum swings all the way around once
and then is a damped oscillation about x = 2π (after one
full rotation). For problem 22, this later case is not
damped, so x continues to increase, as shown earlier.

27a. Setting c = 0 in Eq.(10) of Section 9.2 we obtain
$mL^2 d^2\theta/dt^2 + mgL\sin\theta = 0$. Considering dθ/dt as a function
of θ and using the chain rule we have
$$\frac{d}{dt}\left(\frac{d\theta}{dt}\right) = \frac{d}{d\theta}\left(\frac{d\theta}{dt}\right)\frac{d\theta}{dt} = \frac{1}{2}\frac{d}{d\theta}\left(\frac{d\theta}{dt}\right)^2.$$ Thus
$(1/2)mL^2 d[(d\theta/dt)^2]/d\theta = -mgL\sin\theta$. Now integrate both
sides from α to θ where dθ/dt = 0 at θ = α:
$(1/2)mL^2(d\theta/dt)^2 = mgL(\cos\theta - \cos\alpha)$. Thus
$(d\theta/dt)^2 = (2g/L)(\cos\theta - \cos\alpha)$. Since we are releasing
the pendulum with zero velocity from a positive angle α,

the angle θ will initially be decreasing so $d\theta/dt < 0$.
If we restrict attention to the range of θ from $\theta = \alpha$ to
$\theta = 0$, we can assert $d\theta/dt = -\sqrt{2g/L}\ \sqrt{\cos\theta - \cos\alpha}$.
Solving for dt gives $dt = -\sqrt{L/2g}\ d\theta/\sqrt{\cos\theta - \cos\alpha}$.

27b. Since there is no damping, the pendulum will swing from
its initial angle α through 0 to $-\alpha$, then back through 0
again to the angle α in one period. It follows that
$\theta(T/4) = 0$. Integrating the last equation and noting
that as t goes from 0 to $T/4$, θ goes from α to 0 yields
$T/4 = -\sqrt{L/2g} \displaystyle\int_\alpha^0 (1/\sqrt{\cos\theta - \cos\alpha})\,d\theta$.

28a. If $\dfrac{dx}{dt} = y$, then $\dfrac{d^2x}{dt^2} = \dfrac{dy}{dt} = -g(x) - c(x)y$.

28b. Under the given assumptions we have $g(x) = g(0) + g'(0)x$
$+ g''(\xi_1)x^2/2$ and $c(x) = c(0) + c'(\xi_2)x$, where
$0 < \xi_1,\ \xi_2 < x$ and $g(0) = 0$. Hence
$\dfrac{dy}{dt} = (-g(0) - g'(0)x) - c(0)y - [g''(\xi_1)x^2/2 - c'(\xi_2)xy]$
and thus the system can be written as
$$\frac{d}{dt}\begin{pmatrix} x \\ y \end{pmatrix} = \begin{pmatrix} 0 & 1 \\ -g'(0) & -c(0) \end{pmatrix}\begin{pmatrix} x \\ y \end{pmatrix} - \begin{pmatrix} 0 \\ -g''(\xi_1)x^2/2 - c'(\xi_2)xy \end{pmatrix},$$
from which the results follow.

Section 9.4, Page 501

3b. $x(1.5 - .5x - y) = 0$ and $y(2 - y - 1.125x) = 0$ yield $(0,0)$,
$(0,2)$ and $(3,0)$ very easily. The fourth critical point is
the intersection of $.5x + y = 1.5$ and $1.125x + y = 2$, which
is $(.8,1.1)$.

3c. From Eq (5) we get $\dfrac{d}{dt}\begin{pmatrix} u \\ v \end{pmatrix} = \begin{pmatrix} 1.5-x_0-y_0 & -x_0 \\ -1.125y_0 & 2-2y_0-1.125x_0 \end{pmatrix}\begin{pmatrix} u \\ v \end{pmatrix}$. For
$(0,0)$ we get $u' = 1.5u$ and $v' = 2v$, so $r = 3/2$ and $r = 2$, and
thus $(0,0)$ is an unstable node. For $(0,2)$ we have $u' = -.5u$
and $v' = -2.25u-2v$, so $r = -.5$, -2 and thus $(0,2)$ is an
asymptotically stable node. For $(3,0)$ we get $u' = -1.5u-3v$
and $v' = -1.375v$, so $r = -1.5$, -1.375 and hence $(3,0)$ is an
symptotically stable node. For $(.8,1.1)$ we have
$u' = -.4u -.8v$ and $v' = -1.2375u - 1.1v$ which give
$r = -1.80475$, $.30475$ and thus $(.8,1.1)$ is an unstable saddle
point.

3e.

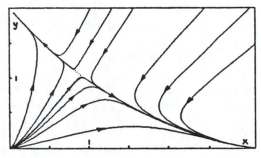

5b. The critical points are found by setting dx/dt = 0 and
    dy/dt = 0 and thus we need to solve x(1 - x - y) = 0 and
    y(1.5 - y - x) = 0. The first yields x = 0 or y = 1 - x
    and the second yields y = 0 or y = 1.5 - x. Thus (0,0),
    (0,3/2) and (1,0) are the only critical points since the
    two straight lines do not intersect in the first quadrant
    (or anywhere in this case). This is an example of one of
    the cases shown in Figure 9.4.5 a or b.

5e.

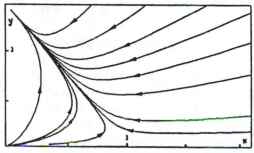

6b. The critical points are found by setting dx/dt = 0 and
    dy/dt = 0 and thus we need to solve x(1-x + y/2) = 0 and
    y(5/2 - 3y/2 + x/4) = 0. The first yields x = 0 or
    y = 2x - 2 and the second yields y = 0 or y = x/6 + 5/3.
    Thus we find the critical points (0,0), (1,0), (0,5/3)
    and (2,2). The last point is the intersection of the two
    straight lines, which will be used again in part d.

6c. For (0,0) the linearized system is x′ = x and y′ = 5y/2,
    which has the eigenvalues $r_1 = 1$ and $r_2 = 5/2$. Thus the
    origin is an unstable node. For (2,2) we let
    x = u + 2 and y = v + 2 in the given system to find
    (since x′ = u′ and y′ = v′) that
    du/dt = (u+2)[1 - (u+2) + (v+2)/2] = (u+2)(-u+v/2) and
    dv/dt = (v+2)[5/2 - 3(v+2)/2 + (u+2)/4] = (v+2)(u/4 - 3v/2).

    Hence the linearized equations are $\begin{pmatrix} u \\ v \end{pmatrix}' = \begin{pmatrix} -2 & 1 \\ 1/2 & -3 \end{pmatrix} \begin{pmatrix} u \\ v \end{pmatrix}$

    which has the eigenvalues $r_{1,2} = (-5 \pm \sqrt{3})/2$. Since
    these are both negative we conclude that (2,2) is an
    asymptotically stable node. In a similar fashion for

(1,0) we let x = u + 1 and y = v to obtain the linearized

system $\begin{pmatrix} u \\ v \end{pmatrix}' = \begin{pmatrix} -1 & 1/2 \\ 0 & 11/4 \end{pmatrix} \begin{pmatrix} u \\ v \end{pmatrix}$.   This has

$r_1 = -1$ and $r_2 = 11/4$ as eigenvalues and thus (1,0) is an
unstable saddle point.   Likewise, for (0,5/3) we let

x = u, y = v + 5/3 to find $\begin{pmatrix} u \\ v \end{pmatrix}' = \begin{pmatrix} 11/6 & 0 \\ 5/12 & -5/2 \end{pmatrix} \begin{pmatrix} u \\ v \end{pmatrix}$ as the

corresponding linear system.   Thus $r_1 = 11/6$ and
$r_2 = -5/2$ and thus (0,5/3) is an unstable saddle point.

6d. To sketch the required trajectories, we must find the
     eigenvectors for each of the linearized systems and then
     analyze the behavior of the linear solution near the
     critical point. Using this approach we find that the

     solution near (0,0) has the form $\begin{pmatrix} x \\ y \end{pmatrix} = c_1 \begin{pmatrix} 1 \\ 0 \end{pmatrix} e^t +$

$c_2 \begin{pmatrix} 0 \\ 1 \end{pmatrix} e^{5t/2}$ and thus the origin is approached only for

large negative values of t.   In this case e^t dominates
$e^{5t/2}$ and hence in the neighborhood of the origin all
trajectories are tangent to the x-axis except for one
pair ($c_1 = 0$) that lies along the y-axis.
For (2,2) we find the eigenvector corresponding to
$r = (-5 + \sqrt{3})/2 = -1.63$ is given by $(1-\sqrt{3})\xi_1/2 + \xi_2 = 0$

and thus $\begin{pmatrix} 1 \\ (\sqrt{3}-1)/2 \end{pmatrix} = \begin{pmatrix} 1 \\ .37 \end{pmatrix}$ is one eigenvector.   For

$r = (-5 -\sqrt{3})/2 = -3.37$ we have $(1 +\sqrt{3})\xi_1/2 + \xi_2 = 0$ and

thus $\begin{pmatrix} 1 \\ -(\sqrt{3}+1)/2 \end{pmatrix} = \begin{pmatrix} 1 \\ -1.37 \end{pmatrix}$ is the second eigenvector.

Hence the linearized solution is
$\begin{pmatrix} u \\ v \end{pmatrix} = c_1 \begin{pmatrix} 1 \\ .37 \end{pmatrix} e^{-1.63t} + c_2 \begin{pmatrix} 1 \\ -1.37 \end{pmatrix} e^{-3.37t}$.   For large

positive values of t the first term is the dominant one
and thus we conclude that all trajectories but two
approach (2,2) tangent to the straight line with slope
.37.   If $c_1 = 0$, we see that there are exactly two
($c_2 > 0$ and $c_2 < 0$) trajectories that lie on the straight
line with slope -1.37.
In similar fashion, we find the linearized solutions near
(1,0) and (0,5/3) to be, respectively,

$$\begin{pmatrix} u \\ v \end{pmatrix} = c_1 \begin{pmatrix} 1 \\ 0 \end{pmatrix} e^{-t} + c_2 \begin{pmatrix} 1 \\ 15/2 \end{pmatrix} e^{11t/4}$$

and

$$\begin{pmatrix} u \\ v \end{pmatrix} = c_1 \begin{pmatrix} 0 \\ 1 \end{pmatrix} e^{-5t/2} + c_2 \begin{pmatrix} 1 \\ 5/52 \end{pmatrix} e^{11t/6},$$

which, along with the above analysis, yields the sketch shown.

6e. From the above sketch, it appears that $(x,y) \rightarrow (2,2)$ as
6f. $t \rightarrow \infty$ as long as (x,y) starts in the first quadrant. To ascertain this, we need to prove that x and y cannot become unbounded as $t \rightarrow \infty$. From the given system, we can observe that, since $x > 0$ and $y > 0$, that dx/dt and dy/dt have the same sign as the quantities $1 - x + y/2$ and $5/2 - 3y/2 + x/4$ respectively. If we set these quantities equal to zero we get the straight lines $y = 2x - 2$ and $y = x/6 + 5/3$, which divide the first quadrant into the four sectors shown. The signs of x' and y' are indicated, from which it can be concluded that x and y must remain bounded [and in fact approach $(2,2)$] as $t \rightarrow \infty$. The discussion leading up to Fig.9.4.4 is also useful here.

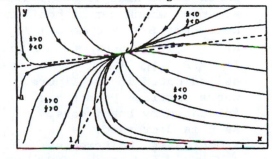

8a. Setting the right sides of the equations equal to zero gives the critical points $(0,0)$, $(0, \varepsilon_2/\sigma_2)$, $(\varepsilon_1/\sigma_1, 0)$, and possibly
$([\varepsilon_1\sigma_2 - \varepsilon_2\alpha_1]/[\sigma_1\sigma_2 - \alpha_1\alpha_2], [\varepsilon_2\sigma_1 - \varepsilon_1\alpha_2]/[\sigma_1\sigma_2 - \alpha_1\alpha_2])$.
(The last point can be obtained from Eq.(36) also). The conditions $\varepsilon_2/\alpha_2 > \varepsilon_1/\sigma_1$ and $\varepsilon_2/\sigma_2 > \varepsilon_1/\alpha_1$ imply that $\varepsilon_2\sigma_1 - \varepsilon_1\alpha_2 > 0$ and $\varepsilon_1\sigma_2 - \varepsilon_2\alpha_1 < 0$. Thus either the x coordinate or the y coordinate of the last critical point is negative so a mixed state is not possible. The linearized system for $(0,0)$ is $x' = \varepsilon_1 x$ and $y' = \varepsilon_2 y$ and thus $(0,0)$ is an unstable equilibrium point. Similarly, it can be shown [by linearizing the given system or by using Eq.(35)] that $(0, \varepsilon_2/\sigma_2)$ is an asymptotically stable critical point and that $(\varepsilon_1\sigma_1, 0)$ is an unstable critical point. Thus the fish represented by y(redear) survive.

8b. The conditions $\varepsilon_1/\sigma_1 > \varepsilon_2/\alpha_2$ and $\varepsilon_1/\alpha_1 > \varepsilon_2/\sigma_2$ imply that
     $\varepsilon_2\sigma_1 - \varepsilon_1\alpha_2 < 0$ and $\varepsilon_1\sigma_2 - \varepsilon_2\alpha_1 > 0$ so again one of the
     coordinates of the fourth point in 8a. is negative and
     hence a mixed state is not possible. An analysis similar
     to that in part(a) shows that $(0,0)$ and $(0,\varepsilon_2/\sigma_2)$ are
     unstable while $(\varepsilon_1/\sigma_1,0)$ is stable. Hence the bluegill
     (represented by x) survive in this case.

9a. $x' = \varepsilon_1 x(1 - \dfrac{\sigma_1}{\varepsilon_1}x - \dfrac{\alpha_1}{\varepsilon_1}y) = \varepsilon_1 x(1 - \dfrac{1}{B}x - \dfrac{\gamma_1}{B}y)$

     $y' = \varepsilon_2 y(1 - \dfrac{\sigma_2}{\varepsilon_2}y - \dfrac{\alpha_2}{\varepsilon_2}x) = \varepsilon_2 y(1 - \dfrac{1}{R}y - \dfrac{\gamma_2}{R}x)$.   The coexistence

     equilibrium point is given by $\dfrac{1}{B}x + \dfrac{\gamma_1}{B}y = 1$ and $\dfrac{\gamma_2}{R}x + \dfrac{1}{R}y = 1$.
     Solving these (using determinants) yields
     $X = (B - \gamma_1 R)/(1 - \gamma_1\gamma_2)$ and $Y = (R - \gamma_2 B)/(1-\gamma_1\gamma_2)$.

9b. If B is reduced, it is clear from the answer to part(a)
     that X is reduced and Y is increased. To determine
     whether the bluegill will die out, we give an intuitive
     argument which can be confirmed by doing the analysis.
     Note that $B/\gamma_1 = \varepsilon_1/\alpha_1 > \varepsilon_2/\sigma_2 = R$ and
     $R/\gamma_2 = \varepsilon_2/\alpha_2 > \varepsilon_1/\sigma_1 = B$ so that the graph of the lines
     $1 - x/B - \gamma_1 y/B = 0$ and $1 - y/R - \gamma_2 x/R = 0$ must appear
     as indicated in the figure,
     where critical points are
     inidcated by heavy dots.
     As B is decreased,
     X decreases, Y increases
     (as indicated above) and
     the point of intersection
     moves closer to $(0,R)$. If
     $B/\gamma_1 < R$ coexistence is not
     possible, and the only
     critical points are $(0,0),(0,R)$ and $(B,0)$.
     It can be shown that $(0,0)$ and $(B,0)$ are unstable and
     $(0,R)$ is asymptotically stable. Hence we conlcude, when
     coexistence is no longer possible, that $x \to 0$ and $y \to R$
     and thus the bluegill population will die out.

12a. Setting each equation equal to zero, we obtain $x = 0$ or
     $(4 - x - y) = 0$ and $y = 0$ or $(2 + 2\alpha - y - \alpha x) = 0$. Thus we
     have $(0,0)$, $(4,0)$, $(0,2 + 2\alpha)$, and the intersection of
     $x + y = 4$ and $\alpha x + y = 2 + 2\alpha$. If $\alpha \neq 1$, this yields $(2,2)$
     as the fourth critical point.

12b. For α = .75 the linear system is $\begin{pmatrix} u \\ v \end{pmatrix}' = \begin{pmatrix} -2 & -2 \\ -1.5 & -2 \end{pmatrix}\begin{pmatrix} u \\ v \end{pmatrix}$, which

has the characteristic equation $r^2 + 4r + 1 = 0$ so that
$r = -2 \pm \sqrt{3}$. Thus the critical point is an asymptotically

stable node. For α = 1.25, we have $\begin{pmatrix} u \\ v \end{pmatrix}' = \begin{pmatrix} -2 & -2 \\ -2.5 & -2 \end{pmatrix}\begin{pmatrix} u \\ v \end{pmatrix}$, so

$r^2 + 4r -1 = 0$ and $r = -2 \pm \sqrt{5}$. Thus (2,2) is an unstable
saddle point.

12c. Letting x = u+2 and y = v+2 yields
     u' = (u+2)(4 - u-2 - v-2) = -2u - 2v - u² - uv and
     v' = (v+2)(2 + 2α - v-2 - αu - 2α) = -2αu - 2v - v² - αuv.
     Thus the approximate linear system is u' = -2u - 2v and
     v' = -2αu - 2v.

12d. The eigenvalues are given by
     $\begin{vmatrix} -2-r & -2 \\ -2\alpha & -2-r \end{vmatrix} = r^2 + 4r + 4 - 4\alpha = 0$, or $r = -2 \pm 2\sqrt{\alpha}$.
     Thus for $0 < \alpha < 1$ there are 2 negative real roots
     (asymptotially stable node) and for $\alpha > 1$ the real roots
     differ in sign, yielding an unstable saddle point. α = 1
     is the bifurcation point.

Section 9.5, Page 509

3b. We have x = 0 or (1 - .5x - .5y) = 0 and y = 0 or
     (-.25 + .5x) = 0 and thus we have three critical points:
     (0,0), (2,0) and (1/2,3/2).

3c. For (0,0) the linear system is dx/dt = x and

     dy/dt = -.25y and hence A = $\begin{pmatrix} 1 & 0 \\ 0 & -1/4 \end{pmatrix}$ which has

     eigenvalues r_1 = 1 and r_2 = -1/4 and corresponding

     eigenvectors $\begin{pmatrix} 1 \\ 0 \end{pmatrix}$ and $\begin{pmatrix} 0 \\ 1 \end{pmatrix}$. Thus (0,0) is an unstable

     saddle point.
     For (2,0), we let x = 2 + u and y = v in the given

     equations and obtain $\dfrac{du}{dt} = -(u+v) - \dfrac{1}{2}u(u+v)$ and

     $\dfrac{dv}{dt} = \dfrac{3}{4}v + \dfrac{1}{2}uv$. The linear portion of this has matrix

$A = \begin{pmatrix} -1 & -1 \\ 0 & 3/4 \end{pmatrix}$, which has the eigenvalues $r_1 = -1$,

$r_2 = 3/4$ and corresponding eigenvectors $\begin{pmatrix} 1 \\ 0 \end{pmatrix}$ and $\begin{pmatrix} -4 \\ 7 \end{pmatrix}$.

Thus $(2,0)$ is also an unstable saddle point.

For $\left(\dfrac{1}{2}, \dfrac{3}{2}\right)$ we let $x = 1/2 + u$ and $y = 3/2 + v$ in the

given equations, which yields $\dfrac{du}{dt} = -\dfrac{1}{4}u - \dfrac{1}{4}v$, $\dfrac{dv}{dt} = \dfrac{3}{4}u$

as the linear portion. Thus $A = \begin{pmatrix} -\dfrac{1}{4} & -\dfrac{1}{4} \\ \dfrac{3}{4} & 0 \end{pmatrix}$, which has

eigenvalues $r_{1,2} = (-1 \pm \sqrt{11}\,i)/8$. Thus $\left(\dfrac{1}{2}, \dfrac{3}{2}\right)$ is an

asymptotically stable spiral point since the eigenvalues
are complex with negative real part. Using
$r_1 = (-1 + \sqrt{11}\,i)/8$ we find that one eigenvector is
$\begin{pmatrix} -2 \\ 1 + \sqrt{11}\,i \end{pmatrix}$ and by Section 7.6 the second eigenvector is
$\begin{pmatrix} -2 \\ 1 - \sqrt{11}\,i \end{pmatrix}$.

3e.

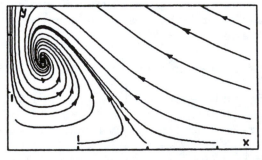

3f. For (x,y) above the line $x + y = 2$ we see that $x' < 0$ and
thus x must remain bounded. For (x,y) to the right of
$x = 1/2$, $y' > 0$ so it appears that y could grow large
asymptotic to $x =$ constant. However, this implies a
contradiction ($x =$ constant implies $x' = 0$, but as y gets
larger, x' gets increasingly negative) and hence we
conclude y must remain bounded and hence $(x,y) \to$
$(1/2,3/2)$ as $t \to \infty$, again assuming they start in the
first quadrant.

7a. The amplitude ratio is $(cK/\gamma)/(\sqrt{ac}\ K/\alpha) = \alpha\sqrt{c}/\gamma\sqrt{a}$.

7b. From Eq (2) $\alpha = .5$, $a = 1$, $\gamma = .25$ and $c = .75$, so the ratio is $.5\sqrt{.75}/.25\sqrt{1} = 2\sqrt{.75} = \sqrt{3} \cong 1.732$.

7c. A rough measurement of the amplitudes is $(6.1 - 1)/2 = 2.55$ and $(3.8 - .9)/2 = 1.45$ and thus the ratio is approximately 1.76. In this case the linear approximation is a good predictor.

11. The presence of a trapping company actually would require a modification of the equations, either by altering the coefficients or by including nonhomogeneous terms on the right sides of the D.E. The effects of indiscreminate trapping could decrease the populations of both rabbits and fox significantly or decrease the fox population which could possibly lead to a large increase in the rabbit population. Over the long run it makes sense for a trapping company to operate in such a way that a consistent supply of pelts is available and to disturb the predator-prey system as little as possible. Thus, the company should trap fox only when their population is increasing, trap rabbits only when their population is increasing, trap rabbits and fox only during the time when both their populations are increasing, and trap neither during the time both their populations are decreasing. In this way the trapping company can have a moderating effect on the population fluctuations, keeping the trajectory close to the center.

13. The critical points of the system are the solutions of the algebraic equations $x(a - \sigma x - \alpha y) = 0$, and $y(-c + \gamma x) = 0$. the critical points are $x = 0$, $y = 0$; $x = a/\sigma$, $y = 0$; and $x = c/\gamma$, $y = a/\alpha - c\sigma/\alpha\gamma = \sigma A/\alpha$ where $A = a/\sigma - c/\gamma > 0$.
To study the critical point $(0,0)$ we discard the nonlinear terms in the system of D.E. to obtain the corresponding linear system $dx/dt = ax$, $dy/dt = -cy$. The characteristic equation is $r^2 - (a+c)r - ac = 0$ so $r_1 = a$, $r_2 = -c$. Thus the critical point $(0,0)$ is an unstable saddle point.
To study the critical point $(a/\sigma,0)$ we let $x = (a/\sigma) + u$, $y = 0 + v$ and substitute in the D.E. to obtain the almost linear system $du/dt = -au - (a\alpha/\sigma)v - \sigma u^2 - \alpha uv$, $dv/dt = \gamma Av + \gamma uv$. The corresponding linear system is $du/dt = -au - (a\alpha/\sigma)v$, $dv/dt = \gamma Av$. The characteristic equation is $r^2 + (a - \gamma A)r - a\gamma A = 0$ so $r_1 = -a$, $r_2 = \gamma A$.

Thus the critical point $(a/\sigma, 0)$ is an unstable saddle point.

To study the critical point $(c/\gamma, \sigma A/\alpha)$ we let $x = (c/\gamma) + u$, $y = (\sigma A/\alpha) + v$ and substitute in the D.E. to obtain the almost linear system

$$\frac{du}{dt} = -(c\sigma/\gamma)u - (ac/\gamma)v - \sigma u^2 - \alpha uv$$

$$\frac{dv}{dt} = (\sigma A\gamma/\alpha)u + \gamma uv$$

The corresponding linear system is $du/dt = -(c\sigma/\gamma)u - (\alpha c/\gamma)v$, $dv/dt = (\sigma A\gamma/\alpha)u$. The characteristic equation is $r^2 + (c\sigma/\gamma)r + c\sigma A = 0$, so $r_1, r_2 = [-(c\sigma/\gamma) \pm \sqrt{(c\sigma/\gamma)^2 - 4c\sigma A}]/2$. Thus, depending on the sign of the discriminant we have that $(c/\gamma, \sigma A/\alpha)$ is either an asymptotically stable spiral point or an asymptotically stable node. Thus for nonzero initial data $(x,y) \to (c/\gamma, \sigma A/\alpha)$ as $t \to \infty$.

Section 9.6, Page 519

1. Assuming that $V(x,y) = ax^2 + cy^2$ we find $V_x(x,y) = 2ax$,

    $V_y = 2cy$ and thus Eq.(7) yields $\dot{V}(x,y) = 2ax(-x^3 + xy^2) + 2cy(-2x^2y - y^3) = -[2ax^4 + 2(2c-a)x^2y^2 + 2cy^4]$. If we choose a and c to be any positive real numbers with

    $2c > a$, then $\dot{V}$ is a negative definite. Also, V is positive definite by Theorem 9.6.4. Thus by Theorem 9.6.1 the origin is an asymptotically stable critical point.

3. Assuming the same form for $V(x,y)$ as in Problem 1, we have

    $\dot{V}(x,y) = 2ax(-x^3 + 2y^3) + 2cy(-2xy^2) = -2ax^4 + 4(a-c)xy^3$.

    If we choose $a = c > 0$, then $\dot{V}(x,y) = -2ax^4 \leq 0$ in any

    neighborhood containing the origin and thus $\dot{V}$ is negative semidefinite and V is positive definite. Theorem 9.6.1 then concludes that the origin is a stable critical point. Note that the origin may still be asymptotically stable, however, the $V(x,y)$ used here is not sufficient to prove that.

6a. The correct system is $dx/dt = y$ and $dy/dt = -g(x)$. Since $g(0) = 0$, we conclude that $(0,0)$ is a critical point.

6b. From the given conditions, the graph of g must be positive for $0 < x < k$ and negative for $-k < x < 0$. Thus

if $\quad 0 < x < k \quad$ then $\displaystyle\int_0^x g(s)\,ds > 0$,

if $\quad -k < x < 0 \quad$ then $\displaystyle\int_0^x g(s)\,ds = -\int_x^0 g(s)\,ds > 0$.

Since $V(0,0) = 0$ it follows that $V(x,y) = y^2/2 + \displaystyle\int_0^x g(s)\,ds$ is positive definite for $-k < x < k$, $-\infty < y < \infty$. Next, we have $\dot{V}(x,y) = V_x\dfrac{dx}{dt} + V_y\dfrac{dy}{dt} = g(x)y + y[-g(x)] = 0$.

Since $\dot{V}(x,y)$ is never positive, we may conclude that it is negative semidefinite and hence by Theorem 9.6.1 $(0,0)$ is at least a stable critical point.

7b. V is positive definite by Theorem 9.6.4. Since $V_x(x,y) = 2x$, $V_y(x,y) = 2y$, we obtain

$\dot{V}(x,y) = 2xy - 2y^2 - 2y\sin x = 2y[-y + (x - \sin x)]$. If $x < 0$, then $\dot{V}(x,y) < 0$ for all $y > 0$. If $x > 0$, choose y so that $0 < y < x - \sin x$. Then $\dot{V}(x,y) > 0$. Hence V is not a Liapunov function.

7c. Since $V(0,0) = 0$, $1 - \cos x > 0$ for $0 < |x| < 2\pi$ and $y^2 > 0$ for $y \neq 0$, it follows that $V(x,y)$ is positive definite in a neighborhood of the origin. Next $V_x(x,y) = \sin x$, $V_y(x,y) = y$, so

$\dot{V}(x,y) = (\sin x)(y) + y(-y - \sin x) = -y^2$. Hence $\dot{V}$ is negative semidefinite and $(0,0)$ is a stable critical point by Theorem 9.6.1.

7d. $V(x,y) = (x+y)^2/2 + x^2 + y^2/2 = 3x^2/2 + xy + y^2$ is positive definite by Theorem 9.6.4. Next $V_x(x,y) = 3x + y$, $V_y(x,y) = x + 2y$ so

$$
\begin{aligned}
\dot{V}(x,y) &= (3x+y)y - (x+2y)(y+\sin x) \\
&= 2xy - y^2 - (x+2y)\sin x \\
&= 2xy - y^2 - (x+2y)(x - \alpha x^3/6) \\
&= -x^2 - y^2 + \alpha(x+2y)x^3/6 \\
&= -r^2 + \alpha r^4 (\cos\theta + 2\sin\theta)(\cos^3\theta)/6 < -r^2 + r^4/2 \\
&= -r^2(1-r^2/2).
\end{aligned}
$$

Thus $\dot{V}$ is negative definite for $r < \sqrt{2}$. From Theorem 9.6.1 it follows that the origin is an asymptotically stable critical point.

8.   Let $x = u$ and $y = du/dt$ to obtain the system $dx/dt = y$
     and $dy/dt = -c(x)y - g(x)$. Now consider
     $V(x,y) = y^2/2 + \int_0^x g(s)ds$, which yields

     $\dot{V} = g(x)y + y[-c(x)y - g(x)] = -y^2 c(x)$.

10b. Since $V_x(x,y) = 2Ax + By$, $V_y(x,y) = Bx + 2Cy$, we have

     $\dot{V}(x,y) = (2Ax + By)(a_{11}x + a_{12}y) + (Bx + 2Cy)(a_{21}x + a_{22}y) =$
     $(2Aa_{11} + Ba_{21})x^2 + [2(Aa_{12} + Ca_{21}) + B(a_{11}+a_{22})]xy +$
     $(2Ca_{22} + Ba_{12})y^2$. We choose A, B, and C so that
     $2Aa_{11} + Ba_{21} = -1$, $2(Aa_{12} + Ca_{21}) + B(a_{11}+a_{22}) = 0$, and
     $2Ca_{22} + Ba_{12} = -1$. The first and third equations give us A
     and C in terms of B, respectively. We substitute in the
     second equation to find B and then calculate A and C. The
     result is given in the text.

10c. Since $a_{11}a_{22} - a_{12}a_{21} > 0$ and $a_{11} + a_{22} < 0$, we see that
     $\Delta < 0$ and so $A > 0$. Using the expressions for A, B, and
     C found in part (b) we obtain
     $(4AC-B^2)\Delta^2 = [a_{21}^2+a_{22}^2 + (a_{11}a_{22}-a_{12}a_{21})][a_{11}^2+a_{12}^2 + (a_{11}a_{22}-a_{12}a_{21})]$
     $\qquad\qquad\qquad - (a_{12}a_{22}+a_{11}a_{21})^2$
     $\quad = (a_{11}^2+a_{12}^2+a_{21}^2+a_{22}^2)(a_{11}a_{22}-a_{12}a_{21}) + (a_{11}^2+a_{12}^2)(a_{21}^2+a_{22}^2)$
     $\qquad\qquad\qquad + (a_{11}a_{22}-a_{12}a_{21})^2 - (a_{12}a_{22}+a_{11}a_{21})^2$
     $\quad = (a_{11}^2+a_{12}^2+a_{21}^2+a_{22}^2)(a_{11}a_{22}-a_{12}a_{21}) + 2(a_{11}a_{22}-a_{12}a_{21})^2$.
     Since $a_{11}a_{22} - a_{12}a_{21} > 0$ it follows that $4AC - B^2 > 0$.

11a. For $V(x,y) = Ax^2 + Bxy + Cy^2$ we have
     $\dot{V} = (2Ax + By)(a_{11}x+a_{12}y + F_1(x,y)) + (Bx+2Cy)(a_{21}x+a_{22}y + G_1(x,y))$
     $\quad = (2Ax+By)(a_{11}x+a_{12}y)+(Bx+2Cy)(a_{21}x+a_{22}y)$
     $\qquad\qquad\qquad + (2Ax+By)F_1(x,y) + (Bx+2Cy)G_1(x,y)$
     $\quad = -x^2-y^2 + (2Ax+By)F_1(x,y) + (Bx+2Cy)G_1(x,y)$, if A,B and C are
     chosen as in Problem 10.

11b. Substituting $x = r\cos\theta$, $y = r\sin\theta$ we find that

     $\dot{V}[x(r,\theta), y(r,\theta)] = -r^2+r(2A\cos\theta+B\sin\theta)F_1[x(r,\theta), y(r,\theta)]$
     $+ r(B\cos\theta + 2C\sin\theta)G_1[x(r,\theta), y(r,\theta)]$. Now we make use
     of the facts that (1) there exists an M such that
     $|2A| \leq M$, $|B| \leq M$, and $|2C| \leq M$; and (2) given any
     $\varepsilon > 0$ there exists a circle $r = R$ such that
     $|F_1(x,y)| < \varepsilon r$ and $|G_1(x,y)| < \varepsilon r$ for $0 \leq r < R$. We have
     $|2A\cos\theta + B\sin\theta| \leq 2M$ and $|B\cos\theta + 2C\sin\theta| \leq 2M$. Hence

     $\dot{V}[x(r,\theta), y(r,\theta)] \leq -r^2 + 2Mr(\varepsilon r)+2Mr(\varepsilon r) = -r^2(1 - 4M\varepsilon)$.

If we choose $\varepsilon = M/8$ we obtain $\dot{V}[x(r,\theta), y(r,\theta)] \leq -r^2/2$

for $0 \leq r < R$. Hence $\dot{V}$ is negative definite in $0 \leq r < R$ and from Problem 10c V is positive definite and thus V is a Liapunov function for the almost linear system.

Section 9.7, Page 530

1.   Note that $r = 1$, $\theta = t + t_0$ satisfy the two equations for all t and is thus a periodic solution. If $r < 1$, then $dr/dt > 0$, and the direction of motion on a trajectory is outward. If $r > 1$, then the direction of motion is inward. It follows that the periodic solution $r = 1$, $\theta = t + t_0$ is a stable limit cycle.

2. $r = 1$, $\theta = -t + t_0$ is a periodic solution. If $r < 1$, then $dr/dt > 0$, and the direction of motion on a trajectory is outward. If $r > 1$, the $dr/dt > 0$, and the direction of motion is still outward. It follows that the solution $r = 1$, $\theta = -t + t_0$ is a semistable limit cycle.

4. $r = 1$, $\theta = -t + t_0$ and $r = 2$, $\theta = -t + t_0$ are periodic solutions. If $r < 1$, then $dr/dt < 0$, and the direction of motion on a trajectory is inward. If $1 < r < 2$, then $dr/dt > 0$, and the direction of motion is outward. Similarly, if $r > 2$, the direction of motion is inward. It follows that the periodic solution $r = 1$, $\theta = -t + t_0$ is unstable and the periodic solution $r = 2$, $\theta = -t + t_0$ is a stable limit cycle.

7.   Differentiating x and y with respect to t we find that $dx/dt = (dr/dt)\cos\theta - (r\sin\theta)d\theta/dt$ and $dy/dt = (dr/dt)\sin\theta + (r\cos\theta)d\theta/dt$. Hence
$ydx/dt - xdy/dt = (r\sin\theta\cos\theta)dr/dt - (r^2\sin^2\theta)d\theta/dt -$
$(r\cos\theta\sin\theta)dr/dt - (r^2\cos^2\theta)d\theta/dt$
$= - r^2d\theta/dt.$

8a. Multiplying the first equation by x and the second by y and adding yields $xdx/dt + ydy/dt = (x^2+y^2)f(r)/r$, or $rdr/dt = rf(r)$, as in the derivation of Eq.(8), and thus $dr/dt = f(r)$. To obtain an equation for θ multiply the first equation by y, the second by x and substract to obtain $ydx/dt - xdy/dt = -x^2-y^2$, or $-r^2d\theta/dt = -r^2$, using the results of Problem 7. Thus $d\theta/dt = 1$. It follows that periodic solutions are given by $r = c$, $\theta = t + t_0$ where $f(c) = 0$. Since $\theta = t + t_0$, the motion is counterclockwise.

8b. First note that $f(r) = r(r-2)^2(r-3)(r-1)$. Thus $r = 1$,
     $\theta = t + t_0$; $r = 2$, $\theta = t + t_0$; and $r = 3$, $\theta = t + t_0$ are
     periodic solutions. If $r < 1$, then $dr/dt > 0$, and the
     direction of motion on a trajectory is outward. If
     $1 < r < 2$, then $dr/dt < 0$ and the direction of motion is
     inward. Thus the periodic solution $r = 1$, $\theta = t + t_0$ is
     a stable limit cycle. If $2 < r < 3$, then $dr/dt < 0$, and
     the direction of motion is inward. Thus the periodic
     solution $r = 2$, $\theta = t + t_0$ is a semistable limit cycle.
     If $r > 3$, then $dr/dt > 0$, and the direction of motion is
     outward. Thus the periodic solution $r = 3$, $\theta = t + t_0$ is
     unstable.

9.   Setting $x = r\cos\theta$, $y = r\sin\theta$ and using the techniques of
     Problem 8 the equations transform to $dr/dt = r^2 - 2$,
     $d\theta/dt = -1$. This system has a periodic solution $r = \sqrt{2}$,
     $\theta = -t + t_0$. If $r < \sqrt{2}$, then $dr/dt < 0$, and the
     direction of motion along a trajectory is inward. If
     $r > \sqrt{2}$, then $dr/dt > 0$, and the direction of motion is
     outward. Thus the periodic solution $r = \sqrt{2}$, $\theta = -t + t_0$
     is unstable.

11. If $F(x,y) = x+y+x^3-y^2$, $G(x,y) = -x+2y+x^2y+y^3/3$, then
     $F_x(x,y) + G_y(x,y) = 1+3x^2+2+x^2+y^2 = 3+4x^2+y^2$. Since the
     conditions of Theorem 9.7.2 are satisfied for all x and
     y, and since $F_x + G_y > 0$ for all x and y, it follows that
     the system has no periodic nonconstant solution.

13. Since $x = \phi(t)$, $y = \psi(t)$ is a solution of Eqs.(15), we
     have $d\phi/dt = F[\phi(t),\psi(t)]$, $d\psi/dt = G[\phi(t),\psi(t)]$. Hence
     on the curve C,
     $F(x,y)dy - G(x,y)dx = \phi'(t)\psi'(t)dt -\psi'(t)\phi'(t)dt = 0$. It
     follows that the line integral around C is zero.
     However, if $F_x + G_y$ has the same sign throughout D, then
     the double integral cannot be zero. This gives a
     contradiction. Thus either the solution of Eqs.(15) is
     not periodic or if it is, it cannot lie entirely in D.

16a. Setting $x' = 0$ and solving for y yields $y = x^3/3 - x + k$.
     Substituting this into $y'= 0$ then gives
     $x + .8(x^3/3 - x + k) - .7 = 0$ Using an equation solver we
     obtain $x = 1.1994$, $y = -.62426$ for $k = 0$ and $x = .80485$,
     $y = -.13106$ for $k = .5$. To determine the type of critical
     points these are, we use Eq.(13) of Section 9.3 to find the

linear coefficient matrix to be $\mathbf{A} = \begin{pmatrix} 3(1-x_c^2) & 3 \\ -1/3 & -.8/3 \end{pmatrix}$, where x_c

is the critical point. For $x_c = 1.1994$ we obtain complex conjugate eigenvalues with a negative real part, and therefore $k = 0$ yields an asymptotically stable spiral point. For $x_c = .80485$ the eigenvalues are also complex conjugates, but with positive real parts, so $k = .5$ yields an unstable spiral point.

16b. Letting $k = .1, .2, .3, .4$ in the cubic equation of part (a) and finding the corresponding eigenvalues from the matrix in part (a), we find that the real part of the eigenvalues change sign between $k = .3$ and $k = .4$. Continuing to iterate in this fashion we find that for $k = .3464$ that the real part of the eigenvalue is $-.0002$ while for $k = .3465$ the real part is $.00005$, which indicates $k_0 = .3465$ is the critical point for which the system changes from stable to unstable.

16d. You must plot the solution for values of k slightly less than k_0, found in part (c), to determine whether a limit cycle exists.

Section 9.8, Page 538

1a. From Eq (6), $\lambda = -8/3$ is clearly one eigenvalue and the other two may be found from $\lambda^2 + 11\lambda - 10(r-1) = 0$ using the quadratic formula.

1b. For $\lambda = \lambda_1$ we have

$$\begin{pmatrix} -10+8/3 & -10 & 0 \\ r & -1+8/3 & 0 \\ 0 & 0 & 0 \end{pmatrix} \begin{pmatrix} \xi_1 \\ \xi_2 \\ \xi_3 \end{pmatrix} = \begin{pmatrix} 0 \\ 0 \\ 0 \end{pmatrix},$$ which requires $\xi_1 = \xi_2 = 0$

and ξ_3 arbitrary and thus $\xi^{(1)} = (0,0,1)^T$.

For $\lambda = \lambda_3 = (-11 + \alpha)/2$, where $\alpha = \sqrt{81+40r}$, we have

$$\begin{pmatrix} -10+(11-\alpha)/2 & 10 & 0 \\ r & -1+(11-\alpha)/2 & 0 \\ 0 & 0 & -8/3+(11-\alpha)/2 \end{pmatrix} \begin{pmatrix} \xi_1 \\ \xi_2 \\ \xi_3 \end{pmatrix} = \begin{pmatrix} 0 \\ 0 \\ 0 \end{pmatrix}.$$

The last line implies $\xi_3 = 0$ and multiplying the first line by

$(-9+\alpha)/2$ we obtain $\begin{pmatrix} (81-\alpha^2)/4 & 10(-9+\alpha)/2 \\ r & (9-\alpha)/2 \end{pmatrix} \begin{pmatrix} \xi_1 \\ \xi_2 \end{pmatrix} = \begin{pmatrix} 0 \\ 0 \end{pmatrix}.$

Substituting $\alpha^2 = 81+40r$ we have

$\begin{pmatrix} -10r & -10(9-\alpha)/2 \\ r & (9-\alpha)/2 \end{pmatrix} \begin{pmatrix} \xi_1 \\ \xi_2 \end{pmatrix} = \begin{pmatrix} 0 \\ 0 \end{pmatrix}.$ Thus $\xi^{(3)} = \begin{pmatrix} 9 - \sqrt{81+40r} \\ -2r \\ 0 \end{pmatrix},$

which is proportional to the answer given in the text. Similar calculations give $\xi^{(2)}$.

1c. Simply substitute $r = 28$ into the answers in parts (a) and (b).

2a. The calculations are somewhat simplified if you let
     $x = \beta + u$, $y = \beta + v$, and $z = (r-1)+w$, where
     $\beta = \sqrt{8(r-1)/3}$. An alternate approach is to extend
     Eq.(13) of Section 9.3, which is:

$$\begin{pmatrix} u \\ v \\ w \end{pmatrix}' = \begin{pmatrix} F_x & F_y & F_z \\ G_x & G_y & G_z \\ H_x & H_y & H_z \end{pmatrix}_{(x_0,y_0,z_0)} \begin{pmatrix} u \\ v \\ w \end{pmatrix}.$$

In this example $F = -10x + 10y$, $G = rx - y - xz$ and
$H = -8z/3 + xy$ and thus

$$\begin{pmatrix} u \\ v \\ w \end{pmatrix}' = \begin{pmatrix} -10 & 10 & 0 \\ r & -1 & -x_0 \\ y_0 & x_0 & -8/3 \end{pmatrix} \begin{pmatrix} u \\ v \\ w \end{pmatrix},$$ which is Eq.(8) for P_2 since
$x_0 = y_0 = \sqrt{8(r-1)/3}$.

2b. Eq.(9) is found by evaluating $\begin{vmatrix} -10-\lambda & 10 & 0 \\ 1 & -1-\lambda & -\beta \\ \beta & \beta & -8/3-\lambda \end{vmatrix} = 0.$

2c. If $r = 28$, then Eq.(9) is $3\lambda^3 + 41\lambda^2 + 304\lambda + 4320 = 0$,
     which has the roots -13.8546 and $.093956 \pm 10.1945i$.

3b. If r_1, r_2, r_3 are the three roots of a cubic polynomial,
     then the polynomial can be factored as
     $(x-r_1)(x-r_2)(x-r_3)$. Expanding this and equating to the
     given polynomial we have $A = -(r_1+r_2+r_3)$,
     $B = r_1r_2 + r_1r_3 + r_2r_3$ and $C = -r_1r_2r_3$. We are interested
     in the case when the real part of the complex conjugate
     roots changes sign. Thus let $r_2 = \alpha+i\beta$ and $r_3 = \alpha-i\beta$,
     which yields
     $A = -(r_1+2\alpha)$, $B = 2\alpha r_1 + \alpha^2 + \beta^2$ and $C = -r_1(\alpha^2+\beta^2)$.
     Hence, if $AB = C$, we have

$-(r_1+2\alpha)(2\alpha r_1+\alpha^2+\beta^2) = -r_1(\alpha^2+\beta^2)$ or
$-2\alpha[r_1^2 + 2\alpha r_1 + (\alpha^2+\beta^2)] = 0$ or $-2\alpha[(r_1+\alpha)^2+\beta^2] = 0$.
Since the square bracket term is positive, we conclude
that if AB = C, then $\alpha = 0$. That is, the conjugate
complex roots are pure imaginary. Note that the converse
is also true. That is, if the conjugate complex roots
are pure imaginary then AB = C.

3c. Comparing Eq.(9) to that of part b, we have A = 41/3,
     B = 8(r+10)/3 and C = 160(r-1)/3. Thus AB = C yields
     r = 470/19.

4.   We have $\dot{V} = 2x[\sigma(-x+y)] + 2\sigma y[rx-y-xz] + 2\sigma z[-bz+xy]$
         $= -2\sigma x^2 + 2\sigma xy + 2\sigma rxy - 2\sigma y^2 - 2\sigma bz^2$
         $= 2\sigma\{-[x^2-(r+1)xy+y^2]-bz^2\}$. For r < 1, the
     term in the square brackets remains positive for all

     values of x and y, by Theorem 9.6.4, and thus $\dot{V}$ is
     negative definite. Thus, by the extension of Theorem
     9.6.1 to three equations, we conclude that the origin is
     an asymptotically stable critical point.

5a. $V = rx^2 + \sigma y^2 + \sigma(z-2r)^2 = c > 0$ yields
     $\dfrac{dv}{dt} = 2rx[\sigma(-x+y)] + 2\sigma y(rx-y-xz) + 2\sigma(z-2r)(-bz+xy)$. Thus

$\dot{V} = -2\sigma[rx^2+y^2 + b(z^2 - 2rz)] = -2\sigma[rx^2 + y^2 + b(z-r)^2 - br^2]$.

5b. From the proof of Theorem 9.6.1, we find that we need to
     show that $\dot{V}$, as found in part a, is always negative as it
     crosses V(x,y,z) = c. (Actually, we need to use the
     extension of Theorem 9.6.1 to three equations, but the
     proof is very similar using the vector calculus
     approach.) From part a we see that
     $\dot{V} < 0$ if $rx^2 + y^2 + b(z-r)^2 > br^2$, which holds if (x,y,z)
     lies outside the ellipsoid $\dfrac{x^2}{br} + \dfrac{y^2}{br^2} + \dfrac{(z-r)^2}{r^2} = 1$. (i)
     Thus we need to choose c such that V = c lies outside
     Eq.(i). Writing V = c in the form of Eq.(i) we obtain
     the ellipsoid $\dfrac{x^2}{c/r} + \dfrac{y^2}{c/\sigma} + \dfrac{(z-2r)^2}{c/\sigma} = 1$. (ii) Now let
     $M = \max(\sqrt{br}, r\sqrt{b}, r)$, then the ellipsoid (i) is
     contained inside the sphere S1: $\dfrac{x^2}{M^2} + \dfrac{y^2}{M^2} + \dfrac{(z-r)^2}{M^2} = 1$.
     Let S2 be a sphere centered at (0,0,2r) with radius

M+r: $\dfrac{x^2}{(M+r)^2} + \dfrac{y^2}{(M+r)^2} + \dfrac{(z-2r)}{(M+r^2)} = 1$, then S1 is contained

in S2.   Thus, if we choose c, in Eq.(ii), such that

$\dfrac{c}{r} > (M+r)^2$ and $\dfrac{c}{\sigma} > (M+r)^2$, then $\dot{V} < 0$ as the trajectory

crosses $V(x,y,z) = c$.   Note that this is a sufficient
condition and there may be many other "better" choices
using different techniques.

8b.   Several cases are shown. Results may vary, particularly
      for r = 24, due to the closeness of r to $r_3 \cong 24.06$.

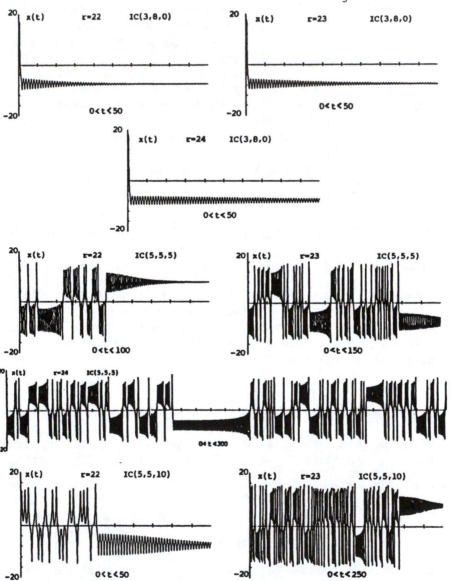

CHAPTER 10

Section 10.1 Page 547

2. $y(x) = c_1\cos\sqrt{2}\,x + c_2\sin\sqrt{2}\,x$ is the general solution of the D.E. Thus $y'(x) = -\sqrt{2}\,c_1\sin\sqrt{2}\,x + \sqrt{2}\,c_2\cos\sqrt{2}\,x$ and hence $y'(0) = \sqrt{2}\,c_2 = 1$, which gives $c_2 = 1/\sqrt{2}$. Now, $y'(\pi) = -\sqrt{2}\,c_1\sin\sqrt{2}\,\pi + \cos\sqrt{2}\,\pi = 0$ then yields $c_1 = \dfrac{\cos\sqrt{2}\,\pi}{\sqrt{2}\,\sin\sqrt{2}\,\pi} = \cot\sqrt{2}\,\pi/\sqrt{2}$. Thus the desired solution is $y = (\cot\sqrt{2}\,\pi\cos\sqrt{2}\,x + \sin\sqrt{2}\,x)/\sqrt{2}$.

3. We have $y(x) = c_1\cos x + c_2\sin x$ as the general solution and hence $y(0) = c_1 = 0$ and $y(L) = c_2\sin L = 0$. If $\sin L \neq 0$, then $c_2 = 0$ and $y(x) = 0$ is the only solution. If $\sin L = 0$, then $y(x) = c_2\sin x$ is a solution for arbitrary c_2.

7. $y(x) = c_1\cos 2x + c_2\sin 2x$ is the solution of the related homogeneous equation and $y_p(x) = \dfrac{1}{3}\cos x$ is a particular solution, yielding $y(x) = c_1\cos 2x + c_2\sin 2x + \dfrac{1}{3}\cos x$ as the general solution of the D.E. Thus $y(0) = c_1 + \dfrac{1}{3} = 0$ and $y(\pi) = c_1 - \dfrac{1}{3} = 0$ and hence there is no solution since there is no value of c_1 that will satisfy both boundary conditions.

11. If $\lambda < 0$, the general solution of the D.E. is $y = c_1\sinh\sqrt{\mu}\,x + c_2\cosh\sqrt{\mu}\,x$ where $-\lambda = \mu$. The two B.C. require that $c_2 = 0$ and $c_1 = 0$ so $\lambda < 0$ is not an eigenvalue. If $\lambda = 0$, the general solution of the D.E. is $y = c_1 + c_2 x$. The B.C. require that $c_1 = 0$, $c_2 = 0$ so again $\lambda = 0$ is not an eigenvalue. If $\lambda > 0$, the general solution of the D.E. is $y = c_1\sin\sqrt{\lambda}\,x + c_2\cos\sqrt{\lambda}\,x$. The B.C. require that $c_2 = 0$ and $\sqrt{\lambda}\,c_1\cos\sqrt{\lambda}\,\pi = 0$. The second condition is satisfied for $\lambda \neq 0$ and $c_1 \neq 0$ if $\sqrt{\lambda}\,\pi = (2n-1)\pi/2$, $n = 1,2,\ldots$. Thus the eigenvalues are $\lambda_n = (2n-1)^2/4$, $n = 1,2,3\ldots$ with the corresponding eigenfunctions $y_n(x) = \sin[(2n-1)x/2]$, $n = 1,2,3\ldots$.

15. For $\lambda < 0$ there are no eigenvalues, as shown in Problem 11. For $\lambda = 0$ we have $y(x) = c_1 + c_2 x$, so $y'(0) = c_2 = 0$ and

$y'(\pi) = c_2 = 0$, and thus $\lambda = 0$ is an eigenvalue, with $y_0(x) = 1$ as the eigenfunction. For $\lambda > 0$ we again have $y(x) = c_1\sin\sqrt{\lambda}\,x + c_2\cos\sqrt{\lambda}\,x$, so $y'(0) = \sqrt{\lambda}\,c_1 = 0$ and $y'(L) = -c_2\sqrt{\lambda}\,\sin\sqrt{\lambda}\,L = 0$. We know $\lambda > 0$, in this case, so the eigenvalues are given by $\sin\sqrt{\lambda}\,L = 0$ or $\sqrt{\lambda}\,L = n\pi$. Thus $\lambda_n = (n\pi/L)^2$ and $y_n(x) = \cos(n\pi x/L)$ for $n = 1,2,3\ldots$.

Section 10.2, Page 555

3. We look for values of T for which $\sinh2(x+T) = \sinh2x$ for all x. Expanding the left side of this equation gives $\sinh2x\cosh2T + \cosh2x\sinh2T = \sinh2x$, which will be satisfied for all x if we can choose T so that $\cosh2T = 1$ and $\sinh2T = 0$. The only value of T satisfying these two constraints is $T = 0$. Since T is not positive we conclude that the function $\sinh2x$ is not periodic.

5. We look for values of T for which $\tan\pi(x+T) = \tan\pi x$. Expanding the left side gives $\tan\pi(x+T) = (\tan\pi x + \tan\pi T)/(1-\tan\pi x\tan\pi T)$ which is equal to $\tan\pi x$ only for $\tan\pi T = 0$. The only positive solutions of this last equation are $T = 1,2,3\ldots$ and hence $\tan\pi x$ is periodic with fundamental period $T = 1$.

7. To start, let $n = 0$, then $f(x) = \begin{cases} 0 & -1 \le x < 0 \\ 1 & 0 \le x < 1 \end{cases}$; for $n = 1$, $f(x) = \begin{cases} 0 & 1 \le x < 2 \\ 1 & 2 \le x < 3 \end{cases}$; and for $n = 2$, $f(x) = \begin{cases} 0 & 3 \le x < 4 \\ 1 & 4 \le x < 5 \end{cases}$. By continuing in this fashion, and drawing a graph, it can be seen that $T = 2$.

10. The graph of f(x) is: We note that f(x) is a straight line with a slope of one in any

interval. Thus f(x) has the form x+b in any interval for the correct value of b. Since $f(x+2) = f(x)$, we may set $x = -1/2$ to obtain $f(3/2) = f(-1/2)$. Noting that 3/2 is on the interval $1 < x < 2[f(3/2) = 3/2 + b]$ and that $-1/2$ is on the interval $-1 < x < 0[f(-1/2) = -1/2 + 1]$, we conclude that $3/2 + b = -1/2 + 1$, or $b = -1$ for the interval $1 < x < 2$. For the interval $8 < x < 9$ we have $f(x+8) = f(x+6) = \ldots = f(x)$ by successive applications of the periodicity condition. Thus for $x = 1/2$ we have $f(17/2) = f(1/2)$ or $17/2 + b = 1/2$ so $b = -8$ on the interval $8 < x < 9$.

In Problems 13 through 18 it is often necessary to use integration by parts to evaluate the coefficients, although all the details will not be shown here.

13a. The function represents a sawtooth wave. It is periodic with period 2L.

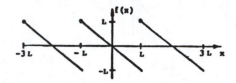

13b. The Fourier series is of the form

$$f(x) = a_0/2 + \sum_{m=1}^{\infty} (a_m \cos m\pi x/L + b_m \sin m\pi x/L), \text{ where the}$$

coefficients are computed from Eqs. (13) and (14). Substituting for $f(x)$ in these equations yields

$$a_0 = (1/L)\int_{-L}^{L}(-x)\,dx = 0 \text{ and } a_m = (1/L)\int_{-L}^{L}(-x)\cos(m\pi x/L)\,dx = 0,$$

$m = 1,2...$ (these can be shown by direct integration, or using the fact that $\int_{-a}^{a} g(x)\,dx = 0$ when $g(x)$ is an odd function). Finally,

$$b_m = (1/L)\int_{-L}^{L}(-x)\sin(m\pi x/L)\,dx$$

$$= (x/m\pi)\cos(m\pi x/L)\Big|_{-L}^{L} - (1/m\pi)\int_{-L}^{L}\cos(m\pi x/L)\,dx$$

$$= (2L\cos m\pi)/m\pi - (L/m^2\pi^2)\sin(m\pi x/L)\Big|_{-L}^{L} = 2L(-1)^m/m\pi$$

Substituting these terms in the above Fourier series for $f(x)$ yields the desired answer.

15a. See the next page.

15b. In this case $f(x)$ is periodic of period 2π and thus $L = \pi$ in Eqs. (9), (13,) and (14). The constant a_0 is found to be $a_0 = (1/\pi)\int_{-\pi}^{0} x\,dx = -\pi/2$ since $f(x)$ is zero on the interval $[0,\pi]$. Likewise $a_n = (1/\pi)\int_{-\pi}^{0} x\cos nx\,dx = [1 - (-1)^n]/n^2\pi$, using integration by parts and recalling that $\cos n\pi = (-1)^n$. Thus $a_n = 0$ for n even and $a_n = 2/n^2\pi$ for n odd, which may be written as $a_{2n-1} = 2/(2n-1)^2\pi$ since $2n-1$ is always an odd number. In a similar fashion $b_n = (1/\pi)\int_{-\pi}^{0} x\sin nx\,dx = (-1)^{n+1}/n$ and thus the desired solution is obtained. Notice that in this case both cosine and sine terms appear in the Fourier series for the given $f(x)$.

15a.

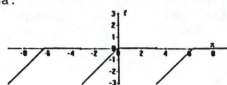

21a.

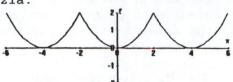

21b. $a_0 = \dfrac{1}{2}\displaystyle\int_{-2}^{2}\dfrac{x^2}{2}dx = \dfrac{1}{12}x^3 \Big|_{-2}^{2} = \dfrac{4}{3}$, so $\dfrac{a_0}{2} = \dfrac{2}{3}$ and

$a_n = \dfrac{1}{2}\displaystyle\int_{-2}^{2}\dfrac{x^2}{2}\cos\dfrac{n\pi x}{2}dx$

$\quad = \dfrac{1}{4}[\dfrac{2x^2}{n\pi}\sin\dfrac{n\pi x}{2} + \dfrac{8x}{n^2\pi^2}\cos\dfrac{n\pi x}{2} - \dfrac{16}{n^3\pi^3}\sin\dfrac{n\pi x}{2}]\Big|_{-2}^{2}$

$\quad = (8/n^2\pi^2)\cos(n\pi) = (-1)^n 8/n^2\pi^2$

where the second line for a_n is found by integration by parts or a computer algebra system. Similarly,

$b_n = \dfrac{1}{2}\displaystyle\int_{-2}^{2}\dfrac{x^2}{2}\sin\dfrac{n\pi x}{2}dx = 0$, since $x^2\sin\dfrac{n\pi x}{2}$ is an odd

function. Thus $f(x) = \dfrac{2}{3} + \dfrac{8}{\pi^2}\displaystyle\sum_{n=1}^{\infty}\dfrac{(-1)^n}{n^2}\cos\dfrac{n\pi x}{2}$.

21c. As in Eq. (27), we have $s_m(x) = \dfrac{2}{3} + \dfrac{8}{\pi^2}\displaystyle\sum_{n=1}^{m}\dfrac{(-1)^n}{n^2}\cos\dfrac{n\pi x}{2}$

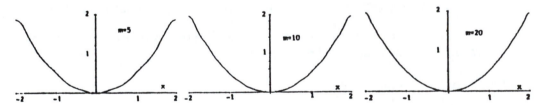

21d. Observing the graphs we see that the Fourier series
    converges quite rapidly, except, at $x = -2$ and $x = 2$,
    where there is a sharp "point" in the periodic function.

25.

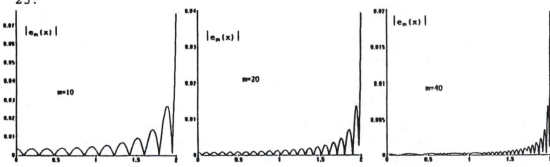

27a. First we have $\int_{T}^{a+T} g(x)\,dx = \int_{0}^{a} g(s)\,ds$ by letting x = s + T
in the left integral. Now, if $0 \leq a \leq T$, then from
elementary calculus we know that
$$\int_{a}^{a+T} g(x)\,dx = \int_{a}^{T} g(x)\,dx + \int_{T}^{a+T} g(x)\,dx = \int_{a}^{T} g(x)\,dx + \int_{0}^{a} g(x)\,dx$$
using the equality derived above. This last sum is
$\int_{0}^{T} g(x)\,dx$ and thus we have the desired result.

Section 10.3, Page 562

2a. Substituting for f(x) in Eqs.(2) and (3) with L = π
yields $a_0 = (1/\pi)\int_{0}^{\pi} x\,dx = \pi/2$;

$a_m = (1/\pi)\int_{0}^{\pi} x\cos mx\,dx = (\cos m\pi - 1)/\pi m^2 = 0$ for m even and

$= -2/\pi m^2$ for m odd; and

$b_m = (1/\pi)\int_{0}^{\pi} x\sin mx\,dx = -(\pi\cos m\pi)/m\pi = (-1)^{m+1}/m$,

m = 1,2... . Substituting these values into Eq.(1) with
L = π yields the desired solution.

2b. The function to which the series converges is indicated
in the figure and is periodic with period 2π. Note that

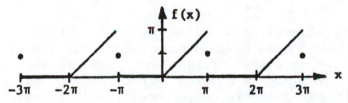

the Fourier series converges to $\pi/2$ at x = $-\pi$, π, etc.,
even though the function is defined to be zero there.
This value ($\pi/2$) represents the mean value of the left
and right hand limits at those points. In $(-\pi, 0)$,
f(x) = 0 and f'(x) = 0 so both f and f' are continuous and
have finite limits as x → $-\pi$ from the right and as
x → 0 from the left. In $(0, \pi)$, f(x) = x and f'(x) = 1
and again both f and f' are continuous and have limits as
x → 0 from the right and as x → π from the left. Since
f and f' are piecewise continuous on $[-\pi, \pi]$ the
conditions of the Fourier theorem are satisfied.

4a. Substituting for f(x) in Eqs.(2) and (3), with L = 1

yields $a_0 = \int_{-1}^{1} (1-x^2)\,dx = 4/3$;

$$a_n = \int_{-1}^{1} (1-x^2)\cos n\pi x\, dx = (2/n\pi)\int_{-1}^{1} x\sin n\pi x\, dx$$

$$= (-2/n^2\pi^2)[x\cos n\pi x\Big|_{-1}^{1} - \int_{-1}^{1}\cos n\pi x\, dx]$$

$$= 4(-1)^{n+1}/n^2\pi^2; \text{ and}$$

$$b_n = \int_{-1}^{1} (1-x^2)\sin n\pi x\, dx = 0. \quad \text{Substituting these values}$$

into Eq.(1) gives the desired series.

4b. The function to which the series converges is shown in
the figure and is periodic of fundamental period 2. In
$[-1,1]$ $f(x)$ and $f'(x) = -2x$ are both continuous and have
finite limits as the endpoints of the interval are
approached from within the interval.

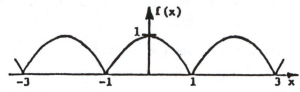

7a. As in Problem 15, Section 10.2, we have
$$f(x) = -\frac{\pi}{4} + \sum_{n=1}^{\infty}[\frac{2\cos(2n-1)x}{\pi(2n-1)^2} + \frac{(-1)^{n+1}\sin nx}{n}].$$

7b. $$e_n(x) = f(x) + \frac{\pi}{4} - \sum_{k=1}^{n}[\frac{2\cos(2k-1)x}{\pi(2k-1)^2} + \frac{(-1)^{k+1}\sin kx}{k}].$$

Using a computer algebra system, we find that for
$n = 5$, 10 and 20 the maximum error occurs at $x = -\pi$ in
each case and is 1.6025, 1.5867 and 1.5787 respectively.
Note that the author's n values are 10, 20 and 40, since
he has included the zero cosine coefficient terms and the
sine terms are all zero at $x = -\pi$.

7c. It's not possible in this case, due to Gibb's phenomenon,
to satisfy $|e_n(x)| \leq 0.01$ for all x.

12a. $$a_0 = \int_{-1}^{1} (x-x^3)\, dx = 0 \text{ and } a_n = \int_{-1}^{1} (x-x^3)\cos n\pi x\, dx = 0 \text{ since}$$
$(x-x^3)$ and $(x-x^3)\cos n\pi x$ are odd functions.
$$b_n = \int_{-1}^{1} (x-x^3)\sin n\pi x\, dx$$

$$= [\frac{x^3}{n\pi}\cos n\pi x - \frac{3x^2}{n^2\pi^2}\sin n\pi x - \frac{(n^2\pi^2+6)}{n^3\pi^3}x\cos n\pi x + \frac{(n^2\pi^2+6)}{n^4\pi^4}\sin n\pi x]_{-1}^{1}$$

$$= \frac{-12}{n^3\pi^3}\cos n\pi, \text{ so } f(x) = -\frac{12}{\pi^3}\sum_{n=1}^{\infty}\frac{(-1)^n}{n^3}\sin n\pi x.$$

12b. $e_n(x) = f(x) + \dfrac{12}{\pi^3} \displaystyle\sum_{k=1}^{n} \dfrac{(-1)^k}{k^3} \sin k\pi x$.   These errors will be

much smaller than in the earlier problems due to the n^3 factor in the denominator. Convergence is much more rapid in this case.

14. The solution to the corresponding homogeneous equation is found by methods presented in Section 3.4 and is $y(t) = c_1 \cos\omega t + c_2 \sin\omega t$. For the nonhomogeneous terms we use the method of superposition and consider the sequence of equations $y_n'' + \omega^2 y_n = b_n \sin nt$ for $n = 1,2,3...$. If $\omega > 0$ is not equal to an integer, then the particular solution to this last equation has the form $Y_n = a_n \cos nt + d_n \sin nt$, as previously discussed in Section 3.6. Substituting this form for Y_n into the equation and solving, we find $a_n = 0$ and $d_n = b_n/(\omega^2 - n^2)$. Thus the formal general solution of the original nonhomogeneous D.E. is

$$y(t) = c_1 \cos\omega t + c_2 \sin\omega t + \sum_{n=1}^{\infty} b_n (\sin nt)/(\omega^2 - n^2), \text{ where}$$

we have superimposed all the Y_n terms to obtain the infinite sum. To evalute c_1 and c_2 we set $t = 0$ to obtain $y(0) = c_1 = 0$ and

$$y'(0) = \omega c_2 + \sum_{n=1}^{\infty} nb_n/(\omega^2 - n^2) = 0 \text{ where we have formally}$$

differentiated the infinite series term by term and evaluated it at $t = 0$. (Differentiation of a Fourier Series has not been justified yet and thus we can only consider this method a formal solution at this point).

Thus $c_2 = -(1/\omega) \displaystyle\sum_{n=1}^{\infty} nb_n/(\omega^2 - n^2)$, which when substituted into the above series yields the desired solution.

If $\omega = m$, a positive integer, then the particular solution of $y_m'' + m^2 y_m = b_m \sin mt$ has the form $Y_m = t(a_m \cos mt + d_m \sin mt)$ since $\sin mt$ is a solution of the related homogeneous D.E. Substituting Y_m into the D.E. yields $a_m = -b_m/2m$ and $d_m = 0$ and thus the general solution of the D.E. (when $\omega = m$) is now $y(t) = c_1 \cos mt$

$$+ c_2 \sin mt - b_m t(\cos mt)/2m + \sum_{n=1, n \neq m}^{\infty} b_n (\sin nt)/(m^2 - n^2).$$

To evaluate c_1 and c_2 we set $y(0) = 0 = c_1$ and

$$y'(0) = c_2 m - b_m/2m + \sum_{n=1, n \neq m}^{\infty} b_n n/(m^2 - n^2) = 0. \quad \text{Thus}$$

$$c_2 = b_m/2m^2 - \sum_{n=1, n \neq m}^{\infty} b_n n/m(m^2 - n^2), \text{ which when substituted}$$

into the equation for $y(t)$ yields the desired solution.

15. In order to use the results of Problem 14, we must first find the Fourier series for the given $f(t)$. Using Eqs.(2) and (3) with $L = \pi$, we find that
$$a_0 = (1/\pi) \int_0^{\pi} dx - (1/\pi) \int_{\pi}^{2\pi} dx = 0;$$

$$a_n = (1/\pi) \int_0^{\pi} \cos nx \, dx - (1/\pi) \int_{\pi}^{2\pi} \cos nx \, dx = 0; \text{ and}$$

$$b_n = (1/\pi) \int_0^{\pi} \sin nx \, dx - (1/\pi) \int_{\pi}^{2\pi} \sin nx \, dx = 0 \text{ for } n \text{ even and}$$
$$= 4/n\pi \text{ for } n \text{ odd. Thus}$$

$$f(t) = (4/\pi) \sum_{n=1}^{\infty} \sin(2n-1)t/(2n-1). \quad \text{Comparing this to the}$$

forcing function of Problem 14 we see that b_n of Problem 14 has the specific values $b_{2n} = 0$ and $b_{2n-1} = (4/\pi)/(2n-1)$ in this example. Substituting these into the answer to Problem 14 yields the desired solution. Note that we have asumed ω is not a positive integer. Note also, that if the solution to Problem 14 is not available, the procedure for solving this problem would be exactly the same as shown in Problem 14.

16. From Problem 8, the Fourier series for $f(t)$ is given by
$$f(t) = 1/2 + (4/\pi^2) \sum_{n=1}^{\infty} \cos(2n-1)\pi t/(2n-1)^2 \text{ and thus we may}$$
not use the form of the answer in Problem 14. The procedure outlined there, however, is applicable and will yield the desired solution.

18a. We will assume $f(x)$ is continuous for this part. For the case where $f(x)$ has jump discontinuities, a more detailed proof can be developed, as shown in part b. From Eq.(3) we have $b_n = \dfrac{1}{L} \int_{-L}^{L} f(x) \sin \dfrac{n\pi x}{L} dx$. If we let $u = f(x)$ and $dv = \sin \dfrac{n\pi x}{L} dx$, then $du = f'(x) dx$ and $v = \dfrac{-L}{n\pi} \cos \dfrac{n\pi x}{L}$.
Thus

$$b_n = \frac{1}{L}\left[\frac{-L}{n\pi}f(x)\cos\frac{n\pi x}{L}\Big|_{-L}^{L} + \frac{L}{n\pi}\int_{-L}^{L}f'(x)\cos\frac{n\pi x}{L}dx\right]$$

$$= -\frac{1}{n\pi}[f(L)\cos n\pi - f(-L)\cos(-n\pi)] + \frac{1}{n\pi}\int_{-L}^{L}f'(x)\cos\frac{n\pi x}{L}dx$$

$$= \frac{1}{n\pi}\int_{-L}^{L}f'(x)\cos\frac{n\pi x}{L}dx, \text{ since } f(L) = f(-L) \text{ and}$$

$\cos(-n\pi) = \cos n\pi$. Hence $nb_n = \frac{1}{\pi}\int_{-L}^{L}f'(x)\cos\frac{n\pi x}{L}dx$, which

exists for all n since f'(x) is piecewise continuous.
Thus nb_n is bounded as $n \to \infty$. Likewise, for a_n, we

obtain $na_n = -\frac{1}{\pi}\int_{-L}^{L}f'(x)\sin\frac{n\pi x}{L}dx$ and hence na_n is also

bounded as $n \to \infty$.

18b. Note that f and f' are continuous at all points where f"
is continuous. Let $x_1, \ldots, x_m$ be the points in $(-L,L)$
where f" is not continuous. By splitting up the interval
of integration at these points, and integrating Eq.(3) by
parts twice, we obtain

$$n^2 b_n = \frac{n}{\pi}\sum_{i=1}^{m}[f(x_i+)-f(x_i-)]\cos\frac{n\pi x_i}{L} - \frac{n}{\pi}[f(L-)-f(-L+)]\cos n\pi$$

$$-\frac{L}{\pi^2}\sum_{i=1}^{m}[f'(x_i+)-f'(x_1-)]\sin\frac{n\pi x_i}{L} - \frac{L}{\pi^2}\int_{-L}^{L}f''(x)\sin\frac{n\pi x}{L}dx, \text{ where}$$

we have used the fact that cosine is continuous. We want
the first two terms on the right side to be zero, for
otherwise they grow in magnitude with n. Hence f must be
continuous throughout the closed interval [-L,L]. The
last two terms are bounded, by the hypotheses on f' and
f". Hence $n^2 b_n$ is bounded; similarly $n^2 a_n$ is bounded.
Convergence of the Fourier series then follows by

comparison with $\displaystyle\sum_{n=1}^{\infty}n^{-2}$.

Section 10.4, Page 570

3.   Let $f(x) = \tan 2x$, then

$$f(-x) = \tan(-2x) = \frac{\sin(-2x)}{\cos(-2x)} = \frac{-\sin(2x)}{\cos(2x)} = -\tan 2x = -f(x)$$

and thus tan2x is an odd function.

6.   Let $f(x) = e^{-x}$, then $f(-x) = e^x$ so that $f(-x) \neq f(x)$ and
$f(-x) \neq -f(x)$ and thus e^{-x} is neither even nor odd.

7.

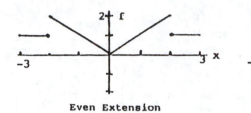

Even Extension

Odd Extension

10.

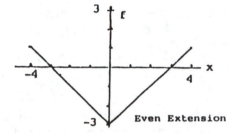

Even Extension

Odd Extension

13. By the hint $f(-x) = g(-x) + h(-x) = g(x) - h(x)$, since g
is an even function and f is an odd function. Thus
$f(x) + f(-x) = 2g(x)$ and hence $g(x) = [f(x) + f(-x)]/2$
defines $g(x)$. Likewise
$f(x) - f(-x) = g(x) - g(-x) + h(x) - h(-x) = 2h(x)$ and thus
$h(x) = [f(x) - f(-x)]/2$.

All functions and their derivatives in Problems 14 through 30
are piecewise continuous on the given intervals and their
extensions. Thus the Fourier Theorem applies in all cases.

14. For the cosine series we use the even extension of the
function given in Eq.(13) and hence
$$f(x) = \begin{cases} 0 & -2 \le x < -1 \\ 1+x & -1 \le x < 0 \end{cases} \quad \text{on the interval } -2 \le x < 0.$$
However, we don't really need this, as the coefficients
in this case are given by Eqs.(7), which just use the
original values for $f(x)$ on $0 < x \le 2$. Applying Eqs.(7)
we have $L = 2$ and thus
$$a_0 = (2/2)\int_0^1 (1-x)dx + (2/2)\int_1^2 0dx = 1/2. \quad \text{Similarly,}$$

$$a_n = (2/2)\int_0^1 (1-x)\cos(n\pi x/2)dx = 4[1-\cos(n\pi/2)]/n^2\pi^2 \text{ and}$$

$b_n = 0$. Substituting these values in the Fourier series
yields the desired results.
For the sine series, we use Eqs.(8) with $L = 2$. Thus
$a_n = 0$ and
$$b_n = (2/2)\int_0^1 (1-x)\sin(n\pi x/2)dx = 4[n\pi/2 - \sin(n\pi/2)]/n^2\pi^2.$$

15. The graph of the function to which the series converges
is shown in the figure. Using Eqs.(7) with $L = 2$ we have
$a_0 = \int_0^1 dx = 1$ and $a_n = \int_0^1 \cos(n\pi x/2)\,dx = 2\sin(n\pi/2)/n\pi$.
Thus $a_n = 0$ for n even, $a_n = 2/n\pi$ for $n = 1,5,9,\ldots$ and
$a_n = -2/n\pi$ for $n = 3,7,11,\ldots$. Hence we may write
$a_{2n} = 0$ and $a_{2n-1} = 2(-1)^{n+1}/(2n-1)\pi$, which when
substituted into the series gives the desired answer.

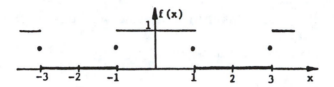

18. The graph of the function to which the series converges
is indicated in the figure.

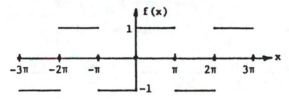

Since we want a sine series, we use Eqs.(8) to find, with
$L = \pi$, that $b_n = (2/\pi)\int_0^\pi \sin nx\,dx = 2[1-(-1)^n]/n\pi$ and thus
$b_n = 0$ for n even and $b_n = 4/n\pi$ for n odd.

20. The graph of the function to which the series converges
is shown in the figure.

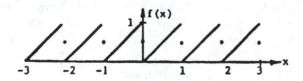

We note that $f(x)$ is specified over its entire
fundamental period ($T = 1$) and hence we cannot extend f
to make it either an odd or an even function. Using
Eqs.(2) and (3) from Section 10.3 we have ($L = 1/2$)
$a_0 = 2\int_0^1 x\,dx = 1$, $a_n = 2\int_0^1 x\cos(2n\pi x)\,dx = 0$ and

$b_n = 2\int_0^1 x\sin(2n\pi x)\,dx = -1/n\pi$. [Note: We have used the results of Problem 27 of Section 10.2 in writing these integrals. That is, if f(x) is periodic of period T, then every integral of f over an interval of length T has the same value. Thus we integrate from 0 to 1 here, rather than −1/2 to 1/2.] Substituting the above values into Eq.(1) of Section 10.3 yields the desired solution. It can also be observed from the above graph that g(x) = f(x) − 1/2 is an odd function. If Eqs.(8) are used with g(x), then it is found that

$$g(x) = (-1/\pi) \sum_{n=1}^{\infty} \sin(2n\pi x)/n \text{ and thus we obtain the same}$$

series for f(x) as found above.

25a. $b_n = \dfrac{2}{2} \displaystyle\int_0^2 (2-x^2)\sin\dfrac{n\pi x}{2}\,dx$

$= [\dfrac{-2}{n\pi}(2-x^2)\cos\dfrac{n\pi x}{2} - \dfrac{8x}{n^2\pi^2}\sin\dfrac{n\pi x}{2} - \dfrac{16}{n^3\pi^3}\cos\dfrac{n\pi x}{2}]\Big|_0^2$

$= \dfrac{4}{n\pi}(1+\cos n\pi) + \dfrac{16}{n^3\pi^3}(1-\cos n\pi)$ and thus

$f(x) = \displaystyle\sum_{n=1}^{\infty} (\dfrac{4n^2\pi^2(1+\cos n\pi) + 16(1-\cos n\pi)}{n^3\pi^3})\sin\dfrac{n\pi x}{2}$

25b.

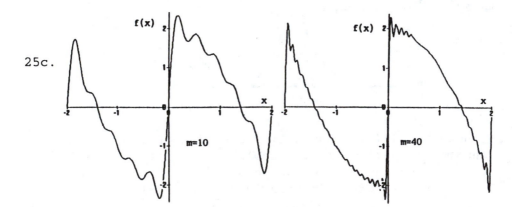

25c.

28b. For the cosine series(even extension) we have

$a_0 = \dfrac{2}{2}\displaystyle\int_0^1 x\,dx = \dfrac{1}{2}$

$$a_n = \frac{2}{2}\int_0^1 x\cos\frac{n\pi x}{2}dx = [\frac{2x}{n\pi}\sin\frac{n\pi x}{2} + \frac{4}{n^2\pi^2}\cos\frac{n\pi x}{2}]\Big|_0^1$$

$$= \frac{2}{n\pi}\sin\frac{n\pi}{2} + \frac{4}{n^2\pi^2}\cos\frac{n\pi}{2} - \frac{4}{n^2\pi^2}, \text{ so}$$

$$g(x) = \frac{1}{4} + \sum_{n=1}^{\infty}\frac{4\cos(n\pi/2) + 2n\pi\sin(n\pi/2) - 4}{n^2\pi^2}\cos\frac{n\pi x}{2}.$$

For the sine series (odd extension) we have

$$b_n = \frac{2}{2}\int_0^1 x\sin\frac{n\pi x}{2}dx = [\frac{-2x}{n\pi}\cos\frac{n\pi x}{2} + \frac{4}{n^2\pi^2}\sin\frac{n\pi x}{2}]\Big|_0^1$$

$$= \frac{-2}{n\pi}\cos\frac{n\pi}{2} + \frac{4}{n^2\pi^2}\sin\frac{n\pi}{2}, \text{ so}$$

$$h(x) = \sum_{n=1}^{\infty}\frac{4\sin(n\pi/2) - 2n\pi\cos(n\pi/2)}{n^2\pi^2}\sin\frac{n\pi x}{2}.$$

28c.

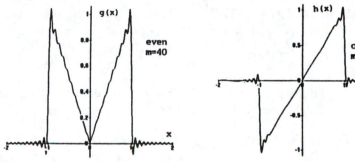

28d. The maximum error does not approach zero in either case, due to Gibb's phenomenon. Note that the coefficients in both series behave like 1/n as n → ∞ since there is an n in the numerator.

31. We have $\int_{-L}^{L}f(x)dx = \int_{-L}^{0}f(x)dx + \int_0^L f(x)dx$. Now, if we let x = -y in the first integral on the right, then $\int_{-L}^0 f(x)dx = \int_L^0 f(-y)(-dy) = \int_0^L f(-y)dy = -\int_0^L f(y)dy$. Thus $\int_{-L}^L f(x)dx = -\int_0^L f(y)dy + \int_0^L f(x)dx = 0$.

32. To prove property 2 let f_1 and f_2 be odd functions and let $f(x) = f_1(x) \pm f_2(x)$. Then $f(-x) = f_1(-x) \pm f_2(-x) = -f_1(x) \pm [-f_2(x)] = -f_1(x) \mp f_2(x) = -f(x)$, so $f(x)$ is odd. Now let $g(x) = f_1(x)f_2(x)$, then $g(-x) = f_1(-x)f_2(-x) = [-f_1(x)][-f_2(x)] = f_1(x)f_2(x) = g(x)$ and thus $g(x)$ is even. Finally, let $h(x) = f_1(x)/f_2(x)$ and hence $h(-x) = f_1(-x)/f_2(-x) = [-f_1(x)]/[-f_2(x)] = f_1(x)/f_2(x) = h(x)$, which says $h(x)$ is also even. Property 3 is proven in a similar manner.

34. Since $F(x) = \int_0^x f(t)dt$ we have

    $F(-x) = \int_0^{-x} f(t)dt = -\int_0^x f(-s)ds$ by letting $t = -s$. If f
    is an even function, $f(-s) = f(s)$, we then have
    $F(-x) = -\int_0^x f(s)ds = -F(x)$ from the original definition of
    F. Thus $F(x)$ is an odd function. The argument is
    similar if f is odd.

35. Set $x = L/2$ in Eq.(6) of Section 10.3. Since we know f
    is continuous at L/2, we may conclude, by the Fourier
    theorem, that the series will converge to
    $f(L/2) = L$ at this point. Thus we have

    $L = L/2 + (2L/\pi) \sum_{n=1}^{\infty} (-1)^{n+1}/(2n-1)$, since

    $\sin[(2n-1)\pi/2] = (-1)^{n+1}$. Dividing by L and simplifying

    yields $\dfrac{\pi}{4} = \sum_{n=1}^{\infty} \dfrac{(-1)^{n+1}}{(2n-1)} = \sum_{n=0}^{\infty} \dfrac{(-1)^n}{2n+1}$.

37a. Multiplying both sides of the equation by $f(x)$ and
     integrating from 0 to L gives

    $\int_0^L [f(x)]^2 dx = \int_0^L [f(x) \sum_{n=1}^{\infty} b_n \sin(n\pi x/L)]dx$

    $= \sum_{n=1}^{\infty} b_n \int_0^L f(x)\sin(n\pi x/L)dx = (L/2) \sum_{n=1}^{\infty} b_n^2$, by Eq.(8).

    This result is identical to that of Problem 17 of Section
    10.3 if we set $a_n = 0$, $n = 0,1,2,\ldots$, since

    $\dfrac{1}{L}\int_{-L}^L [f(x)]^2 dx = \dfrac{2}{L}\int_0^L [f(x)]^2 dx$. In a similar manner, it
    can be shown that

    $(2/L)\int_0^L [f(x)]^2 dx = a_0^2/2 + \sum_{n=1}^{\infty} a_n^2$.

37b. Since $f(x) = x$ and $b_n = 2L(-1)^{n+1}/n\pi$ (from Eq.(9)), we
     obtain

    $(2/L)\int_0^L [f(x)]^2 dx = (2/L)\int_0^L x^2 dx = 2L^2/3 = \sum_{n=1}^{\infty} b_n^2 =$

    $\sum_{n=1}^{\infty} 4L^2/n^2\pi^2 = 4L^2/\pi^2 \sum_{n=1}^{\infty} (1/n^2)$ or $\pi^2/6 = \sum_{n=1}^{\infty} (1/n^2)$.

38. We assume that the extensions of f and f′ are piecewise
     continuous on [-2L,2L]. Since f is an odd periodic
     function of fundamental period 4L it follows from
     properties 2 and 3 that $f(x)\cos(n\pi x/2L)$ is odd and
     $f(x)\sin(n\pi x/2L)$ is even. Thus the Fourier coefficients
     of f are given by Eqs.(8) with L replaced by 2L; that is
     $a_n = 0$, n = 0,1,2,... and

     $$b_n = (2/2L)\int_0^{2L} f(x)\sin(n\pi x/2L)\,dx, \quad n = 1,2,... \ . \quad \text{The}$$

     Fourier sine series for f is $f(x) = \sum_{n=1}^{\infty} b_n \sin(n\pi x/2L)$.

39. From Problem 38 we have $b_n = (1/L)\int_0^{2L} f(x)\sin(n\pi x/2L)\,dx$

     $$= (1/L)\int_0^L f(x)\sin(n\pi x/2L)\,dx + (1/L)\int_L^{2L} f(2L-x)\sin(n\pi x/2L)\,dx$$

     $$= (1/L)\int_0^L f(x)\sin(n\pi x/2L)\,dx - (1/L)\int_L^0 f(s)\sin[n\pi(2L-s)/2L]\,ds$$

     $$= (1/L)\int_0^L f(x)\sin(n\pi x/2L)\,dx - (1/L)\int_0^L f(s)\cos(n\pi)\sin(n\pi s/2L)\,ds$$

     and thus $b_n = 0$ for n even and

     $$b_n = (2/L)\int_0^L f(x)\sin(n\pi x/2L)\,dx \text{ for n odd.}$$ The Fourier

     series for f is given in Problem 38, where the b_n are
     given above.

Section 10.5, Page 579

3.   We seek solutions of the form $u(x,t) = X(x)T(t)$.
     Substituting into the P.D.E. yields $X''T + X'T' + XT' =$
     $X''T + (X' + X)T' = 0$. Formally dividing by the quantity
     $(X' + X)T$ gives the equation $X''/(X' + X) = -T'/T$ in which
     the variables are separated. In order for this equation
     to be valid on the domain of u it is necessary that both
     sides be equal to the same constant λ. Hence
     $X''/(X' + X) = -T'/T = \lambda$ or equivalently,
     $X'' - \lambda(X' + X) = 0$ and $T' + \lambda T = 0$.

5.   We seek solutions of the form $u(x,y) = X(x)Y(y)$.
     Substituting into the P.D.E. yields $X''Y + (x+y)XY'' =$
     $X''Y + xXY'' + yXY'' = 0$. Formally dividing by XY yields
     $X''/X + xY''/Y + yY''/Y = 0$. From this equation we see that
     the presence of the independent variable x multiplying
     the term u_{yy} in the original equation leads to the term
     xY''/Y when we attempt to separate the variables. It
     follows that the argument for a separation constant does
     not carry through and we cannot replace the P.D.E. by two
     O.D.E.

8. Following the procedures of Eqs.(5) through (8), we set
    $u(x,y) = X(x)T(t)$ in the P.D.E. to obtain $X''T = 4XT'$, or
    $X''X = 4T'/T$, which must be a constant. As stated in the text
    this separation constant must be $-\lambda^2$ (we choose $-\lambda^2$ so that
    when a square root is used later, the symbols are simpler)
    and thus $X'' + \lambda^2 X = 0$ and $T' + (\lambda^2/4)T = 0$. Now $u(0,t) =$
    $X(0)T(t) = 0$, for all $t > 0$, yields $X(0) = 0$, as discussed
    after Eq.(11) and similarly $u(2,t) = X(2)T(t) = 0$, for all
    $t > 0$, implies $X(2) = 0$. The D.E. for X has the solution
    $X(x) = C_1\cos\lambda x + C_2\sin\lambda x$ and $X(0) = 0$ yields $C_1 = 0$. Setting
    $x = 2$ in the remaining form of X yields $X(2) = C_2\sin 2\lambda = 0$,
    which has the solutions $2\lambda = n\pi$ or $\lambda = n\pi/2$, $n = 1,2,\ldots$.
    Note that we exclude $n = 0$ since then $\lambda = 0$ would yield
    $X(x) = 0$, which is unacceptable. Hence
    $X(x) = \sin(n\pi x/2)$, $n = 1,2,\ldots$. Finally, the solution of the
    D.E. for T yields $T(t) = \exp(-\lambda^2 t/4) = \exp(-n^2\pi^2 t/16)$. Thus we
    have found $u_n(x,t) = \exp(-n^2\pi^2 t/16)\sin(n\pi x/2)$. Setting $t = 0$
    in this last expression indicates that $u_n(x,0)$ has, for the
    correct choices of n, the same form as the terms in $u(x,0)$,
    the initial condition. Using the principle of superposition
    we know that
    $u(x,t) = c_1 u_1(x,t) + c_2 u_2(x,t) + c_4 u_4(x,t)$ satisfies the
    P.D.E. and the B.C. and hence we let $t = 0$ to obtain
    $u(x,0) = c_1 u_1(x,0) + c_2 u_2(x,0) + c_4 u_4(x,0) =$
    $c_1\sin\pi x/2 + c_2\sin\pi x + c_4\sin 2\pi x$. If we choose $c_1 = 2$,
    $c_2 = -1$ and $c_4 = 4$ then $u(x,0)$ here will match the given
    initial condition, and hence substituting these values in
    $u(x,t)$ above then gives the desired solution.

10. Since the B.C. for this heat conduction problem are
    $u(0,t) = u(40,t) = 0$, $t > 0$, the solution $u(x,t)$ is given
    by Eq.(19) with $\alpha^2 = 1$ cm^2/sec, $L = 40$ cm, and the
    coefficients c_n determined by Eq.(21) with the I.C.
    $u(x,0) = f(x) = x$, $0 \le x \le 20$; $= 40 - x$, $20 \le x \le 40$.

$$c_n = \frac{1}{20}\left[\int_0^{20} x\sin\frac{n\pi x}{40}\,dx + \int_{20}^{40}(40-x)\sin\frac{n\pi x}{40}\,dx\right]$$

$$= \frac{160}{n^2\pi^2}\sin\frac{n\pi}{2}.\quad\text{It follows that}$$

$$u(x,t) = \frac{160}{\pi^2}\sum_{n=1}^{\infty}\frac{\sin(n\pi/2)}{n^2}e^{-n^2\pi^2 t/1600}\sin\frac{n\pi x}{40}.$$

15a.

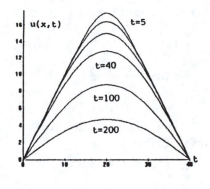

15b.

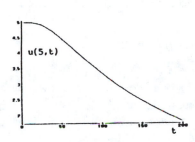

15c.

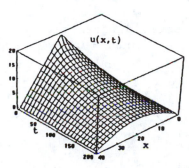

15d. As in Example 1, the maximum temperature will be at the
     midpoint, x = 20, and we use just the first term, since
     the others will be negligible for this temperature, since
     t is so large. Thus

$$u(20,t) = 1 = \frac{160}{\pi^2}\sin(\pi/2)e^{-\pi^2 t/1600}\sin(20\pi/40).$$ Solving

for t, we obtain $e^{-\pi^2 t/1600} = \pi^2/160$, or

$$t = \frac{1600}{\pi^2}\ln\frac{160}{\pi^2} = 451.60 \text{ sec.}$$

18a. Since the B.C. for this heat conduction problem are
     $u(0,t) = u(20,t) = 0$, t > 0, the solution $u(x,t)$ is given by
     Eq.(19) with L = 20 cm, and the coefficients c_n determined by
     Eq.(21) with the I.C. $u(x,0) = f(x) = 100°C$. Thus
     $c_n = (1/10)\int_0^{20}(100)\sin(n\pi x/20)dx = -200[(-1)^n-1]/n\pi$ and hence

$c_{2n} = 0$ and $c_{2n-1} = 400/(2n-1)\pi$. Substituting these values into Eq.(19) yields

$$u(x,t) = \frac{400}{\pi}\sum_{n=1}^{\infty}\frac{e^{-(2n-1)^2\pi^2\alpha^2 t/400}}{2n-1}\sin\frac{(2n-1)\pi x}{20}$$

18b. For aluminum, we have $\alpha^2 = .86$ cm^2/sec (from Table 10.5.1) and thus the first two terms give

$$u(10,30) = \frac{400}{\pi}\{e^{-\pi^2(.86)30/400} - \frac{1}{3}e^{-9\pi^2(.86)30/400}\}$$

$$= 67.228°C.$$ If an additional term is used, the temperature is increased by

$$\frac{80}{\pi}e^{-25\pi^2(.86)30/400} = 3\times 10^{-6} \text{ degrees C.}$$

19b. Using only one term in the series for $u(x,t)$, we must solve the equation $5 = (400/\pi)\exp[-\pi^2(.86)t/400]$ for t. Taking the logarithm of both sides and solving for t yields $t \cong 400\ln(80/\pi)/\pi^2(.86) = 152.56$ sec.

20. Applying the chain rule to partial differentiation of u with respect to x we see that $u_x = u_\xi\xi_x = u_\xi(1/L)$ and $u_{xx} = u_{\xi\xi}(1/L)^2$. Substituting $u_{\xi\xi}/L^2$ for u_{xx} in the heat equation gives $\alpha^2 u_{\xi\xi}/L^2 = u_t$ or $u_{\xi\xi} = (L^2/\alpha^2)u_t$. In a similar manner, $u_t = u_\tau\tau_t = u_\tau(\alpha^2/L^2)$ and hence $\frac{L^2}{\alpha^2}u_t = u_\tau$ and thus $u_{\xi\xi} = u_\tau$.

22. Substituting $u(x,y,t) = X(x)Y(y)T(t)$ in the P.D.E. yields $\alpha^2(X''YT + XY''T) = XYT'$, which is equivalent to $\frac{X''}{X} + \frac{Y''}{Y} = \frac{T'}{\alpha^2 T}$. By keeping the independent variables x and y fixed and varying t we see that $T'/\alpha^2 T$ must equal some constant σ_1 since the left side of the equation is fixed. Hence, $X''/X + Y''/Y = T'/\alpha^2 T = \sigma_1$, or $X''/X = \sigma_1 - Y''/Y$ and $T' - \sigma_1\alpha^2 T = 0$. By keeping x fixed and varying y in the equation involving X and Y we see that $\sigma_1 - Y''/Y$ must equal some constant σ_2 since the left side of the equation is fixed. Hence, $X''/X = \sigma_1 - Y''/Y = \sigma_2$ so $X'' - \sigma_2 X = 0$ and $Y'' - (\sigma_1 - \sigma_2)Y = 0$. For $T' - \sigma_1\alpha^2 T = 0$ to have solutions that remain bounded as $t \to \infty$ we must have $\sigma_1 < 0$. Thus, setting $\sigma_1 = -\lambda^2$, we have $T' + \alpha^2\lambda^2 T = 0$. For $X'' - \sigma_2 X = 0$ and homogeneous B.C., we conclude, as in Sect. 10.1, that $\sigma_2 < 0$ and, if we let

$\sigma_2 = -\mu^2$, then $X'' + \mu^2 X = 0$. With these choices for σ_1 and σ_2 we then have $Y'' + (\lambda^2 - \mu^2)Y = 0$.

Section 10.6, Page 588

3. The steady-state temperature distribution $v(x)$ must satisfy Eq.(9) and also satsify the B.C. $v_x(0) = 0$, $v(L) = 0$. The general solution of $v'' = 0$ is $v(x) = Ax + B$. The B.C. $v_x(0) = 0$ implies $A = 0$ and then $v(L) = 0$ implies $B = 0$, so the steady state solution is $v(x) = 0$.

7. Again, $v(x)$ must satisfy $v'' = 0$, $v'(0) - v(0) = 0$ and $v(L) = T$. The general solution of $v'' = 0$ is $v(x) = ax + b$, so $v(0) = b$, $v'(0) = a$ and $v(L) = T$. Thus $a - b = 0$ and $aL + b = T$, which give $a = b = T/(1+L)$. Hence $v(x) = T(x+1)/(L+1)$.

9a. Since the B.C. are not homogeneous, we must first find the steady state solution. Using Eqs.(9) and (10) we have $v'' = 0$ with $v(0) = 0$ and $v(20) = 60$, which has the solution $v(x) = 3x$. Thus the transient solution $w(x,t)$ satisfies the equations $\alpha^2 w_{xx} = w_t$, $w(0,t) = 0$, $w(20,t) = 0$ and $w(x,0) = 25 - 3x$, which are obtained from Eqs.(13) to (15). The solution of this problem is given by Eq.(4) with the c_n given by Eq.(6):

$$c_n = \frac{1}{10}\int_0^{20}(25-3x)\sin\frac{n\pi x}{20}dx = (70\cos n\pi + 50)/n\pi, \text{ and thus}$$

$$u(x,t) = 3x + \sum_{n=1}^{\infty}\frac{70\cos n\pi + 50}{n\pi}e^{-0.86n^2\pi^2 t/400}\sin\frac{n\pi x}{20} \text{ since}$$

$\alpha^2 = .86$ for aluminum.

9b.   9c.

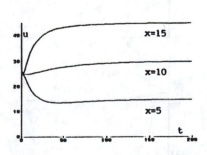

9d. Using just the first term of the sum, we have
$$u(5,t) = 15 - \frac{20}{\pi}e^{-0.86\pi^2 t/400}\sin\frac{\pi}{4} = 15 \pm .15. \text{ Thus}$$

$$\frac{20}{\pi} e^{-0.86\pi^2 t/400} \sin\frac{\pi}{4} = .15, \text{ which yields } t = 160.30 \text{ sec.}$$

To obtain the answer in the text, the first two terms of
the sum must be used, which requires an equation solver
to solve for t. Note that this reduces t by only .01
seconds.

12a. Since the B.C. are $u_x(0,t) = u_x(L,t) = 0$, $t > 0$, the
solution $u(x,t)$ is given by Eq.(35) with the coefficients
c_n determined by Eq.(37). Substituting the I.C.
$u(x,0) = f(x) = \sin(\pi x/L)$ into Eq.(37) yields

$$c_0 = (2/L)\int_0^L \sin(\pi x/L)\,dx = 4/\pi \text{ and}$$

$$c_n = (2/L)\int_0^L \sin(\pi x/L)\cos(n\pi x/L)\,dx$$

$$= (1/L)\int_0^L \{\sin[(n+1)\pi x/L] - \sin[(n-1)\pi x/L]\}\,dx$$

$$= (1/\pi)\{[1 - \cos(n+1)\pi]/(n+1) - [1 - \cos(n-1)\pi]/(n-1)\}$$

$$= 0, n \text{ odd}; = -4/(n^2-1)\pi, n \text{ even.} \quad \text{Thus}$$

$$u(x,t) = 2/\pi - (4/\pi)\sum_{n=1}^{\infty} \exp[-4n^2\pi^2\alpha^2 t/L^2]\cos(2n\pi x/L)/(4n^2-1)$$

where we are now summing over even terms by setting
n = 2n.

12b. As $t \to \infty$ we see that all terms in the series decay to
zero except the constant term, $2/\pi$. Hence
$$\lim_{t\to\infty} u(x,t) = 2/\pi.$$

12c.

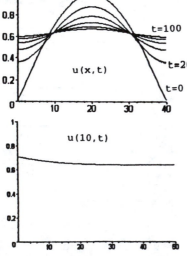

12d. The original heat in the rod is redistributed to give the
final temperature distribution, since no heat is lost.

14a. Since the ends are insulated, the solution to this
     problem is given by Eq.(35), with $\alpha^2 = 1$ and L = 30, and

     Eq.(37). Thus $u(x,t) = \dfrac{c_0}{2} + \sum\limits_{n=1}^{\infty} c_n \exp(-n^2\pi^2 t/900)\cos(n\pi x/30)$,

     where

$$c_0 = \frac{2}{30}\int_0^{30} f(x)\,dx = \frac{1}{15}\int_5^{10} 25\,dx = \frac{25}{3} \text{ and}$$

$$c_n = \frac{2}{30}\int_0^3 f(x)\cos\frac{n\pi x}{30}\,dx = \frac{1}{15}\int_5^{10} 25\cos\frac{n\pi x}{30}\,dx = \frac{50}{n\pi}[\sin\frac{n\pi}{3} - \sin\frac{n\pi}{6}].$$

14b.

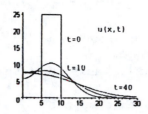
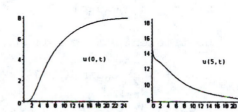
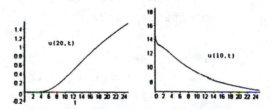

14c.

Although x = 4 and x = 1 are symmetrical to the initial
temperature pulse, they are not symmetrical to the insulated
end points.

15a. Substituting u(x,t) = X(x)T(t) into Eq.(1) leads to the
     two O.D.E. $X'' - \sigma X = 0$ and $T' - \alpha^2\sigma T = 0$. An argument
     similar to the one in the text implies that we must have
     X(0) = 0 and X'(L) = 0. Also, by assuming σ is real and
     considering the three cases $\sigma < 0$, $\sigma = 0$, and $\sigma > 0$ we
     can show that only the case $\sigma < 0$ leads to nontrivial
     solutions of $X'' - \sigma X = 0$ with X(0) = 0 and X'(L) = 0.
     Setting $\sigma = -\lambda^2$, we obtain $X(x) = k_1\sin\lambda x + k_2\cos\lambda x$.
     Now, X(0) = 0 $\rightarrow$ $k_2 = 0$ and thus $X(x) = k_1\sin\lambda x$.

Differentiating and setting x = L yields $\lambda k_1 \cos\lambda L = 0$.
Since $\lambda = 0$ and $k_1 = 0$ lead to $u(x,t) = 0$, we must choose
λ so that $\cos\lambda L = 0$, or $\lambda = (2n-1)\pi/2L$, n = 1,2,3,... .
These values for λ imply that $\sigma = -(2n-1)^2\pi^2/4L^2$ so the
solutions T(t) of $T' - \alpha^2\sigma T = 0$ are proportional to
$\exp[-(2n-1)^2\pi^2\alpha^2 t/4L^2]$. Combining the above results
leads to the desired set of fundamental solutions.

15b. In order to satisfy the I.C. u(x,0) = f(x) we assume that
u(x,t) has the form

$$u(x,t) = \sum_{n=1}^{\infty} c_n \exp[-(2n-1)^2\pi^2\alpha^2 t/4L^2]\sin[(2n-1)\pi x/2L].$$ The

coefficients c_n are determined by the requirement that

$$u(x,0) = \sum_{n=1}^{\infty} c_n \sin[(2n-1)\pi x/2L] = f(x).$$ Referring to

Problem 39 of Section 10.4 reveals that such a
representation for f(x) is possible if we choose the
coefficients $c_n = (2/L)\int_0^L f(x)\sin[(2n-1)\pi x/2L]dx$.

19. We must solve $v_1''(x) = 0$, $0 \le x \le a$ and $v_2''(x) = 0$,
$a \le x \le L$ subject to the B.C. $v_1(0) = 0$, $v_2(L) = T$ and
the continuity conditions at x = a. For the temperature
to be continuous at x = a we must have $v_1(a) = v_2(a)$ and
for the rate of heat flow to be continuous we must have
$-\kappa_1 A_1 v_1'(a) = -\kappa_2 A_2 v_2'(a)$, from Eq.(2) of Appendix A. The
general solutions to the two O.D.E. are $v_1(x) = C_1 x + D_1$
and $v_2(x) = C_2 x + D_2$. By applying the boundary and
continuity conditions we may solve for C_1, D_1, and C_2 and
D_2 to obtain the desired solution.

Section 10.7, Page 600

1a. Since the initial velocity is zero, the solution is given
by Eq.(20) with the coefficients c_n given by Eq.(22).
Substituting f(x) into Eq.(22) yields

$$c_n = \frac{2}{L}[\int_0^{L/2}\frac{2x}{L}\sin\frac{n\pi x}{L}dx + \int_{L/2}^L\frac{2(L-x)}{L}\sin\frac{n\pi x}{L}dx]$$

$$= \frac{8}{n^2\pi^2}\sin\frac{n\pi}{2}.$$ Thus Eq. (20) becomes

$$u(x,t) = \frac{8}{\pi^2}\sum_{n=1}^{\infty}\frac{1}{n^2}\sin\frac{n\pi}{2}\sin\frac{n\pi x}{L}\cos\frac{n\pi a t}{L}.$$

1b.

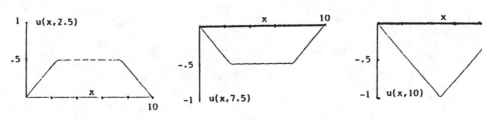

1c.

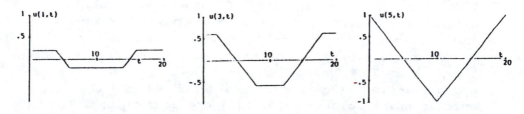

1e. The graphs in part b can best be understood using Eq.(28)
    (or the results of Problems 13 and 14). The original
    triangular shape is composed of two similar triangles of
    1/2 the height, one of which moves to the right, h(x-at),
    and the other to the left, h(x+at). Recalling that the
    series are periodic then gives the results shown. The
    graphs in part c can then be visualized from those in
    part b.

6a. The motion is governed by Eqs.(1), (3) and (31), and thus
    the solution is given by Eq.(34) where the k_n are given
    by Eq.(36):

$$k_n = \frac{2}{n\pi a}\left[\int_0^{L/4}\frac{4x}{L}\sin\frac{n\pi x}{L}dx + \int_{L/4}^{3L/4}\sin\frac{n\pi x}{L}dx + \int_{3L}^{L}\frac{4(L-x)}{L}\sin\frac{n\pi x}{L}dx\right]$$

$$= \frac{8L}{n^3\pi^3 a}\left(\sin\frac{n\pi}{4} + \sin\frac{3n\pi}{4}\right).$$ Substituting this in Eq.(34)

yields $u(x,t) = \dfrac{8L}{a\pi^3}\displaystyle\sum_{n=1}^{\infty}\dfrac{\sin\dfrac{n\pi}{4} + \sin\dfrac{3n\pi}{4}}{n^3}\sin\dfrac{n\pi x}{L}\sin\dfrac{n\pi a t}{L}.$

6b.

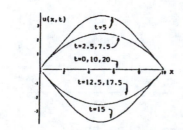

6c.

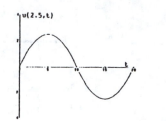

9.   Assuming that $u(x,t) = X(x)T(t)$ and substituting for u in
     Eq.(1) leads to the pair of O.D.E. $X'' + \sigma X = 0$,
     $T'' + a^2\sigma T = 0$. Applying the B.C. $u(0,t) = 0$ and
     $u_x(L,t) = 0$ to $u(x,t)$ we see that we must have $X(0) = 0$
     and $X'(L) = 0$. By considering the three cases $\sigma < 0$,
     $\sigma = 0$, and $\sigma > 0$ it can be shown that nontrivial solutions of
     the problem $X'' + \sigma X = 0$, $X(0) = 0$, $X'(L) = 0$ are possible if
     and only if $\sigma = (2n-1)^2\pi^2/4L^2$, $n = 1,2,...$ and the
     corresponding solutions for $X(x)$ are proportional to
     $\sin[(2n-1)\pi x/2L]$. Using these values for σ we find that $T(t)$
     is a linear combination of $\sin[(2n-1)\pi at/2L]$ and
     $\cos[(2n-1)\pi at/2L]$. Now, the I.C. $u_t(x,0)$ implies that
     $T'(0) = 0$ and thus functions of the form
     $u_n(x,t) = \sin[(2n-1)\pi x/2L]\cos[(2n-1)\pi at/2L]$, $n = 1,2,...$
     satsify the P.D.E. (1), the B.C. $u(0,t) = 0$, $u_x(L,t) = 0$,
     and the I.C. $u_t(x,0) = 0$. We now seek a superposition of
     these fundamental solutions u_n that also satisfies the
     I.C. $u(x,0) = f(x)$. Thus we assume that

$$u(x,t) = \sum_{n=1}^{\infty} c_n\sin[(2n-1)\pi x/2L]\cos[(2n-1)\pi at/2L].$$ The

     I.C. now implies that we must have

$$f(x) = \sum_{n=1}^{\infty} c_n\sin[(2n-1)\pi x/2L].$$ From Problem 39 of Section

     10.4 we see that $f(x)$ can be represented by such a series
     and that

$$c_n = (2/L)\int_0^L f(x)\sin[(2n-1)\pi x/2L]dx, \quad n = 1,2,... \; .$$

     Substituting these values into the above series for
     $u(x,t)$ yields the desired solution.

10a. From Problem 9 we have

$$c_n = \frac{2}{L}\int_{(L-2)/2}^{(L+2)/2}\sin\frac{(2n-1)\pi x}{2L}dx$$

$$= \frac{-4}{(2n-1)\pi}[\cos(\frac{(2n-1)\pi(L+2)}{4L}) - \cos(\frac{(2n-1)\pi(L-2)}{4L})]$$

$$= \frac{8}{(2n-1)\pi}\sin\frac{(2n-1)\pi}{4}\sin\frac{(2n-1)\pi}{2L} \quad \text{using the}$$

trigonometric relations for cos(A ± B). Substituting
this value of c_n into u(x,t) in Problem 9 yields the
desired solution.

10b.

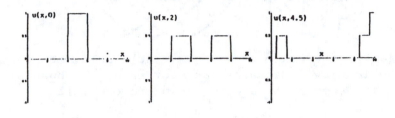

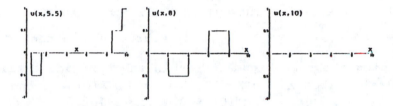

10c.

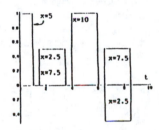

13. Using the chain rule we obtain $u_x = u_\xi \xi_x + u_\eta \eta_x =$
     $u_\xi + u_\eta$ since $\xi_x = \eta_x = 1$. Differentiating a second time
     gives $u_{xx} = u_{\xi\xi} + 2u_{\xi\eta} + u_{\eta\eta}$. In a similar way we obtain
     $u_t = u_\xi \xi_t + u_\eta \eta_t = -au_\xi + au_\eta$, since $\xi_t = -a$, $\eta_t = a$. Thus
     $u_{tt} = a^2(u_{\xi\xi} - 2u_{\xi\eta} + u_{\eta\eta})$. Substituting for u_{xx} and u_{tt}
     in the wave equation, we obtain $u_{\xi\eta} = 0$. Integrating
     both sides of $u_{\xi\eta} = 0$ with respect to η yields
     $u_\xi(\xi,\eta) = \gamma(\xi)$ where γ is an arbitrary function of ξ.
     Integrating both sides of $u_\xi(\xi,\eta) = \gamma(\xi)$ with respect to ξ
     yields $u(\xi,\eta) = \int\gamma(\xi)d\xi + \psi(\eta) = \phi(\xi) + \psi(\eta)$ where $\psi(\eta)$
     is an arbitrary function of η and $\int\gamma(\xi)d\xi$ is some
     function of ξ denoted by $\phi(\xi)$. Thus
     $u(x,t) = u(\xi(x,t),\eta(x,t)) = \phi(x - at) + \psi(x + at)$.

14. The graph of y = sin(x-at) for the various values of t is
     indicated in the figure on the next page. Note that the
     graph of y = sinx is displaced to the right by the
     distance "at" for each value of t.

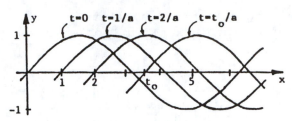

Similarly, the graph of $y = \phi(x + at)$ would be displaced to the left by a distance "at" for each t. Thus $\phi(x + at)$ represents a wave moving to the left.

16. Write the equation as $a^2 u_{xx} = u_{tt} + \alpha^2 u$ and assume $u(x,t) = X(x)T(t)$. This gives $a^2 X''T = XT'' + \alpha^2 XT$, or $\dfrac{X''}{X} = \dfrac{1}{a^2}\left(\dfrac{T''}{T} + \alpha^2\right) = \sigma$. The separation constant σ is $-\lambda^2$ using the same arguments as in the text and earlier problems. Thus $X'' + \lambda^2 X = 0$, $X(0) = 0$, $X(L) = 0$ and $T'' + (\alpha^2 + z^2\lambda^2)T = 0$, $T'(0) = 0$. If we let $\beta_n^2 = \lambda_n^2 a^2 + \alpha^2$, we then have $u_n(x,t) = \cos\beta_n t \sin\dfrac{n\pi x}{L}$, where $\lambda_n = \dfrac{n\pi x}{L}$.

Using superposition we obtain $u(x,t) = \displaystyle\sum_{n=1}^{\infty} c_n \cos\beta_n t \sin\dfrac{n\pi x}{L}$

and thus $u(x,0) = \displaystyle\sum_{n=1}^{\infty} c_n \sin\dfrac{n\pi x}{L} = f(x)$. Hence c_n are given by Eq. (22).

17a. We have $u(x,t) = \phi(x-at) + \psi(x+at)$ and thus $u_t(x,t) = -a\phi'(x-at) + a\chi'(x+at)$. Hence $u(x,0) = \phi(x) + \psi(x) = f(x)$ and $u_t(x,0) = -a\phi'(x) + a\psi'(x) = 0$. Dividing the last equation by a yields the desired result.

17b. Using the hint and the first equation obtained in part (a) leads to $\phi(x) + \psi(x) = 2\phi(x) + c = f(x)$ so $\phi(x) = (1/2)f(x) - c/2$ and $\psi(x) = (1/2)f(x) + c/2$. Hence $u(x,t) = \phi(x - at) + \psi(x + at) = (1/2)[f(x - at) - c] + (1/2)[f(x + at) + c] = (1/2)[f(x - at) + f(x + at)]$.

17c. Substituting $x + at$ for x in $f(x)$ yields
$$f(x + at) = \begin{cases} 2 & -1 < x + at < 1 \\ 0 & \text{otherwise} \end{cases}.$$
Subtracting "at" from both sides of the inequality then yields

$$f(x + at) = \begin{cases} 2 & -1 - at < x < 1 - at \\ 0 & \text{otherwise} \end{cases}.$$

18a. As in Problem 17a, we have $u(x,0) = \phi(x) + \psi(x) = 0$ and $u_t(x,0) = -a\phi'(x) + a\psi'(x) = g(x)$.

18b. From part (a) we have $\psi(x) = -\phi(x)$ which yields $-2a\phi(x) = g(x)$ from the second equation in part a. Integration then yields $\phi(x) - \phi(x_0) = \dfrac{-1}{2a}\int_{x_0}^{x} g(\xi)\,d\xi$ and hence

$$\psi(x) = (1/2a)\int_{x_0}^{x} g(\xi)\,d\xi - \phi(x_0).$$

18c. $u(x,t) = \phi(x-at) + \psi(x+at)$

$$= -(1/2a)\int_{x_0}^{x-at} g(\xi)\,d\xi + \phi(x_0) + (1/2a)\int_{x_0}^{x+at} g(\xi)\,d\xi - \phi(x_0)$$

$$= (1/2a)[\int_{x_0}^{x+at} g(\xi)\,d\xi - \int_{x_0}^{x-at} g(\xi)\,d\xi]$$

$$= (1/2a)[\int_{x_0}^{x+at} g(\xi)\,d\xi + \int_{x-at}^{x_0} g(\xi)\,d\xi]$$

$$= (1/2a\int_{x-at}^{x+at} g(\xi)\,d\xi.$$

24. Substituting $u(r,\theta,t) = R(r)\Theta(\theta)T(t)$ into the P.D.E. yields $R''\Theta T + R'\Theta T/r + R\Theta''T/r^2 = R\Theta T''/a^2$ or equivalently $R''/R + R'/rR + \Theta''/\Theta r^2 = T''/a^2 T$. In order for this equation to be valid for $0 < r < r_0$, $0 \le \theta \le 2\pi$, $t > 0$, it is necessary that both sides of the equation be equal to the same constant $-\sigma$. Otherwise, by keeping r and θ fixed and varying t, one side would remain unchanged while the other side varied. Thus we arrive at the two equations $T'' + \sigma a^2 T = 0$ and $r^2 R''/R + rR'/R + \sigma r^2 = -\Theta''/\Theta$. By an argument similar to the one above we conclude that both sides of the last equation must be equal to the same constant δ. This leads to the two equations $r^2 R'' + rR' + (\sigma r^2 - \delta)R = 0$ and $\Theta'' + \delta\Theta = 0$. Since the circular membrane is continuous, we must have $\Theta(2\pi) = \Theta(0)$, which requires $\delta = \mu^2$, μ a non-negative integer. The condition $\Theta(2\pi) = \Theta(0)$ is also known as the periodicity condition. Since we also desire solutions which vary periodically in time, it is clear that the separation constant σ should be positive, $\sigma = \lambda^2$. Thus we arrive at the three equations $r^2 R'' + rR' + (\lambda^2 r^2 - \mu^2)R = 0$, $\Theta'' + \mu^2\Theta = 0$, and $T'' + \lambda^2 a^2 T = 0$.

Section 10.8, Page 611

1a. Assuming that $u(x,y) = X(x)Y(y)$ leads to the two O.D.E.
    $X'' - \sigma X = 0$, $Y'' + \sigma Y = 0$. The B.C. $u(0,y) = 0$,
    $u(a,y) = 0$ imply that $X(0) = 0$ and $X(a) = 0$. Thus
    nontrivial solutions to $X'' - \sigma X = 0$ which satisfy these
    boundary conditions are possible only if $\sigma = -(n\pi/a)^2$,
    $n = 1,2...$; the corresponding solutions for $X(x)$ are
    proportional to $\sin(n\pi x/a)$. The B.C. $u(x,0) = 0$ implies
    that $Y(0) = 0$. Solving $Y'' - (n\pi/a)^2 Y = 0$ subject to this
    condition we find that $Y(y)$ must be proportional to
    $\sinh(n\pi y/a)$. The fundamental solutions are then
    $u_n(x,y) = \sin(n\pi x/a)\sinh(n\pi y/a)$, $n = 1,2,\ldots$, which
    satisfy Laplace's equation and the homogeneous B.C. We

    assume that $u(x,y) = \sum\limits_{n=1}^{\infty} c_n\sin(n\pi x/a)\sinh(n\pi y/a)$, where

    the coefficients c_n are determined from the B.C.

    $u(x,b) = g(x) = \sum\limits_{n=1}^{\infty} c_n\sin(n\pi x/a)\sinh(n\pi b/a)$. It follows

    that
    $c_n\sinh(n\pi b/a) = (2/a)\int_0^a g(x)\sin(n\pi x/a)dx$, $n = 1,2,\ldots$.

1b. Substituting for $g(x)$ in the equation for c_n we have
    $c_n\sinh(n\pi b/a) = (2/a)[\int_0^{a/2} x\sin(n\pi x/a)dx +$

    $\int_{a/2}^a (a-x)\sin(n\pi x/a)dx] = [4a\sin(n\pi/2)]/n^2\pi^2$, $n = 1,2,\ldots$,

    so $c_n = [4a\sin(n\pi/2)]/[n^2\pi^2\sinh(n\pi b/a)]$. Substituting
    these values for c_n in the above series yields the
    desired solution.

1c.

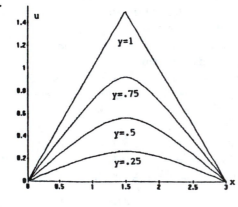
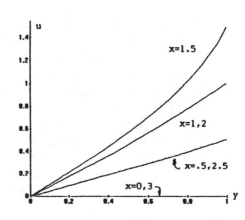

1d.

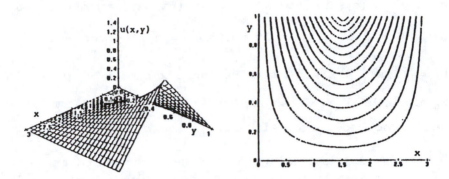

2. In solving the D.E. $Y'' - \lambda^2 Y = 0$, one normally writes
    $Y(y) = c_1\sinh\lambda y + c_2\cosh\lambda y$. However, since we have
    $Y(b) = 0$, it is advantageous to rewrite Y as
    $Y(y) = d_1\sinh\lambda(b-y) + d_2\cosh\lambda(b-y)$, where d_1, d_2 are also
    arbitrary constants and can be related to c_1, c_2 using
    the appropriate hyperbolic trigonometric identities. The
    important thing, however, is that the second form also
    satisfies the D.E. and thus $Y(y) = d_1\sinh\lambda(b-y)$ satisfies
    the D.E. and the homogeneous B.C. $Y(b) = 0$. The rest of
    the problem follows the pattern of Problem 1.

3a. Let $u(x,y) = v(x,y) + w(x,y)$, where u, v and w all
    satisfy Laplace's Eq., $v(x,y)$ satisfies the conditions in
    Eq. (4) and $w(x,y)$ satisfies the conditions of Problem 2.

4. Following the pattern of Problem 3, one could consider
    adding the solutions of four problems, each with only one
    non-homogeneous B.C. It is also possible to consider
    adding the solutions of only two problems, each with only
    two non-homogeneous B.C., as long as they involve the
    same variable. For instance, one such problem would be
    $u_{xx} + u_{yy} = 0$, $u(x,0) = 0$, $u(x,b) = 0$, $u(0,y) = k(y)$,
    $u(a,y) = f(y)$, which has the fundamental solutions
    $u_n(x,y) = [c_n\sinh(n\pi x/b) + d_n\cosh(n\pi x/b)]\sin(n\pi y/b)$.

    Assuming $u(x,y) = \sum_{n=1}^{\infty} u_n(x,y)$ and using the B.C.

    $u(0,y) = k(y)$ we obtain $d_n = (2/b)\int_0^b k(y)\sin(n\pi y/b)\,dy$.

    Using the B.C. $u(a,y) = f(y)$ we obtain
    $c_n\sinh(n\pi a/b) + d_n\cosh(n\pi a/b) = (2/b)\int_0^b f(y)\sin(n\pi y/b)\,dy$,

    which can be solved for c_n, since d_n is already known. The
    second problem, in this approach, would be $u_{xx} + u_{yy} = 0$,
    $u(x,0) = h(x)$, $u(x,b) = g(x)$, $u(0,y) = 0$ and $u(a,y) = 0$. This
    has the fundamental solutions

$u_n(x,y) = [a_n \sinh(n\pi y/a) + b_n \cosh(n\pi y/a)]\sin(n\pi x/a$, so that

$u(x,y) = \sum_{n=1}^{\infty} u_n(x,y)$. Thus $u(x,0) = h(x)$ gives

$b_n = (2/a)\int_0^a h(x)\sin(n\pi x/a)dx$ and $u(x,b) = g(x)$ gives

$a_n \sinh(n\pi b/a) + b_n \cosh(n\pi b/a) = (2/a)\int_0^a g(x)\sin(n\pi x/a)dx$, which

can be solved for a_n since b_n is known.

5. Using Eq.(20) and following the same arguments as
    presented in the text, we find that $R(r) = k_1 r^n + k_2 r^{-n}$
    and $\Theta(\theta) = c_1 \sin n\theta + c_2 \cos n\theta$, for n a positive integer,
    and $u_0(r,\theta) = 1$ for $n = 0$. Since we require that $u(r,\theta)$
    be bounded as $r \to \infty$, we conclude that $k_1 = 0$. The
    fundamental solutions are therefore
    $u_n(r,\theta) = r^{-n}\cos n\theta$, $v_n(r,\theta) = r^{-n}\sin n\theta$, $n = 1,2,\ldots$
    together with $u_0(r,\theta) = 1$. Assuming that u can be
    expressed as a linear combination of the fundamental
    solutions we have

    $u(r,\theta) = c_0/2 + \sum_{n=1}^{\infty} r^{-n}(c_n\cos n\theta + k_n\sin n\theta)$. The B.C.

    requires that

    $u(a,\theta) = c_0/2 + \sum_{n=1}^{\infty} a^{-n}(c_n\cos n\theta + k_n\sin n\theta) = f(\theta)$ for

    $0 \le \theta < 2\pi$. This is precisely the Fourier series
    representation for $f(\theta)$ of period 2π and thus
    $a^{-n}c_n = (1/\pi)\int_0^2 f(\theta)\cos n\theta d\theta$, $n = 0,1,2,\ldots$ and

    $a^{-n}k_n = (1/\pi)\int_0^2 f(\theta)\sin n\theta d\theta$, $n = 1,2\ldots$.

7. Again we let $u(r,\theta) = R(r)\Theta(\theta)$ and thus we have
    $r^2 R'' + rR' - \sigma R = 0$ and $\Theta'' + \sigma\Theta = 0$, with $R(0)$ bounded and
    the B.C. $\Theta(0) = \Theta(\alpha) = 0$. For $\sigma \le 0$ we find that
    $\Theta(0) \equiv 0$, so we let $\sigma = \lambda^2$ (λ^2 real) and thus
    $\Theta(\theta) = c_1\cos\lambda\theta + c_2\sin\lambda\theta$. The B.C. $\Theta(0) = 0 \to c_1 = 0$ and
    the B.C. $\Theta(\alpha) = 0 \to \lambda = n\pi/\alpha$, $n = 1,2,\ldots$.
    Substituting these values into Eq.(31) we obtain
    $R(r) = k_1 r^{n\pi/\alpha} + k_2 r^{-n\pi/\alpha}$. However $k_2 = 0$ since $R(0)$ must
    be bounded, and thus the fundamental solutions are
    $u_n(r,\theta) = r^{n\pi/\alpha}\sin(n\pi\theta/\alpha)$. The desired solution may now
    be formed using previously discussed procedures.

8a. Separating variables, as before, we obtain
$X'' + \lambda^2 X = 0$, $X(0) = 0$, $X(a) = 0$ and $Y'' - \lambda^2 Y = 0$, $Y(y)$ bounded
as $y \to \infty$. Thus $X(x) = \sin(n\pi x/a)$, and $\lambda^2 = (n\pi/a)^2$.
However, since neither $\sinh y$ nor $\cosh y$ are bounded as $y \to \infty$,
we must write the solution to $Y'' - (n\pi/a)^2 Y = 0$ as
$Y(y) = c_1 \exp[n\pi y/a] + c_2 \exp[-n\pi y/a]$. Thus we must choose
$c_1 = 0$ so that $u(x,y) = X(x)Y(y) \to 0$ as $y \to \infty$. The
fundamental solutions are then $u_n(x,t) = e^{-n\pi y/a}\sin(n\pi x/a)$.

$$u(x,y) = \sum_{n=1}^{\infty} c_n u_n(x,y) \text{ then gives}$$

$$u(x,0) = \sum_{n=1}^{\infty} c_n \sin(n\pi x/a) = f(x) \text{ so that } c_n = \frac{2}{a}\int_0^1 f(x)\sin\frac{n\pi x}{a}dx.$$

8b. $$c_n = \frac{2}{a}\int_0^a x(a-x)\sin\frac{n\pi x}{a}dx = \frac{4a^2}{n^3\pi^3}(1-\cos n\pi)$$

8c. Using just the first term and letting $a = 5$, we have
$u(x,y) = \dfrac{200}{\pi^3}e^{-\pi y/5}\sin\dfrac{\pi x}{5}$, which, for a fixed y, has a maximum
at $x = 5/2$ and thus we need to find y such that
$u(5/2,y) = \dfrac{200}{\pi^3}e^{-\pi y/5} = .1$. Taking the logarithm of both
sides and solving for y yields $y_0 = 6.6315$. With an equation
solver, more terms can be used. However, to four decimal
places, three terms yield the same result as above.

13a. Assuming that $u(x,y) = X(x)Y(y)$ and substituting into
Eq.(1) leads to the two O.D.E. $X'' - \sigma X = 0$, $Y'' + \sigma Y = 0$.
The B.C. $u(x,0) = 0$, $u_y(x,b) = 0$ imply that $Y(0) = 0$ and
$Y'(b) = 0$. For nontrivial solutions to exist for
$Y'' + \sigma Y = 0$ with these B.C. we find that σ must take the
values $(2n-1)^2\pi^2/4b^2$, $n = 1,2,\ldots$; the corresponding
solutions for $Y(y)$ are proportional to $\sin[(2n-1)\pi y/2b]$.
Solutions to $X'' - [(2n-1)^2\pi^2/4b^2]X = 0$ are of the form
$X(x) = A\sinh[(2n-1)\pi x/2b] + B\cosh[(2n-1)\pi x/2b]$. However,
the boundary condition $u(0,y) = 0$ implies that
$X(0) = B = 0$. It follows that the fundamental solutions
are $u_n(x,y) = c_n\sinh[(2n-1)\pi x/2b]\sin[(2n-1)\pi y/2b]$,
$n = 1,2,\ldots$. To satisfy the remaining B.C. at $x = a$ we
assume that we can represent the solution $u(x,y)$ in the

form $$u(x,y) = \sum_{n=1}^{\infty} c_n\sinh[(2n-1)\pi x/2b]\sin[(2n-1)\pi y/2b].$$
The coefficients c_n are determined by the B.C.

$$u(a,y) = \sum_{n=1}^{\infty} c_n \sinh[(2n-1)\pi a/2b] \sin[(2n-1)\pi y/2b] = f(y).$$

By properly extending f as a periodic function of period
4b as in Problem 39, Section 10.4, we find that the
coefficients c_n are given by

$$c_n \sinh[(2n-1)\pi a/2b] = (2/b) \int_0^b f(y) \sin[(2n-1)\pi y/2b] dy,$$

$$n = 1,2,\ldots .$$

CHAPTER 11

Section 11.1, Page 626

2.   Since the B.C. at $x = 1$ is nonhomogeneous, the B.V.P. is nonhomogeneous.

4.   The D.E. may be written $y'' + (\lambda - x^2)y = 0$ and is thus homogeneous, as are both B.C.

5.   Since the D.E. contains the nonhomogeneous term 1, the B.V.P. is nonhomogeneous.

9.   If $\lambda = 0$, then $y(x) = c_1 x + c_2$ and thus $y(0) = c_2$, $y(1) = c_1 + c_2$, $y'(0) = c_1$ and $y'(1) = c_1$. Hence the B.C. yield the two equations $c_2 - c_1 = 0$ and $c_1 + c_2 + c_1 = 0$ which give $c_1 = c_2 = 0$ and thus $\lambda = 0$ is not an eigenvalue.

If $\lambda > 0$, the general solution of the D.E. is $y = c_1 \sin\sqrt{\lambda}\, x + c_2 \cos\sqrt{\lambda}\, x$. The B.C. require that $c_2 - \sqrt{\lambda}\, c_1 = 0$, and $(\sin\sqrt{\lambda} + \sqrt{\lambda} \cos\sqrt{\lambda})c_1 + (\cos\sqrt{\lambda} - \sqrt{\lambda} \sin\sqrt{\lambda})c_2 = 0$. In order to have nontrivial solutions λ must satisfy $(\lambda - 1)\sin\sqrt{\lambda} - 2\sqrt{\lambda} \cos\sqrt{\lambda} = 0$. In this case $c_2 = \sqrt{\lambda}\, c_1$ and thus $\phi_n = \sin\sqrt{\lambda_n}\, x + \sqrt{\lambda_n} \cos\sqrt{\lambda_n}\, x$. If $\lambda \neq 1$, the eigenvalue equation is equivalent to $\tan\sqrt{\lambda} = 2\sqrt{\lambda}/(\lambda - 1)$ and thus by graphing $f(\sqrt{\lambda}) = \tan\sqrt{\lambda}$ and $g(\sqrt{\lambda}) = 2\sqrt{\lambda}/(\lambda - 1)$ we can estimate the eigenvalues. Since $g(\sqrt{\lambda})$ has a vertical asymptote at $\lambda = 1$ and $f(\sqrt{\lambda})$ has a vertical asymptote at $\sqrt{\lambda} = \pi/2$, we see that $1 < \sqrt{\lambda_1} < \pi/2$. By interating numerically, we find $\sqrt{\lambda_1} \cong 1.30655$ and thus $\lambda_1 \cong 1.7071$. The second eigenvalue will lie to the right of π, the second zero of $\tan\sqrt{\lambda}$. Again iterating numerically, we find $\sqrt{\lambda_2} \cong 3.6732$ and thus $\lambda_2 \cong 13.4924$. For large values of n, we see from the graph that $\sqrt{\lambda_n} \cong (n-1)\pi$, which are the zeros of $\tan\sqrt{\lambda}$. Thus $\lambda_n \cong (n-1)^2\pi^2$ for large n.

For $\lambda < 0$, the discussion follows the pattern of Example 1 yielding $y(x) = c_1 \sinh\sqrt{\mu}\, x + c_2 \cosh\sqrt{\mu}\, x$. The B.C. then yield $c_2 - \sqrt{\mu}\, c_1 = 0$ and $(\sinh\sqrt{\mu} + \sqrt{\mu} \cosh\sqrt{\mu})c_1 + (\cosh\sqrt{\mu} + \sqrt{\mu} \sinh\sqrt{\mu})c_2 = 0$, which

have a non-zero solution if and only if
$(\mu+1)\sinh\sqrt{\mu} + 2\sqrt{\mu}\cosh\sqrt{\mu} = 0$. By plotting $y = \tanh\sqrt{\mu}$ and
$y = -2\sqrt{\mu}/(\mu+1)$ we see that they intersect only at $\mu = 0$, and
thus there are no negative eigenvalues.

10. If $\lambda = 0$, the general solution of the D.E. is
$y = c_1 + c_2 x$. The B.C. $y(0) + y'(0) = 0$ requires
$c_1 + c_2 = 0$ and the B.C. $y(1) = 0$ requires $c_1 + c_2 = 0$
and thus $\lambda = 0$ is an eigenvalue with corresponding
eigenfunction $\phi_0(x) = 1-x$.

If $\lambda < 0$, set $-\lambda = \mu^2$ to obtain $y = c_1\cos\mu x + c_2\sin\mu x$.
In this case the B.C. require $c_1 + \mu c_2 = 0$ and
$c_1\cos\mu + c_2\sin\mu = 0$ which yields nontrivial solutions for c_1
and c_2 (i.e., $c_1 = -\mu c_2$) if and only if $\tan\mu = \mu$. By
plotting on the same graph $f(\mu) = \mu$ and $g(\mu) = \tan\mu$, we see
that they intersect at $\mu_0 = 0$ ($\mu = 0 \rightarrow \lambda = 0$, which has
already been discussed), $\mu_1 \cong 4.49341$ (which is just to the
left of the vertical asymptote of $\tan\mu$ at $\mu = 3\pi/2$,
$\mu_2 \cong 7.72525$ (which is just to the left of the vertical
asymptote of $\tan\mu$ at $\mu = 5\pi/2$) and for larger values
$\mu_n \cong (2n+1)\pi/2$. Since $\lambda_n = -\mu_n^2$, we have $\lambda_1 \cong -20.1907$,
$\lambda_2 = -59.6795$, $\lambda_n \cong -(2n+1)^2\pi^2/4$ and $\phi_n = \sin\mu_n x - \mu_n\cos\mu_n x$.

If $\lambda > 0$, the general solution of the D.E. is
$y(x) = c_1\cosh\sqrt{\lambda}\,x + c_2\sinh\sqrt{\lambda}\,x$. The B.C. respectively
require that $c_1 + \sqrt{\lambda}\,c_2 = 0$ and $c_1\cosh\sqrt{\lambda} + c_2\sinh\sqrt{\lambda} = 0$ and
thus λ must satisfy $\tanh\sqrt{\lambda} = \sqrt{\lambda}$ in order to have
nontrivial solutions. The only solution of this equation is
$\lambda = 0$ and thus there are no positive eigenvalues.

11a. From Eq.(i) the coefficient of y' is μQ and from Eq.(ii) the
coefficient of y' is $(\mu P)'$. Thus $(\mu P)' = \mu Q$, which gives
Eq.(iii).

11b. Eq.(iii) is both linear and separable. Using the latter
approach we have $d\mu/\mu = [(Q-P')/P]dx$ and thus
$\ln\mu = \int_{x_0}^{x} [Q(s)/P(s)]ds - \ln P$. Taking the exponential of
both sides yields Eq.(iv). The choice of x_0 simply alters
the constant of integration, which is immaterial here.

13. Since $P(x) = x^2$ and $Q(x) = x$, we find that
     $\mu(x) = (1/x^2)\exp[\int_{x_0}^{x}(s/s^2)ds] = k/x$, where k is an

     arbitrary constant which may be set equal to 1. It
     follows that Bessel's equation takes the form
     $(xy')' + (x^2-\upsilon^2/x)y = 0$.

18a. Assuming $y = s(x)u$, we have $y' = s'u + su'$ and
     $y'' = s''u + 2s'u' + su''$ and thus the D.E. becomes
     $su'' + (2s'+4s)u' + [s'' + 4s' + (4+9\lambda)s]u = 0$. Setting
     $2s' + 4s = 0$ we find $s(x) = e^{-2x}$ and the D.E. becomes
     $u'' + 9\lambda u = 0$. The B.C. $y(0) = 0$ yields $s(0)u(0) = 0$, or
     $u(0) = 0$ since $s(0) \neq 0$. The B.C. at L
     is $y'(L) = s'(L)u(L) + s(L)u'(L) = e^{-2L}(-2u(L) + u'(L)) = 0$
     and thus $u'(L) - 2u(L) = 0$. Thus the B.V.P. satisfied by
     $u(x)$ is $u'' + 9\lambda u = 0$, $u(0) = 0$, $u'(L) - 2u(L) = 0$.

     If $\lambda < 0$, the general solution of the D.E. $u'' + 9\lambda u = 0$
     is $u = c_1\sinh3\mu x + c_2\cosh3\mu x$ where $-\lambda = \mu^2$. The B.C.
     require that $c_2 = 0$, $c_1(3\mu\cosh3\mu L - 2\sinh3\mu L) = 0$. In
     order to have nontrivial solutions μ must satisfy the
     equation $3\mu/2 = \tanh3\mu L$. A graphical analysis reveals
     that for $L \leq 1/2$ this equation has no solutions for $\mu \neq 0$
     so there are no negative eigenvalues for $L \leq 1/2$. If
     $L > 1/2$ there is one solution and hence one negative
     eigenvalue with eigenfuction $\phi_{-1}(x) = e^{-2x}\sinh3\mu x$.

     If $\lambda = 0$, the general solution of the D.E. $u'' + 9\lambda u = 0$
     is $u = c_1 + c_2x$. The B.C. require that $c_1 = 0$,
     $c_2(1-2L) = 0$ so nontrivial solutions are possible only if
     $L = 1/2$. In this case the eigenfuction is $\phi_0(x) = xe^{-2x}$.

     If $\lambda > 0$, the general solution of the D.E. $u'' + 9\lambda u = 0$
     is $u = c_1\sin3\sqrt{\lambda}x + c_2\cos\sqrt{\lambda}x$. The B.C. require that
     $c_2 = 0$, $c_1(3\sqrt{\lambda}\cos3\sqrt{\lambda}L - 2\sin3\sqrt{\lambda}L) = 0$. In order to
     have nontrivial solutions λ must satisfy the equation
     $\sqrt{\lambda} = (2/3)\tan3\sqrt{\lambda}L$. A graphical analysis reveals that
     there is an infinite number of solutions to this
     eigenvalue equation. Thus the eigenfunctions are
     $\phi_n(x) = e^{-2x}\sin3\sqrt{\lambda_n}x$ where the eigenvalues λ_n satisfy
     $\sqrt{\lambda_n} = (2/3)\tan3\sqrt{\lambda_n}L$.

20. This is an Euler equation whose characteristic equation
     has roots $r_1 = \lambda$ and $r_2 = 1$. If $\lambda = 1$ the general
     solution of the D.E. is $y = c_1x + c_2x\ln x$ and the B.C.

require that $c_1 = c_2 = 0$ and thus $\lambda = 1$ is not an
eigenvalue. If $\lambda \neq 1$, $y = c_1 x + c_2 x^\lambda$ is the general
solution and the B.C. require that $c_1 + c_2 = 0$ and
$2c_1 + c_2 2^\lambda - (c_1 + \lambda c_2 2^{\lambda-1}) = 0$. Thus nontrivial solutions
exist if and only if $\lambda = 2(1-2^{-\lambda})$. The graphs of
$f(\lambda) = \lambda$ and $g(\lambda) = 2(1-2^{-\lambda})$ intersect only at $\lambda = 1$
(which has already been discussed) and $\lambda = 0$. Thus the
only eigenvalue is $\lambda = 0$ with corresponding eigenfunction
$\phi(x) = x - 1$ (since $c_1 = -c_2$).

22a. For positive λ, the general solution of the D.E. is
$y = c_1 \sin\sqrt{\lambda}\, x + c_2 \cos\sqrt{\lambda}\, x$. The B.C. require that
$\sqrt{\lambda}\, c_1 + \alpha c_2 = 0$, $c_1 \sin\sqrt{\lambda} + c_2 \cos\sqrt{\lambda} = 0$. Nontrivial
solutions exist if and only if $\sqrt{\lambda}\cos\sqrt{\lambda} - \alpha\sin\sqrt{\lambda} = 0$.
If $\alpha = 0$ this equation is satisfied by the sequence
$\lambda_n = [(2n-1)\pi/2]^2$, $n = 1,2,\ldots$. If $\alpha \neq 0$, λ must
satisfy the equation $\sqrt{\lambda}/\alpha = \tan\sqrt{\lambda}$. A plot of the
graphs of $f(\sqrt{\lambda}) = \sqrt{\lambda}/\alpha$ and $g(\sqrt{\lambda}) = \tan\sqrt{\lambda}$ reveals
that there is an infinite sequence of postive eigenvalues
for $\alpha < 0$ and $\alpha > 0$.

22b. By procedures shown previously, the cases $\lambda < 0$ and
$\lambda = 0$, when $\alpha < 1$, lead to only the trivial solution and
thus by part a all real eigenvalues are positive. For
$0 < \alpha < 1$, the graphs of $f(\sqrt{\lambda})$ and $g(\sqrt{\lambda})$ (see part a)
intersect once on $0 < \sqrt{\lambda} < \pi/2$. As α approaches 1 from
below, the slope of $f(\sqrt{\lambda})$ decreases and thus the
intersection point approaches zero.

22c. If $\lambda = 0$, then $y(x) = c_1 + c_2 x$ and the B.C. yield
$\alpha c_1 + c_2 = 0$ and $c_1 + c_2 = 0$, which have a non-zero
solution if and only if $\alpha = 1$.

22d. Let $-\lambda = \mu^2$, then $y(x) = c_1 \cosh\mu x + c_2 \sinh\mu x$ and thus the
B.C. yield $\alpha c_1 + \mu c_2 = 0$ and $(\cosh\mu)c_1 + (\sinh\mu)c_2 = 0$, which
have non-zero solutions if and only if $\tanh\mu = \mu/\alpha$. For $\alpha > 1$,
the straight line $y = \mu/\alpha$ intersects the curve $y = \tanh\mu$ in
one point, which increases as α increases. Thus $\lambda = -\mu^2$
decreases as α increases.

23. Using the D.E. ϕ_m and following the hint yields:
$\lim\int_0^L \phi_m'' \phi_n dx + \lambda_m \int_0^L 1\phi_m \phi_n dx = 0$. Integrating the first term by

parts yields: $\phi'_m\phi_n\big|_0^L - \int_0^L \phi'_m\phi'_n dx = -\lambda_m \int_0^L \phi_m\phi_n dx$. Upon utilizing
he B.C. the first term on the left vanishes and thus
$\int_0^L \phi'_n\phi'_m dx = \lambda_m \int_0^L \phi_m\phi_n dx$. Similarly, the D.E. for ϕ_n yields
$\int_0^L \phi'_m\phi'_n dx = \lambda_n \int_0^L \phi_n\phi_m dx$ and thus $(\lambda_n - \lambda_m) \int_0^L \phi_m\phi_n dx = 0$. If
$\lambda_n \neq \lambda_m$ the desired result follows.

24b. The general solution of the D.E. is
$y = c_1\sin\mu x + c_2\cos\mu x + c_3\sinh\mu x + c_4\cosh\mu x$ where $\lambda = \mu^4$.
The B.C. require that $c_2 + c_4 = 0$, $-c_2 + c_4 = 0$,
$c_1\sin\mu L + c_2\cos\mu L + c_3\sinh\mu L + c_4\cosh\mu L = 0$, and
$c_1\cos\mu L - c_2\sin\mu L + c_3\cosh\mu L + c_4\sinh\mu L = 0$. The first two
equations yield $c_2 = c_4 = 0$, and the last two have nontrivial
solutions if and only if $\sin\mu L\cosh\mu L - \cos\mu L\sinh\mu L = 0$. In
this case the third equation yields $c_3 = -c_1\sin\mu L/\sinh\mu L$ and
thus the desired eigenfunctions are obtained. The quantity
μL can be approximated by finding the intersetion of
$f(x) = \tan x$ and $g(x) = \tanh x$, where $x = \mu L$. The first
intersection is at $x \cong 3.9266$, which gives $\lambda_1 \cong 237.72/L^4$ and
the second intersection is at $x \cong 7.0686$, which gives
$\lambda_2 \cong 2,496.5/L^4$.

Section 11.2, Page 639

1.   We have $y(x) = c_1\cos\sqrt{\lambda}\,x + c_2\sin\sqrt{\lambda}\,x$ and thus $y(0) = 0$
     yields $c_1 = 0$ and $y'(1) = 0$ yields $\sqrt{\lambda}\,c_2\cos\sqrt{\lambda} = 0$. $\lambda = 0$
     gives $y(x) = 0$, so is not an eigenvalue. Otherwise
     $\lambda = (2n-1)^2\pi^2/4$ and the eigenfuctions are $\sin[(2n-1)\pi x/2]$,
     $n = 1,2,\ldots$. Thus, by Eq.(20), we must choose k_n so that
     $\int_0^1 \{k_n\sin[(2n-1)\pi x/2]\}^2 dx = 1$, since the weight function
     $r(x) = 1$ (by comparing the D.E. to Eq.(1)). Evaluating the
     integral yields $k_n^2/2 = 1$ and thus $k_n = \sqrt{2}$ and the desired
     normalized eigenfuctions are obtained.

3.   Note here that $\phi_0(x) = 1$ satisifes Eq.(20) and hence it
     is already normalized.

5.   From Problem 17 of Section 11.1 we have $e^x\sin n\pi x$,
     $n = 1,2,\ldots$ as the eigenfunctions and thus k_n must be chosen
     so that $\int_0^1 r(x)k_n^2 e^{2x}\sin^2 n\pi x\,dx = 1$. To determine $r(x)$, we must
     write the D.E. in the form of Eq.(1). That is, we multiply

the D.E. by $r(x)$ to obtain $ry'' - 2ry' + ry + \lambda ry = 0$. Now choose r so that $(ry')' = ry''-2ry'$, which yields $r' = -2r$ or $r(x) = e^{-2x}$. Thus the above integral becomes $\int_0^1 k_n^2\sin^2 n\pi x\,dx = 1$ and $k_n = \sqrt{2}$. Hence $\phi_n(x) = \sqrt{2}\,e^x\sin n\pi x$ are the normalized eigenfunctions.

7. Using Eq.(34) with $r(x) = 1$, we find that the coefficients of the series (32) are determined by

$$a_n = (f,\phi_n) = \sqrt{2}\int_0^1 x\sin[(2n-1)\pi x/2]\,dx$$

$$= (4\sqrt{2}/(2n-1)^2\pi^2)\sin(2n-1)\pi/2. \text{ Thus Eq.(32) yields}$$

$$f(x) = \frac{4\sqrt{2}}{\pi^2}\sum_{n=1}^{\infty}\frac{(-1)^{n-1}}{(2n-1)^2}\sqrt{2}\sin[(2n-1)\pi x/2],\ 0 \le x \le 1,$$

which agrees with the expansion using the approach developed in Problem 39 of Section 10.4.

10. In this case $\phi_n(x) = (\sqrt{2}/\alpha_n)\cos\sqrt{\lambda_n}\,x$, where $\alpha_n = (1 + \sin^2\sqrt{\lambda_n})^{1/2}$. Thus Eq.(34) yields

$$a_n = (\sqrt{2}/\alpha_n)\int_0^1\cos\sqrt{\lambda_n}\,x\,dx = \sqrt{2}\sin\sqrt{\lambda_n}/\alpha_n\sqrt{\lambda_n}.$$

14. In this case $L[y] = y'' + y' + 2y$ is not of the form shown in Eq.(3) and thus the B.V.P. is not self adjoint.

17. In this case $L[y] = [(1+x^2)y']' + y$ and thus the D.E. has the form shown in Eq.(3). However, the B.C. are not separated and thus we must determine by integration whether Eq.(8) is satisfied. Therefore, for u and v satisfying the B.C., integration by parts yields the following:

$$(L[u],v) = \int_0^1\{[(1+x^2)u']'+u\}v\,dx = vu'(1+x^2)\Big|_0^1 - \int_0^1\{(1+x^2)v'u'+uv\}dx$$

$$= vu'(1+x^2)\Big|_0^1 - uv'(1+x^2)\Big|_0^1 + \int_0^1\{[(1+x^2)v']'+v\}u\,dx$$

$$= (u,L[v])$$

since the integrated terms add to zero with the given B.C. Thus the B.V.P. is self-adjoint.

21a. Substituting ϕ for y in the D.E., multiplying both sides by ϕ, and integrating form 0 to 1 yields

$$\lambda\int_0^1 r\phi^2 dx = \int_0^1\{-[p(x)\phi']'\phi + q(x)\phi^2\}dx. \text{ Integrating the}$$

first term on the right side once by parts, we obtain

$$\lambda \int_0^1 r\phi^2 dx = -p(1)\phi'(1)\phi(1) + p(0)\phi'(0)\phi(0) + \int_0^1 (p\phi'^2 + q\phi^2) dx.$$

If $a_2 \neq 0$, $b_2 \neq 0$, then $\phi'(1) = -b_1\phi(1)/b_2$ and
$\phi'(0) = -a_1\phi(0)/a_2$ and the result follows. If $a_2 = 0$,
then $\phi(0) = 0$ and the boundary term at 0 will be missing.
A similar result is obtained if $b_2 = 0$.

21b. From the text, $p(x) > 0$ and $r(x) > 0$ for $0 \leq x \leq 1$ (following
     Eq.(4)). If $q(x) \geq 0$ and if b_1/b_2 and $-a_1/a_2$ are non-negative,
     then all terms in the final equation of part a are non-negative
     and thus λ must be non-negative. Note that λ could be zero if
     $q(x) = 0$, $b_1 = 0$, $a_1 = 0$ and $\phi(x) = 1$.

21c. If either $q(x) \neq 0$ or $a_1 \neq 0$ or $b_1 \neq 0$, there is at least one
     positive term on the right and thus λ must be positive.

23a. Using $\phi(x) = U(x) + iV(x)$ in Eq.(4) we have
     $L[\phi] = L[U(x) + iV(x)] = \lambda r(x)[U(x)+iV(x)]$. Using the
     linearity of L and the fact that λ and $r(x)$ are real we
     have $L[U(x)] + iL[V(x)] = \lambda r(x)U(x) + i\lambda r(x)V(x)$.
     Equating the real and imaginary parts shows that both U
     and V satisfy Eq.(1). The B.C. Eq.(2) are also satisfied
     by both U and V, using the same arguments, and thus both
     U and V are eigenfunctions.

23b. By Theorem 11.2.3 each eigenvalue λ has only one linearly
     independent eigenfuction. By part a we have U and V
     being eigenfunctions corresponding to λ and thus U and V
     must be linearly dependent.

23c. From part b we have $V(x) = cU(x)$ and thus
     $\phi(x) = U(x) + icuU(x) = (1+ic)U(x)$.

24. This is an Euler equation, so for $y = x^r$ we have
    $r^2 - (\lambda+1)r + \lambda = 0$ or $(r-1)(r-\lambda) = 0$. If $\lambda = 1$, the general
    solution to the D.E. is $y = c_1 x + c_2 x\ln x$. The B.C. require
    that $c_1 = 0$, $2c_1 + 2(\ln 2)c_2 = 0$ so $c_1 = c_2 = 0$ and $\lambda = 1$ is
    not an eigenvalue. If $\lambda \neq 1$, the general solution to the
    D.E. is $y = c_1 x + c_2 x^\lambda$. The B.C. require that $c_1 + c_2 = 0$ and
    $2c_1 + 2^\lambda c_2 = 0$. Nontrivial solutions exist if and only if
    $2^\lambda - 2 = 0$. If λ is real, this equation has no solution
    (other than $\lambda = 1$) and again $y = 0$ is the only solution to
    the boundary value problem. Suppose that $\lambda = a + bi$ with
    $b \neq 0$. Then $2^\lambda = 2^{a+bi} = 2^a 2^{bi} = 2^a \exp(ib\ln 2)$, which upon
    substitution into $2^\lambda = 2$ yields the equation

$\exp(ib\ln 2) = 2^{1-a}$. Since $e^{ix} = \cos x + i\sin x$, it follows that $\cos(b\ln 2) = 2^{1-a}$ and $\sin(b\ln 2) = 0$, which yield $a = 1$ and $b(\ln 2) = 2n\pi$ or $b = 2n\pi/\ln 2$, $n = \pm 1, \pm 2,\ldots$. Thus the only eigenvalues of the problem are $\lambda_n = 1 + i(2n\pi/\ln 2)$, $n = \pm 1, \pm 2,\ldots$.

25b. For $\lambda \leq 0$, there are no eigenfuctions. For $\lambda > 0$ the general solution of the D.E. is

$y = c_1 + c_2 x + c_3\sin\sqrt{\lambda}\,x + c_4\cos\sqrt{\lambda}\,x$. The B.C. require that

$c_1 + c_4 = 0$, $c_4 = 0$, $c_1 + c_2 L + c_3\sin\sqrt{\lambda}\,L + c_4\cos\sqrt{\lambda}\,L = 0$, and

$c_2 + \sqrt{\lambda}\,c_3\cos\sqrt{\lambda}\,L - \sqrt{\lambda}\,c_4\sin\sqrt{\lambda}\,L = 0$. Thus $c_1 = c_4 = 0$ and for nontrivial solutions to exist λ must satisfy the equation $\sqrt{\lambda}\,L\cos\sqrt{\lambda}\,L - \sin\sqrt{\lambda}\,L = 0$. In this case $c_2 = (-\sqrt{\lambda}\cos\sqrt{\lambda}\,L)(c_3)$ and it follows that the eigenfunction ϕ_1 is given by

$\phi_1(x) = \sin(\sqrt{\lambda_1}\,x) - \sqrt{\lambda_1}\,x\cos\sqrt{\lambda_1}\,L$ where λ_1 is the smallest positive solution of the equation $\sqrt{\lambda}\,L = \tan\sqrt{\lambda}\,L$. A graphical or numerical estimate of λ_1 reveals that $\lambda_1 \cong (4.4934)^2/L^2$.

25c. Assuming $\lambda \neq 0$ the eigenvalue equation is $2(1 - \cos x) = x\sin x$, where $x = \sqrt{\lambda}\,L$. Graphing $f(x) = 2(1 - \cos x)$ and $g(x) = x\sin x$ we see that there is an intersection for $6 < x < 7$. Since both $f(x)$ and $g(x)$ are zero for $x = 2\pi$, this then is the precise root and thus $\lambda_1 = (2\pi)^2/L^2$. In addition, it appears there might be an intersection for $0 < x < 1$. Using a Taylor series representation for $f(x)$ and $g(x)$ about $x = 0$, however, shows there is no intersection for $0 < x < 1$. Of course $x = 0$ is also an intersection, which yields $\lambda = 0$, which gives the trivial solution and hence $\lambda = 0$ is not an eigenvalue.

Section 11.3, Page 651

1. We must first find the eigenvalues and normalized eigenfunctions of the associated homogeneous problem $y'' + \lambda y = 0$, $y(0) = 0$, $y(1) = 0$. This problem has the solutions $\phi_n(x) = k_n\sin n\pi x$, for $\lambda_n = n^2\pi^2$, $n = 1, 2,\ldots$.

Choosing k_n so that $\int_0^1 \phi_n^2 dx = 1$ we find $k_n = \sqrt{2}$. Hence the solution of the original nonhomogeneous problem is given by $y = \sum\limits_{n=1}^{\infty} b_n \phi_n(x)$, where the coefficients b_n are found from Eq.(12), $b_n = c_n/(\lambda_n - 2)$ where c_n is given by $c_n = \sqrt{2}\int_0^1 x\sin n\pi x\, dx$ (Eq.9). [Note that the original problem can be written as $-y'' = 2y + x$ and therefore comparison with Eq.(1) yields $r(x) = 1$ and $f(x) = x$]. Integrating the expression for c_n by parts yields $c_n = \sqrt{2}\,(-1)^{n+1}/n\pi$ and thus $y = \sum\limits_{n=1}^{\infty} \dfrac{\sqrt{2}\,(-1)^{n+1}}{(n^2\pi^2 - 2)\,n\pi}\,\sqrt{2}\,\sin n\pi x$.

2. From Problem 1 of Section 11.2 we have $\phi_n = \sqrt{2}\sin[(2n-1)\pi x/2]$ for $\lambda_n = (2n-1)^2\pi^2/4$ and from Problem 7 of that section we have $c_n = 4\sqrt{2}\,(-1)^{n+1}/(2n-1)^2\pi^2$. Substituting these values into $b_n = c_n/(\lambda_n - 2)$ and $y = \sum\limits_{n=1}^{\infty} b_n \phi_n$ yields the desired result.

3. Referring to Problem 3 of Section 11.2 we have $y = b_0 + \sum\limits_{n=1}^{\infty} b_n(\sqrt{2}\cos n\pi x)$, where $b_n = c_n/(\lambda_n - 2)$ for $n = 0,1,2,\ldots$. The rest of the calculations follow those of Problem 1.

5. Note that the associated eigenvalue problem is the same as for Problem 1 and that $|1-2x| = 1-2x$ for $0 \le x \le 1/2$ while $|1-2x| = 2x-1$ for $1/2 \le x \le 1$.

8. Writing the D.E. in the form Eq.(1), we have $-y'' = \mu y + f(x)$, so $r(x) = 1$. The associated eigenvalue problem is $y'' + \lambda y = 0$, $y'(0) = 0$, $y'(1) = 0$ which has the eigenvalues $\lambda_n = n^2\pi^2$, $n = 0,1,2\ldots$ and the normalized eigenfunctions $\phi_0 = 1$, $\phi_n(x) = \sqrt{2}\cos n\pi x$, $n = 1,2\ldots$, as found in Problem 3, Section 11.2. Now, we assume $y(x) = b_0 + \sum\limits_{n=1}^{\infty} b_n\phi_n(x) = \sum\limits_{n=0}^{\infty} b_n\phi_n(x)$, Eq.(5), and thus $b_n = \dfrac{c_n}{\lambda_n - \mu}$, where $c_n = \int_0^1 f(x)\phi_n(x)\,dx$,

$n = 0, 1, 2 \ldots$, Eq.(9).   Thus

$$y(x) = \sum_{n=0}^{\infty} \frac{c_n}{\lambda_n - \mu} \phi_n(x) = -c_0/\mu + \sqrt{2} \sum_{n=1}^{\infty} \frac{c_n \cos n\pi x}{\lambda_n - \mu}, \text{ where we}$$

have assumed $\mu \neq \lambda_n$, $n = 0, 1, 2 \ldots$.

10. Since $\mu = \pi^2$ is an eigenvalue of the corresponding
     homogeneous equation, Theorem 11.3.1 tells us that a
     solution will exist only if $-(a+x)$ is orthogonal to the
     corresponding eigenfunction $\sqrt{2} \sin \pi x$.   Thus we require
     $\int_0^1 (a+x) \sin \pi x \, dx = 0$, which yields $a = -1/2$. With $a = -1/2$,
     we find that $Y = (x-1/2)/\pi^2$ and $y_c = c \sin \pi x + d \cos \pi x$ by
     methods of Chapter 3.   Setting $y = y_c + Y$ and choosing d
     to satisfy the B.C. we obtain the desired family of
     solutions.

11. Note that in this case $\mu = 4\pi^2$ and $\phi_2 = \sqrt{2} \sin 2\pi x$ are the
     eigenvalue and eigenfunction respectively of the
     corresponding homogeneous equation.   However, there is no
     value of a for which $-\sqrt{2} \int_0^1 (a+x) \sin 2\pi x \, dx = 0$, and thus
     there is no solution.

12. In this case a solution will exist only if $-a$ is
     orthogonal to $\sqrt{2} \cos \pi x$., that is if $\int_0^1 a \cos \pi x \, dx = 0$.
     Since this condition is valid for all a, a family of
     solutions exists.

14. Since $\sum_{n=1}^{\infty} c_n \phi_n(x)$ converges to zero we have $\sum_{n=1}^{\infty} c_n \phi_n(x) = 0$.
     Multiplying and integrating as suggested yields
     $$\int_0^1 [\sum_{n=1}^{\infty} c_n \phi_n(x)] r(x) \phi_m(x) \, dx = 0 \text{ or}$$
     $$\sum_{n=1}^{\infty} c_n \int_0^1 r(x) \phi_n(x) \phi_m(x) \, dx = 0. \text{ The integral that multiplies } c_n$$
     is just δ_{nm} [Eq.(22) of Section 11.2].   Thus the infinite sum
     becomes c_m and the last equation yields $c_m = 0$.

18. A twice differentiable function v satisfying the boundary
     conditions can be found by assuming that $v = ax + b$.
     Thus $v(0) = b = 1$ and $v(1) = a + 1$ while $v'(1) = a$.
     Hence $2a + 1 = -2$ or $a = -3/2$ and $v(x) = 1 - 3x/2$.

Assuming $y = u + v$ we have $(u+v)'' + 2(u+v) =$ $u'' + 2u + 2(1-3x/2) = 2 - 4x$ or $u'' + 2u = -x$, $u(0) = 0$, $u(1) + u'(1) = 0$ which is the same as Example 1 of the text.

19. From Eq.(30) we assume $u(x,t) = \sum\limits_{n=1}^{\infty} b_n(t)\phi_n(x)$, where the ϕ_n are the eigenfunctions of the related eigenvalue problem $y'' + \lambda y = 0$, $y(0) = 0$, $y'(1) = 0$ and the $b_n(t)$ are given by Eq.(42). From Problem 2, we have $\phi_n = \sqrt{2}\sin[(2n-1)\pi x/2]$ and $\lambda_n = (2n-1)^2\pi^2/4$. To evaluate Eq.(42) we need to calculate $\alpha_n = \int_0^1 \sin(\pi x/2)\sqrt{2}\sin[(2n-1)\pi x/2\,dx$ [Eq.(41) with $r(x) = 1$ and $f(x) = \sin(\pi x/2)$], which is zero except for $n = 1$ in which case $\alpha_1 = \sqrt{2}/2$, and $\gamma_n = \int_0^1 (-x)\sqrt{2}\sin[(2n-1)\pi x/2]dx$ [Eq.(35) with $F(x,t) = -x$]. This integral is the negative of the c_n in Problem 2 and thus $\gamma_n = -4\sqrt{2}\,(-1)^{n+1}/(2n-1)^2\pi^2 = -c_n$, $n = 1,2,\ldots$. Setting $\gamma_n = -c_n$ in Eq.(42) we then have

$$b_1 = \frac{\sqrt{2}}{2}e^{-\pi^2 t/4} - c_1\int_0^t e^{-\pi^2(t-s)/4}ds$$

$$= \frac{\sqrt{2}}{2}e^{-\pi^2 t/4} - \frac{4c_1}{\pi^2}e^{-\pi^2(t-s)/4}\Big|_0^t$$

$$= \frac{\sqrt{2}}{2}e^{-\pi^2 t/4} - \frac{4c_1}{\pi^2} + \frac{4c_1}{\pi^2}e^{-\pi^2 t/4} \text{ and}$$

similarly $b_n = -c_n\int_0^t e^{-\lambda_n(t-s)}ds = -(c_n/\lambda_n)e^{-\lambda_n(t-s)}\Big|_0^t$

$$= -(c_n/\lambda_n)(1-e^{-\lambda_n t}), \text{ where } \lambda_n = (2n-1)^2\pi^2/4,$$

$n = 2,3,\ldots$. Substituting these values for b_n along with $\phi_n = \sqrt{2}\sin[(2n-1)\pi x/2]$ into the series for $u(x,t)$ yields the solution to the given problem.

22. In this case $\alpha_n = 0$ for all n and γ_n is given by

$$\gamma_n = \int_0^1 e^{-t}(1-x)\sqrt{2}\sin[(2n-1)\pi x/2]dx$$

$$= e^{-t}\int_0^1 (1-x)\sqrt{2}\sin[(2n-1)\pi x/2]dx. \text{ This last integral}$$

can be written as the sum of two integrals, each of which has been evaluated in either Problem 6 or 7 of Section

11.2. Letting c_n denote the value obtained, we then have
$\gamma_n = c_n e^{-t}$ and thus $b_n = c_n \int_0^t e^{-\lambda_n(t-s)} e^{-s} ds =$
$c_n e^{-\lambda_n t} \int_0^t e^{(\lambda_n - 1)s} ds = [c_n/(\lambda_n - 1)](e^{-t} - e^{-\lambda_n t})$, where
$\lambda_n = (2n-1)^2 \pi^2/4$. Substituting these values into Eq.(30)
yields the desired solution.

24. Using the approach of Problem 23 we find that $v(x)$
satisfies $v'' = 2$, $v(0) = 1$, $v(1) = 0$. Thus
$v(x) = x^2 + c_1 x + c_2$ and the B.C. yield $v(0) = c_2 = 1$ and
$v(1) = 1 + c_1 + 1 = 0$ or $c_1 = -2$. Hence
$v(x) = x^2 - 2x + 1$ and $w(x,t) = u(x,t) - v(x)$ where, from
Problem 23, we have $w_t = w_{xx}$, $w(0,t) = 0$, $w(1,t) = 0$ and
$w(x,0) = x^2 - 2x + 2 - v(x) = 1$. This last problem can
be solved by methods of this section or by methods of
Chapter 10. Using the approach of this section we have
$$w(x,t) = \sum_{n=1}^{\infty} b_n(t)\phi_n(x) \text{ where } \phi_n(x) = \sqrt{2}\sin n\pi x$$
[which are the normalized eigenfunctions of the
associated eigenvalue problem $y'' + \lambda y = 0$, $y(0) = 0$,
$y(1) = 0$] and the b_n are given by Eq.(42). Since the
P.D.E. for $w(x,t)$ is homogeneous Eq.(42) reduces to
$b_n = \alpha_n e^{-\lambda_n t}$ ($\lambda_n = n^2\pi^2$ from the above eigenvalue problem),
where
$\alpha_n = \int_0^1 1\cdot\sqrt{2}\sin n\pi x dx = \sqrt{2}[1-(-1)^n]/n\pi$. Thus
$$u(x,t) = x^2 - 2x + 1 + \sum_{n=1}^{\infty} \frac{\sqrt{2}[1-(-1)^n]}{n\pi} e^{-n^2\pi^2 t}\sqrt{2}\sin n\pi x,$$
which simplifies to the desired solution.

28a. Since $y_c = c_1 + c_2 x$, we assume that
$Y(x) = u_1(x) + xu_2(x)$. Then $Y' = u_2$ since we require
$u_1' + xu_2' = 0$. Differentiating again yields $Y'' = u_2'$ and
thus $u_2' = -f(x)$ by substitution into the D.E. Hence
$u_2(x) = -\int_0^x f(s)ds$, $u_1' = xf(x)$, and $u_1(x) = \int_0^x sf(s)ds$.
Therefore $Y = \int_0^x sf(s)ds - x\int_0^x f(s)ds = -\int_0^x (x-s)f(s)ds$ and
$\phi(x) = c_1 + c_2 x - \int_0^x (x-s)f(s)ds$.

28b. From part a we have $y(0) = c_1 = 0$.   Thus
$y(x) = c_2 x - \int_0^x (x-s)f(s)ds$ and hence
$y(1) = c_2 - \int_0^1 (1-s)f(s)ds = 0$, which yields the desired
value of c_2.

28c. From parts a and b we have
$$\phi(x) = x\int_0^1 (1-s)f(s)ds - \int_0^x (x-s)f(s)ds$$
$$= \int_0^x x(1-s)f(s)ds + \int_x^1 x(1-s)f(s)ds - \int_0^x (x-s)f(s)ds$$
$$= \int_0^x (x-xs-x+s)f(s)ds + \int_x^1 x(1-s)f(s)ds$$
$$= \int_0^x s(1-x)f(s)ds + \int_x^1 x(1-s)f(s)ds.$$

28d. We have $\phi(x) = \int_0^1 G(x,s)f(s)ds$
$$= \int_0^x G(x,s)f(s)ds + \int_x^1 G(x,s)f(s)ds$$
$$= \int_0^x s(1-x)ds + \int_x^1 x(1-s)ds,$$ which is the same
as found in part c.

30b. In this case $y_1(x) = \sin x$ and $y_2(x) = \sin(1-x)$ [assume
$y_2(x) = c_1\cos x + c_2\sin x$, let $x = 1$, solve for c_2 in terms
of c_1 using $y(1) = 0$ and then let $c_1 = \sin 1$]. Using
these functions for y_1 and y_2 we find $W(y_1,y_2) = -\sin 1$
and thus $G(x,s) = -\sin s\,\sin(1-x)/(-\sin 1)$, since $p(x) = 1$,
for $0 \le s \le x$. Interchanging the x and s verifies $G(x,s)$
for $x \le s \le 1$.

30c. Since $W(y_1,y_2)(x) = y_1(x)y_2'(x) - y_2(x)y_1'(x)$ we find that
$[p(x)W(y_1,y_2)(x)]' = p'(x)[y_1(x)y_2'(x) - y_2(x)y_1'(x)]$
$+ p(x)[y_1'(x)y_2'(x) + y_1(x)y_2''(x) - y_2'(x)y_1'(x) - y_2(x)y_1''(x)]$
$= y_1[py_2']' - y_2[py_1']' = y_1[q(x)y_2] - y_2[q(x)y_1] = 0.$

30d. Let $c = p(x)W(y_1,y_2)(x)$. If $0 \le s \le x$, then
$G(x,s) = -y_1(s)y_2(x)/c$. Since the first argument in
$G(s,x)$ is less than the second argument, the bottom
expression of formula (iv) must be used to determine
$G(s,x)$. Thus, $G(s,x) = -y_1(s)y_2(x)/c$. A similar argument
holds if $x \le s \le 1$.

30e. We have $\phi(x) = \int_0^1 G(x,s)f(s)ds$

$$= -\int_0^x \frac{y_1(s)y_2(x)f(s)}{c}ds - \int_x^1 \frac{y_1(x)y_2(s)f(s)}{c}ds$$

(where $c = p(x)W(y_1,y_2)$ and thus, by Leibnitz's rule,

$$c\phi'(x) = -y_1(x)y_2(x)f(x) - \int_0^x y_1(s)y_2'(x)f(s)ds + y_1(x)y_2(x)f(x)$$

$$-\int_x^1 y_1'(x)y_2(s)f(s)ds. \quad \text{From this we obtain}$$

$$-c(p\phi')' = (py_2')'\int_0^x y_1(s)f(s)ds + py_2'y_1f(x)$$

$$+ (py_1')'\int_x^1 y_2(s)f(s)ds - py_1'y_2f(x). \quad \text{Dividing by}$$

c and adding $q(x)\phi(x)$ we get

$$(-p\phi')'+q\phi = \frac{(py_2')'}{c}\int_0^x y_1(s)f(s)ds - \frac{qy_2}{c}\int_0^x y_1(s)f(s)ds$$

$$+ \frac{(py_1')'}{c}\int_x^1 y_2(s)f(s)ds - \frac{qy_1}{c}\int_x^1 y_2(s)f(s)ds + f(x)$$

$$= \frac{(py_2')'-qy_2}{c}\int_0^x y_1(s)f(s)ds + \frac{(py_1')'-qy_1}{c}\int_x^1 y_2(s)f(s)ds + f(x)$$

$$= f(x),$$

since y_1 and y_2 satisfy $L[y] = 0$. Using $\phi(x)$ and $\phi'(x)$ as found above, the B.C. are both satisfied since $y_1(x)$ satisfies one B.C. and $y_2(x)$ satisfies the other B.C.

33. In general $y(x) = c_1\cos x + c_2\sin x$. For $y'(0) = 0$ we must choose $c_2 = 0$ and thus $y_1(x) = \cos x$. For $y(1) = 0$ we have $c_1\cos 1 + c_2\sin 1 = 0$, which yields $c_2 = -c_1(\cos 1)/\sin 1$ and thus $y_2(x) = c_1\cos x - c_1(\cos 1)\sin x/\sin 1$

$$= c_1(\sin 1\cos x - \cos 1\sin x)/\sin 1$$

$$= \sin(1-x) \text{ [by setting } c_1 = \sin 1].$$

Furthermore, $W(y_1,y_2) = -\cos 1$ and thus

$$G(x,s) = \begin{cases} \dfrac{\cos s \, \sin(1-x)}{\cos 1} & 0 \le s \le x \\[2mm] \dfrac{\cos x \, \sin(1-s)}{\cos 1} & x \le s \le 1 \end{cases},$$

and hence

$$\phi(x) = \int_0^x [\cos s \, \sin(1-x)f(s)/\cos 1]ds$$

$$+ \int_x^1 [\cos x \, \sin(1-s)f(s)/\cos 1]ds \text{ is the solution of the}$$

given B.V.P.

Section 11.4, Page 661

2a. The D.E. is the same as Eq.(6) and thus, from Eq.(9), the general solution of the D.E. is
$y = c_1 J_0(\sqrt{\lambda}\,x) + c_2 Y_0(\sqrt{\lambda}\,x)$. The B.C. at $x = 0$ requires that $c_2 = 0$, and the B.C. at $x = 1$ requires
$c_1\sqrt{\lambda}\,J_0'(\sqrt{\lambda}) = 0$. For $\lambda = 0$ we have $\phi_0(x) = J_0(0) = 1$
and if λ_n is the nth positive root of $J_0'(\sqrt{\lambda}) = 0$ then
$\phi_n(x) = J_0(\sqrt{\lambda_n}\,x)$. Note that for $\lambda = 0$ the D.E. becomes
$(xy')' = 0$, which has the general solution $y = c_1 \ln x + c_2$.
To satisfy the bounded conditions at $x = 0$ we must choose $c_1 = 0$, thus obtaining the same solution as above.

2b. For $n \neq 0$, set $y = J_0(\sqrt{\lambda_n}\,x)$ in the D.E. and integrate
from 0 to 1 to obtain $-\int_0^1 (xJ_0')'dx = \lambda_n \int_0^1 xJ_0(\sqrt{\lambda_n}\,x)\,dx$.
Integrating the left side of this equation yields
$\int_0^1 (xJ_0')'dx = xJ_0'(\sqrt{\lambda_n}\,x)\Big|_0^1 = J_0'(\sqrt{\lambda_n}) - 0 = 0$ since the λ_n
are eigenvalues from part a. Thus $\int_0^1 xJ_0(\sqrt{\lambda_n}\,x)\,dx = 0$.
For other n and m, we let $L[y] = -(xy')'$. Then
$L[J_0(\sqrt{\lambda_n}\,x)] = \lambda_n xJ_0(\sqrt{\lambda_n}\,x)$ and
$L[J_0(\sqrt{\lambda_m}\,x)] = \lambda_m xJ_0(\sqrt{\lambda_m}\,x)$. Multiply the first equation
by $J_0(\sqrt{\lambda_m}\,x)$, the second by $J_0(\sqrt{\lambda_n}\,x)$, subtract the
second from the first, and integrate from 0 to 1 to obtain
$\int_0^1 \{J_0(\sqrt{\lambda_m}\,x)L[J_0(\sqrt{\lambda_n}\,x)] - J_0(\sqrt{\lambda_n}\,x)L[J_0(\sqrt{\lambda_m}\,x)]\}dx =$
$(\lambda_n - \lambda_m)\int_0^1 xJ_0(\sqrt{\lambda_n}\,x)J_0(\sqrt{\lambda_m}\,x)\,dx$. Again the left side
is zero after each term is integrated by parts once, as
was done above. If $\lambda_n \neq \lambda_m$, the result follows with
$\phi_n(x) = J_0(\sqrt{\lambda_n}\,x)$.

2c. Since $\lambda = 0$ is an eigenvalue we assume that
$y = b_0 + \sum_{n=1}^{\infty} b_n J_0(\sqrt{\lambda_n}\,x)$. Since
$-[xJ_0'(\sqrt{\lambda_n}\,x)]' = \lambda_n xJ_0(\sqrt{\lambda_n}\,x)$, $n = 0, 1, \ldots$, we find that

$-(xy')' = x\sum_{n=1}^{\infty}\lambda_n b_n J_0(\sqrt{\lambda_n}\,x)$ [note that $\lambda_0 = 0$ and b_0 are missing on the right]. Now assume

$f(x)/x = c_0 + \sum_{n=1}^{\infty}c_n J_0(\sqrt{\lambda_n}\,x)$. Multiplying both sides by $xJ_0(\sqrt{\lambda_m}\,x)$, integrating from 0 to 1 and using the orthogonality relations of part b, we find

$c_n = \int_0^1 f(x)J_0(\sqrt{\lambda_n}\,x)\,dx / \int_0^1 xJ_0^2(\sqrt{\lambda_n}\,x)\,dx$, $n = 0,1,2,\ldots$.

[Note that $c_0 = 2\int_0^1 f(x)\,dx$ since the denominator can be integrated.] Substituting the series for y and $f(x)/x$ into the D.E., using the above result for $-(xy')'$, and simplifying we find that

$(\mu b_0 + c_0) + \sum_{n=1}^{\infty}[c_n - b_n(\lambda_n - \mu)]J_0(\sqrt{\lambda_n}\,x) = 0$. Thus

$b_0 = -c_0/\mu$ and $b_n = c_n/(\lambda_n-\mu)$, $n = 1,2,\ldots$, where $\sqrt{\lambda_n}$ are obtained from $J_0'(\sqrt{\lambda_n}) = 0$.

4a. Let $L[y] = -[(1-x^2)y']'$. Then $L[\phi_n] = \lambda_n\phi_n$ and $L[\phi_m] = \lambda_m\phi_m$. Multiply the first equation by ϕ_m, the second by ϕ_n, subtract the second from the first, and integrate from 0 to 1 to obtain $\int_0^1(\phi_m L[\phi_n] - \phi_n L[\phi_m])\,dx = (\lambda_n - \lambda_m)\int_0^1\phi_n\phi_m\,dx$. The integral on the left side can be shown to be 0 by integrating each term once by parts. Since $\lambda_n \neq \lambda_m$ if $m \neq n$, the result follows. Note that the result may also be written as $\int_0^1 P_{2m-1}(x)P_{2n-1}(x)\,dx = 0$, $m \neq n$.

4b. First let $f(x) = \sum_{n=1}^{\infty}c_n\phi_n(x)$, multiply both sides by $\phi_m(x)$, and integrate term by term from $x = 0$ to $x = 1$. The orthogonality condition yields

$c_n = \int_0^1 f(x)\phi_n(x)\,dx / \int_0^1\phi_n^2(x)\,dx$, $n = 1,2,\ldots$ where it is understood that $\phi_n(x) = P_{2n-1}(x)$. Now assume

$y = \sum_{n=1}^{\infty}b_n\phi_n(x)$. As in Problem 2 and in the text

$-[(1-x^2)y']' = \sum_{n=1}^{\infty} \lambda_n b_n \phi_n$ since the ϕ_n are eigenfunctions.

Thus, substitution of the series for y and f into the D.E. and simplification yields $\sum_{n=1}^{\infty} [b_n(\lambda_n - \mu) - c_n]\phi_n(x) = 0$. Hence $b_n = c_n/(\lambda_n - \mu)$, $n = 1, 2, \ldots$ and the desired solution is obtained [after setting $\phi_n(x) = P_{2n-1}(x)$].

Section 11.5, Page 666

1a. Since $u(x,0) = 0$ we have $Y(0) = 0$. However, since the other two boundaries are given by $y = 2x$ and $y = 2(x-2)$ we cannot separate x and y dependence and thus neither X nor Y satisfy homogeneous B.C. at both end points.

1b. The line $y = 2x$ is transformed into $\xi = 0$ and $y = 2(x-2)$ is transformed into $\xi = 2$. The lines $y = 0$ and $y = 2$ are transformed into $\eta = 0$ and $\eta = 2$ respectively, so the parallelogram is transformed into a square of side 2. From the given equations, we have $x = \xi + \eta/2$ and $y = \eta$. Thus
$u_\xi = u_x x_\xi + u_y y_\xi = u_x$ and
$u_\eta = u_x x_\eta + u_y y_\eta = 1/2 u_x + u_y$. Likewise
$u_{\xi\xi} = u_{xx} - x_\xi + u_{xy} y_\xi = u_{xx}$
$u_\eta = u_{xx} x_\eta + u_{xy} y_\eta = 1/2 u_{xx} + u_{xy}$ and
$u_{\eta\eta} = 1/2 u_{xx} x_\eta + 1/2 u_{xy} y_\eta + u_{yx} x_\eta + u_{yy} y_\eta$
$\quad = 1/4 u_{xx} + u_{xy} + u_{yy}$. Therefore,
$5/4 u_{\xi\xi} - u_{\xi\eta} + u_{\eta\eta} = u_{xx} + u_{yy} = 0$.

1c. Substituting $u(\xi,\eta) = U(\xi)V(\eta)$ into the equation of part b yields $5/4 U''V - U'V' + UV'' = 0$ or upon dividing by UV
$\dfrac{5}{4}\dfrac{U''}{U} + \dfrac{V''}{V} = \dfrac{U'V'}{UV}$, which is not separable. The
B.C. become $U(\xi,0) = 0$, $U(\xi,2) = f(\xi+1)$ (since $x = \xi + \eta/2$), $U(0,\eta) = 0$, and $U(2,\eta) = 0$.

2. This problem is very similar to the example worked in the text. The fundamental solutions satisfying the P.D.E.(3), the B.C. $u(1,t) = 0$, $t \geq 0$ and the finiteness condition are given by Eqs.(15) and (16). Thus assume $u(r,t)$ is of the form given by Eq.(17). The I.C. require that $u(r,0) = \sum_{n=1}^{\infty} c_n J_0(\lambda_n r) = 0$ and

$$u_t(r,0) = \sum_{n=1}^{\infty} \lambda_n a k_n J_0(\lambda_n r) = g(r). \quad \text{From Eq.(26) of}$$

Section 11.4 we obtain $c_n = 0$ and

$$\lambda_n k_n a = \int_0^1 rg(r)J_0(\lambda_n r)\,dr \Big/ \int_0^1 rJ_0^2(\lambda_n r)\,dr, \quad n = 1,2,\dots .$$

4.   This problem is the same as Problem 22 of Section 10.7.
      The periodicity condition requires that μ of that problem
      be an integer and thus substituting $\mu^2 = n^2$ into the
      previous results yields the given equations.

5a.   Substituting $u(r,\theta,z) = R(r)\Theta(\theta)Z(z)$ into Laplace's
      equation yields $R''\Theta Z + R'\Theta Z/r + R\Theta''Z/r^2 + R\Theta Z'' = 0$ or
      equivalently $R''/R + R'/rR + \Theta''/r^2\Theta = -Z''/Z = \sigma$. In order
      to satisfy arbitrary B.C. it can be shown that σ must be
      negative, so assume $\sigma = -\lambda^2$, and thus $Z'' - \lambda^2 Z = 0$ and,
      after some algebra, it follows that
      $r^2 R''/R + rR'/R + \lambda^2 r^2 = -\Theta''/\Theta = \alpha$. The periodicity
      condition $\Theta(0) = \Theta(2\pi)$ requires that $\sqrt{\alpha}$ be an integer n
      so $\alpha = n^2$. Thus $r^2 R'' + rR' + (\lambda^2 r^2 - n^2)R = 0$,
      $\Theta'' + n^2\Theta = 0$, and $Z'' - \lambda^2 Z = 0$.

5b.   If $u(r,\theta,z)$ is independent of θ, then the $\Theta''/r^2\Theta$ term
      does not appear in the second equation of part a and thus
      $R''/R + R'/rR = -Z''/Z = -\lambda^2$, from which the desired result
      follows.

6.   Assuming that $u(r,z) = R(r)Z(z)$ it follows from Problem 5
      that $R = c_1 J_0(\lambda r) + c_2 Y_0(\lambda r)$, from Eq.(13), and
      $Z = k_1 e^{-\lambda z} + k_2 e^{\lambda z}$. Since $u(r,z)$ is bounded as $r \to 0$ and
      approaches zero as $z \to \infty$ we require that $c_2 = 0$, $k_2 = 0$.
      The B.C. $u(1,z) = 0$ requires that $J_0(\lambda) = 0$ leading to an
      infinite set of discrete positive eigenvalues
      $\lambda_1, \lambda_2, \dots \lambda_n \dots$. The fundamental solutions of the

      problem are then $u_n(r,z) = J_0(\lambda_n r)e^{-\lambda_n z}$, $n = 1,2,\dots$.

      Thus assume $u(r,z) = \sum_{n=1}^{\infty} c_n J_0(\lambda_n r)e^{-\lambda_n z}$. The B.C.

      $u(r,0) = f(r)$, $0 \le r \le 1$ requires that

      $$u(r,0) = \sum_{n=1}^{\infty} c_n J_0(\lambda_n r) = f(r) \quad \text{so}$$

      $$c_n = \int_0^1 rf(r)J_0(\lambda_n r)\,dr \Big/ \int_0^1 rJ_0^2(\lambda_n r)\,dr, \quad n = 1,2,\dots .$$

7b. Again, Θ periodic of period 2π implies $\lambda^2 = n^2$. Thus the
     solutions to the D.E. are $R(r) = c_1 J_n(kr) + c_2 Y_n(kr)$ (note
     that λ and k here are the reverse of Problem 3 of Section
     11.4) and $\Theta(\theta) = d_1 \cos n\theta + d_2 \sin n\theta$, $n = 0,1,2...$. For the
     solution to remain bounded, $c_2 = 0$ and thus

$$v(r,\theta) = (1/2)c_0 J_0(kr) + \sum_{m=1}^{\infty} J_m(kr)(b_m \sin m\theta + c_m \cos m\theta).$$

     Hence $v(c,\theta)$ is then a Fourier Series of period 2π and
     the coefficients are found as in Section 10.2, Eqs.(13),
     (14) and Problem 27.

9a. Substituting $u(\rho,\theta,\phi) = P(\rho)\Theta(\theta)\Phi(\phi)$ into Laplace's
     equation leads to
     $\rho^2 P''/P + 2\rho P'/P = -(\csc^2\phi)\Theta''/\Theta - \Phi''/\Phi - (\cot\phi)\Phi'/\Phi = \sigma$.
     In order to satisfy arbitrary B.C. it can be shown that σ
     must be positive, so assume $\sigma = \mu^2$.
     Thus $\rho^2 P'' + 2\rho P' - \mu^2 P = 0$. Then we have
     $(\sin^2\phi)\Phi''/\Phi + (\sin\phi\cos\phi)\Phi'/\Phi + \mu^2\sin^2\phi = -\Theta''/\Theta = \alpha$.
     The periodicity condition $\Theta(0) = \Theta(2\pi)$ requires that $\sqrt{\alpha}$
     be an integer λ so $\alpha = \lambda^2$. Hence $\Theta'' + \lambda^2\Theta = 0$ and
     $(\sin^2\phi)\Phi'' + (\sin\phi\cos\phi)\Phi' + (\mu^2\sin^2\phi - \lambda^2)\Phi = 0$.

10. Since u is independent of θ, only the first and third of
     the Eqs. in 9a hold. The general solution to the Euler
     equation is

     $P = c_1\rho^{r_1} + c_2\rho^{r_2}$ where $r_1 = (-1+\sqrt{1+4\mu^2})/2 > 0$ and

     $r_2 = (-1-\sqrt{1+4\mu^2})/2 < 0$. Since we want u to be bounded

     as $\rho \to 0$, we set $c_2 = 0$. As found in Problem 22 of
     Section 5.3, the solutions of Legendre's equation,
     Problem 9c, are either singular at 1, at -1, or at both
     unless $\mu^2 = n(n+1)$, where n is an integer. In this case,
     one of the two linearly independent solutions is a
     polynomial denoted by P_n (Problems 23 and 24 of

     Section 5.3). Since $r_1 = (-1 + \sqrt{1+4n(n+1)})/2 = n$, the

     fundamental solutions of this problem satisfying the

     finiteness condition are $u_n(\rho,\phi) = \rho^n P_n(s) = \rho^n P_n(\cos\phi)$,

     $n = 1,2,...$. It can be shown that an arbitrary
     piecewise continuous function on $[-1,1]$ can be expressed
     as a linear combination of Legendre polynomials. Hence
     we assume that

$u(\rho,\phi) = \sum_{n=1}^{\infty} c_n\rho^n P_n(\cos\phi)$. The B.C. $u(1,\phi) = f(\phi)$ requires

that $u(1,\phi) = \sum_{n=1}^{\infty} c_n P_n(\cos\phi) = f(\phi)$, $0 \le \phi \le \pi$. From

Problem 28 of Section 5.3 we know that $P_n(x)$ are
orthogonal. However here we have $P_n(\cos\phi)$ and thus we
must rewrite the equation in Problem 9b to find
$-[(\sin\phi)\Phi']' = \mu^2(\sin\phi)\Phi$. Thus $P_n(\cos\phi)$ and $P_m(\cos\phi)$ are
orthogonal with weight function $\sin\phi$. Thus we must
multiply the series expansion for $f(\phi)$ by $\sin\phi P_m(\cos\phi)$
and integrate from 0 to π to obtain
$c_m = \int_0^{\pi} f(\phi)\sin\phi P_m(\cos\phi)d\phi / \int_0^{\pi}\sin\phi P_m^2(\cos\phi)d\phi$. To obtain
the answer as given in the text let $s = \cos\phi$.

Section 11.6, Page 675

2a. $b_m = \sqrt{2}\int_0^1 x\sin m\pi x\,dx = \sqrt{2}\,(-1)^{m+1)}/m\pi$ and thus

$S_n = \frac{2}{\pi}\sum_{m=1}^{n}\frac{(-1)^{m+1}\sin m\pi x}{m}$.

2b. We have $R_n = \int_0^1 [x - S_n(x)]^2 dx$, where $S_n(x)$ is given in
part a. Using appropriate computer software we find
$R_1 = .1307$, $R_2 = .0800$, $R_5 = .0367$, $R_{10} = .0193$, $R_{15} = .0131$
and $R_{19} = .0104$.

2c. As in part b, we find $R_{20} = .0099$, and thus $n = 20$ will
insure a mean square error less than $.01$.

4a. Write $S_n(x)$ as $n\sqrt{x}/e^{nx^2/2}$ and use L'Hopital's Rule:

$\lim_{n\to\infty}\frac{n\sqrt{x}}{e^{nx^2/2}} = \lim_{n\to\infty}\frac{\sqrt{x}}{\dfrac{x^2}{2}e^{nx^2/2}} = 0$ for $x \ne 0$. For $x = 0$,

$S_n(0) = 0$ and thus $\lim_{n\to\infty} S_n(x) = 0$ for all x in $[0,1]$.

$R_n > \int_0^1 [0-S_n(x)]^2 dx = n^2\int_0^1 xe^{-nx^2}dx$

$= -\frac{n}{2}e^{nx^2}\Big|_0^1 = \frac{n}{2}(1-e^{-n})$. Since $e^{-n} \to 0$ as $n \to \infty$, we

have that $R_n \to \infty$ as $n \to \infty$.

5. Expanding the integrand we get

$$R_n = \int_0^1 r(x)\,[f(x) - S_n(x)]^2\,dx$$

$$= \int_0^1 r(x)\,f^2(x)\,dx - 2\sum_{i=1}^{n} c_i \int_0^1 r(x)\,f(x)\,\phi_i(x)\,dx$$

$$+ \sum_{i=1}^{n}\sum_{j=1}^{n} c_i c_j \int_0^1 r(x)\,\phi_i(x)\,\phi_j(x)\,dx,$$

where the last term is obtained by calculating $S_n^2(x)$.
Using Eqs.(1) and (9) this becomes

$$R_n = \int_0^1 r(x)\,f^2(x)\,dx - 2\sum_{i=1}^{n} c_i a_i + \sum_{i=1}^{n} c_i^2$$

$$= \int_0^1 r(x)\,f^2(x)\,dx - \sum_{i=1}^{n} a_i^2 + \sum_{i=1}^{n}(c_i - a_i)^2, \text{ by completing}$$

the square. Since all terms involve a real quantity
squared (and $r(x) > 0$) we may conclude R_n is minimized by
choosing $c_i = a_i$. This can also be shown by calculating
$\partial R_n/\partial c_i = 2(c_i - a_i)$ and setting equal to zero.

7b. From part a we have $f_0(x) = 1$ and thus $f_1(x) = c_1 + c_2 x$
must satisfy $(f_0, f_1) = \int_0^1 (c_1 + c_2 x)\,dx = 0$ and
$(f_1, f_1) = \int_0^1 (c_1 + c_2 x)^2\,dx = 1$. Evaluating the integrals
yields $c_1 + c_2/2 = 0$ and $c_1^2 + c_1 c_2 + c_2^2/3 = 1$, which have
the solution $c_1 = \sqrt{3}$, $c_2 = -2\sqrt{3}$ and thus
$f_1(x) = \sqrt{3}\,(1-2x)$.

7c. $f_2(x) = c_1 + c_2 x + c_3 x^2$ must satisfy $(f_0, f_2) = 0$,
$(f_1, f_2) = 0$ and $(f_2, f_2) = 1$.

7d. For $g_2(x) = c_1 + c_2 x + c_3 x^2$ we have $(g_0, g_2) = 0$ and
$(g_1, g_2) = 0$, which yield the same ratio of coefficients
as found in 7c. Thus $g_2(x) = c f_2(x)$, where c may now be
found from $g_2(1) = 1$.

8. This problem follows the pattern of Problem 7 except now
the limits on the orthogonality integral are from -1 to
1. That is $(P_i, P_j) = \int_{-1}^{1} P_i(x) P_j(x)\,dx = 0$, $i \neq j$. For
$i = 0$ we have $P_0(x) = 1$ and for $i = 0$ and $j = 1$ we

have$(P_0, P_1) = \int_{-1}^{1} (c_1 + c_2 x) dx = (c_1 x + c_2 x^2/2) \Big|_{-1}^{1} = 2c_1 = 0$ and

thus $P(1) = 1$ yields $P_1(x) = x$. The others follow in a similar fashion.

9a. This part has essentially been worked in Problem 5 by setting $c_i = a_i$.

9b. Eq.(6) shows that $R_n \geq 0$ since $r(x) \geq 0$ and thus

$$\int_0^1 r(x) f^2(x) dx - \sum_{i=1}^{n} a_i^2 \geq 0. \quad \text{The result follows.}$$

9c. Since f is square integrable, $\int_0^1 r(x) f^2(x) dx = M < \infty$ and therefore the monotone increasing sequence of partial

sums $T_n = \sum_{i=1}^{n} a_i^2$ is bounded above. Thus $\lim_{n \to \infty} T_n$ exists,

which proves the convergence of the given sum.

9d. This result follows from part a and part c.

9e. By definition if $\sum_{i=1}^{\infty} a_i \phi_i(x)$ converges to $f(x)$ in the

mean, then $R_n \to 0$ as $n \to \infty$. Hence $\int_0^1 r(x) f^2(x) dx = \sum_{i=1}^{\infty} a_i^2$.

Conversely, if $\int_0^1 r(x) f^2(x) dx = \sum_{i=1}^{\infty} a_i^2$, $\lim_{n \to \infty} R_n = 0$ and

$\sum_{i=1}^{\infty} a_i \phi_i(x)$ converges to $f(x)$ in the mean.

10. Bessel's inequality implies that $\sum_{i=1}^{\infty} a_i^2$ converges and thus

the nth term $a_n \to 0$ as $n \to \infty$.

12. If the series were the eigenfunction series for a square

integrable function, the series $\sum_{i=1}^{\infty} a_i^2$ would have to

converge. But $a_0 = 1$, $a_1 = 1/\sqrt{2}, \ldots, a_n = 1/\sqrt{n}, \ldots,$

and $\sum_{n=1}^{\infty} a_n^2 = \sum_{n=1}^{\infty} 1/n$ is the well-known harmonic series which

does not converge.